Copyright © 2023, Alimentanima Books

This book is composed almost entirely of materials in the public domain. The combination of materials, formatting and editing as it appears in this work is copyright of Alimentanima Books.

First Printing, 2023

FIELD AND GARDEN VEGETABLES OF
THE LATE NINETEENTH CENTURY

Field and Garden Vegetables of the Late Nineteenth Century

THE HEIRLOOM GARDEN COLLECTION

Alimentanima

Contents

Introduction		1
Preface (from the original 1863 edition)		3
Part One: Esculent Roots		5
1	The Beet (Beta vulgaris)	6
2	The Carrot (Daucus carota)	19
3	Parsnip Chervil (Chærophyllum bulbosum)	28
4	Chinese Potato/Japanese Yam (Dioscorea batatas)	30
5	Chufa/Earth Almond (Cyperus esculentus)	32
6	German Rampion/Evening Primrose (Œnothera biennis)	34
7	Jerusalem Artichoke (Helianthus tuberosus)	35
8	Kohl Rabi (Brassica caulo-rapa)	37
9	Oxalis/Oca (Oxalis crenata)	41
10	The Parsnip (Pastinaca sativa)	45
11	The Potato (Solanum tuberosum)	50
12	The Radish (Raphanus sativus)	69
13	Rampion (Campanula rapunculus)	78
14	Ruta-Baga/Swede Turnip (Brassica campestris ruta-baga)	80
15	Salsify/Oyster Plant (Tragopogon porrifolius)	85
16	Scolymus (Scolymus hispanicus)	88
17	Black Salsify (Scorzonera hispanica)	90
18	Skirrret (Sium sisarum)	92
19	Sweet Potato (Ipomœa batatas)	94

20	Tuberous-Rooted Chickling Vetch (Lathyrus tuberosus)	98
21	Tuberous-Rooted Tropaeolu/Ysano (Tropæolum tuberosum)	100
22	The Turnip (Brassica rapa)	102

Part Two: Alliaceous Plants — 115

23	The Chive (Allium schœnoprasum)	116
24	Common Garlic (Allium sativum)	117
25	The Leek (Allium porrum)	119
26	The Onion (Allium cepa)	123
27	Rocambole (Allium scorodoprasum)	133
28	Shallot (Allium ascalonicum)	134
29	Welsh Onion (Allium fistulosum)	137

Part Three: Asparaginous Plants — 139

30	The Artichoke (Cynarus scolymus)	140
31	Asparagus (Asparagus officinalis)	144
32	Cardoon (Cynara cardunculus)	151
33	The Hop (Humulus lupulus)	155
34	Hoosung/Oosung	156
35	Pokeweed/Pigeon Berry (Phytolacca decandra)	157

Part Four: Curcubitaceous Plants — 159

36	The Cucumber (Cucumis sativus)	160
37	Egyptian Cucumber (Cucumis chate)	168
38	Globe Cucumber (Cucumis prophetarum)	169
39	Calabash/Bottle Gourd (Cucurbita lagenaria)	170
40	The Melon (The Musk, Persian & Water Melons)	172
41	Luffa/Sponge Cucumber (Luffa acutangula)	185
42	West Indian Gherkin (Cucumis anguria)	187
43	The Pumpkin (Curcurbita pepo)	188

44	Snake Cucumber (Cucumis flexuosus)	193
45	Squash and Marrows (Curcurbita pepo)	194

Part Five: Brassicaceous Plants — 211

46	Borecole/Kale (Brassica oleracea sabellica)	212
47	Broccoli (Brassica oleracea var.)	219
48	Brussel Sprouts (Brassica oleracea var.)	228
49	Cabbage (Brassica oleracea capitata)	230
50	Cauliflower (Brassica oleracea var.)	242
51	Colewort/Collards (Brassica oleracea var. viridis.)	247
52	Couve Tronchuda/Portugal Cabbage (Brassica oleracea var.)	248
53	Pak-Choi/Bok-Choy (Brassica rapa chinensi)	251
54	Pe-Tsai/Chinese Cabbage (Brassica rapa pekinensi)	253
55	Savoy Cabbage (Brassica oleracea var. sabauda)	254
56	Sea-Kale (Crambe maritima)	259

Part Six: Spinaceous Plants — 265

57	Amaranth (Amaranthus)	266
58	Black Nightshade (Solanum nigrum)	269
59	Swiss Chard (Beta cicla)	270
60	Malabar Spinach (Basella alba)	273
61	Common Nettle (Urtica dioica)	275
62	New Zealand Spinach (Tetragonia tetragonioides)	277
63	Orach (Atriplex hortensis)	279
64	Patience (Rumex patientia)	283
65	Quinoa (Chenopodium quinoa)	285
66	Sea-Beet (Beta maritima)	287
67	Shephard's Purse (Capsella bursa-pastoris)	289
68	Sorrel (Rumex. sp. et var.)	291
69	Spinach (Spinacia oleracea)	295

70 Wild/Perennial Spinach (Blitum bonus henricus)   299

Part Seven: Salad Plants   301

71 Alexanders (Smyrnium olusatrum)   302
72 Brook-Lime (Veronica beccabunga)   304
73 Buckshorn Plantain (Plantago coronopus)   305
74 Burnet (Poterium sanguisorba)   306
75 Caterpillar Plant (Scorpiurus)   308
76 Celery (Apium graveolens)   310
77 Celeriac/Turnip-Rooted Celery (Apium graveolens var. rapaceum)   319
78 Chervil (Anthriscus cerefolium)   321
79 Chicory (Cichorium intybus)   323
80 Corchorus (Corchorus olitorius)   326
81 Corn-Salad (Valeriana locusta)   327
82 Cress/Peppergrass (Lepidium sativum)   329
83 Cuckoo Flower (Cardamine pratensis)   331
84 The Dandelion (Leontodon taraxacum)   333
85 Endive (Chicorium endivia)   335
86 Horse-Radish (Armoracia rusticana)   341
87 Lettuce (Lactuca sativa)   343
88 Madras Radish (Raphanus caudatus)   361
89 Marrow, Curled-Leaved (Malva crispa)   363
90 Mustard (Sinapis nigra/alba)   364
91 Nasturtium (Tropæolum, sp. et var)   367
92 Picridium (Picridium vulgare)   369
93 Purslain (Portulaca)   370
94 Rape (Brassica napus)   372
95 Rocket (Eruca vesicaria)   374
96 Samphire (Crithmum maritimum)   375
97 Scurvy Grass (Cochlearia officinalis)   377

98	Snail Trefoil (Medicago orbicularis)	378
99	Sweet Cicely (Myrrhis odorata)	379
100	Tarragon (Artemesia dracunculus)	380
101	Valeriana (Valeriana cornucopiæ)	381
102	Water-Cress (Nasturtium officinale)	382
103	Winter-Cress (Barbarea præcox)	384
104	Wood-Sorrel (Oxalis acetocella)	385
105	Worms (Astragalus hamosus)	386

Part Eight: Oleraceaous Plants — 387

106	Angelica (Angelica archangelica)	388
107	Anise (Pimpinella anisum)	389
108	Balm (Melissa officinalis)	391
109	Basil (Ocimum basilicum)	392
110	Borage (Borago officinalis)	394
111	Caraway (Carum carvi)	396
112	Clary Sage (Salvia sclarea)	398
113	Coriander (Coriandrum sativum)	399
114	Costmary/Alecost (Tanacetum balsamita)	401
115	Cumin (Cuminum cyminum)	402
116	Dill (Anethum graveolens)	403
117	Fennel (Fœniculum)	405
118	Lavender (Lavendula spica)	408
119	Lovage (Ligusticum levisticum)	410
120	Calendula (Calendula officinalis)	411
121	Oregano and Marjoram (Origanum vulgare)	413
122	Nigella (Nigella sativa)	416
123	Parsley (Apium petroselinum)	417
124	Peppermint (Mentha piperita)	421
125	Rosemary (Salvia rosmarinus)	423

126	Sage (Salvia officinalis)	425
127	Savory (Satureja)	428
128	Spearmint (Mentha spicata)	430
129	Tansy (Tanacetum vulgare)	431
130	Thyme (Thymus)	433

Part Nine: Leguminous Plants — 435

131	The Common Garden Bean (Phaseolus vulgaris)	436
132	Asparagus Bean (Vigna unguiculata)	467
133	Lima Bean (Phaseolus lunatus)	469
134	Scarlet Runner Bean (Phaseolus coccineus)	473
135	Chick-Pea (Cicer arietinum)	476
136	Chickling Vetch (Lathyrus sativus)	477
137	Broad Bean (Vicia faba)	478
138	The Lentil (Lens culinaris)	484
139	Lupin (Lupinus)	487
140	The Pea (Pisum sativum)	489
141	Sugar/Snap Pea (Pisum macrocarpum)	515
142	Peanut (Arachis hypogaea)	518
143	Vetch /Tare (Vicia sativa)	520
144	Winged Pea (Lotus tetragonolobus)	521

Part Ten: Medicinal Plants — 523

145	Sesame (Sesamum)	524
146	Chamomile (Anthemis nobilis)	526
147	Common Coltsfoot (Tussilago farfara)	528
148	Elecampane (Inula helenium)	529
149	Hoarhound (Marrubium vulgare)	530
150	Hyssop (Hyssopus officinalis)	531
151	Liquorice (Glycyrrhiza glabra)	533

152	Pennyroyal (Hedeoma pulegioides)	534
153	Poppy (Papaver somniferum)	535
154	Palmate-Leaved Rhubarb (Rheum palmatum)	538
155	Rue (Ruta graveolens)	539
156	Safflower (Carthamus tinctorius)	540
157	Southernwood (Artemesia abrotanum)	542
158	Wormwood (Artemesia)	543

Part Eleven: Mushrooms/Esculent Fungi — 545

159	Varied Mushroom Types	546

Part Twelve: Miscellaneous Vegetables — 555

160	Cape Gooseberry/Peruvian Groundcherry (Physalis edulis)	556
161	Corn (Zea mays)	558
162	Eggplant (Solanum melongena)	569
163	Martynie (Martynia proboscidea)	573
164	Oil Radish (Raphanus sativus olifer)	574
165	Okra/Gumbo (Hibiscus esculentus)	575
166	Pepper/Capsicum (Capsicum annuum)	577
167	Rhubarb (Rheum sp. et var)	585
168	Sunflower (Helianthus annuus)	590
169	Tabacco (Nicotiana, sp.)	592
170	Tomato (Solanum lycopersicum)	596

Introduction

This comprehensive book, truly a labour of love, is the first in our *Heirloom Garden Series*. The series aims to republish gardening books from the late nineteenth and early twentieth centuries in order to provide insights not only into practical gardening techniques, but also into attitudes towards plants and gardening in general. This book by Fearing Burr, first published in 1863, was originally titled *The Field and Garden Vegetables of America*. However, as many of the vegetables listed are actually traditional European varieties and the information it provides is valuable well beyond the soil of the United States, we have retitled it as *Field and Garden Vegetables of the Late Nineteenth Century*. It provides a wealth of detailed information about over a thousand varieties of vegetables, some of which will be well-known to contemporary readers and others less so. It is a reminder of the effort that went into the breeding and documentation of productive and tasty vegetable varieties and of the sheer diversity of choices available to us as gardeners, a diversity that is at risk of being lost in the face of industrialised agricultural systems that easily lose sight of the value of localised varieties adjusted to the peculiarities of different regions and also to the novel and attraction of preparing and eating vegetables that don't always look and taste the same.

For those who are interested in reviving older varieties, either for the kitchen garden or for breeding projects, this book is an invaluable source of information. It also provides tips on the growth and storage of each vegetable type and even includes information on taste, texture, and culinary uses. In the section on salad vegetables, for example, many plants often considered weeds are listed, along with ways to prepare or pickle them. You will learn about purslane pickles, dandelion cultivation, and even some curious plants used to trick nineteenth-century gardeners into thinking there were worms and snails in their salads. Practical functions aside, it is wonderful collectors item. For this reason, we have included all the original illustrations as well as additional images from heirloom seed packages from the same era, which are works of art in their own right. Although it has not been possible to include images of every single variety, we have gone to a special effort to include images of less well-known plants, especially in the chapters about herbs and leafy-greens.

Regarding the formatting and editing, since this is a historical book, we have made as few changes as possible. We have changed punctuation in some places, for example by removing full-stops in titles. However, we have left the capitalisation of vegetable names as in the original, despite this being unusual by modern standards. In a few chapters, we have updated the latin names of vegetable varieties where they contained errors or have a more modern version. However, for those interested in historical research, we have also left in the original identifications given. In a few cases, such as with the Asian vegetables, we have added latin names that weren't given, so that the reader can know for certain which vegetable is being referred to. The source of quotes is indicated in footnotes at the end of each chapter, though it should be noted that many of the quotes did not come with indication as to their origin.

Garden Manual for the Southern States (1898), Richard Frotscher Seed Co.

Preface (from the original 1863 edition)

Though embracing all the directions necessary for the successful management of a vegetable garden, the present volume is offered to the public as a manual or guide to assist in the selection of varieties, rather than as a treatise on cultivation. Through the standard works of American authors, as well as by means of the numerous agricultural and horticultural periodicals of our time, all information of importance relative to the various methods of propagation and culture, now in general practice, can be readily obtained.

But, with regard to the characteristics which distinguish the numerous varieties; their difference in size, form, color, quality, and season of perfection; their hardiness, productiveness, and comparative value for cultivation,—these details, a knowledge of which is important as well to the experienced cultivator as to the beginner, have heretofore been obtained only through sources scattered and fragmentary.

To supply this deficiency in horticultural literature, I have endeavored, in the following pages, to give full descriptions of the vegetables common to the gardens of this country. It is not, however, presumed that the list is complete, as many varieties, perhaps of much excellence, are comparatively local: never having been described, they are, of course, little known. Neither is the expectation indulged, that all the descriptions will be found perfect; though much allowance must be made in this respect for the influence of soil, locality, and climate, as well as for the difference in taste of different individuals.

Much time, labor, and expense have been devoted to secure accuracy of names and synonymes; the seeds of nearly all of the prominent varieties having been imported both from England and France, and planted, in connection with American vegetables of the same name, with reference to this object alone.

The delay and patience required in the preparation of a work like the present may be in some degree appreciated from the fact, that in order to obtain some comparatively unimportant particular with regard to the foliage, flower, fruit, or seed, of some obscure and almost unknown plant, it has been found necessary to import the seed or root; to plant, to till, to watch, and wait an entire season.

Though some vegetables have been included which have proved of little value either for the table or for agricultural purposes, still it is believed such descriptions will be found by no means unimportant; as a timely knowledge of that which is inferior, or absolutely worthless, is often as advantageous as a knowledge of that which is of positive superiority.

That the volume may be acceptable to the agriculturist, seedsman, and to all who may possess, cultivate, or find pleasure in, a garden, is the sincere wish of the author.

Fearing Burr

Hingham, March, 1863.

Part One: Esculent Roots

The Beet; Carrot; Parsnip Chervil; Chinese Potato, or Japanese Yam; Chufa, or Earth Almond; German Rampion; Jerusalem Artichoke; Kohl Rabi; Oxalis, Tuberous; Oxalis, Deppes; Parsnip; Potato; Radish; Rampion; Swede, or Ruta-baga Turnip; Salsify, or Oyster Plant; Scolymus; Scorzonera; Skirret; Sweet Potato; Tuberous-rooted Chickling Vetch; Tuberous-rooted Tropæolum; Turnip.

Image of Varied Root Vegetables
Adolphe Millot legume et plante potagères betteraves navets pommes de terre et autres.

1

The Beet (Beta vulgaris)

The Common Beet, sometimes termed the Red Beet, is a half-hardy biennial plant; and is cultivated for its large, succulent, sweet, and tender roots. These attain their full size during the first year, but will not survive the winter in the open ground. The seed is produced the second year; after the ripening of which, the plant perishes.

When fully developed, the beet-plant rises about four feet in height, with an angular, channelled stem; long, slender branches; and large, oblong, smooth, thick, and fleshy leaves. The flowers are small, green, and are either sessile, or produced on very short peduncles. The calyxes, before maturity, are soft and fleshy; when ripe, hard and wood-like in texture. These calyxes, which are formed in small, united, rounded groups, or clusters, are of a brownish color, and about one-fourth of an inch in diameter; the size, however, as well as depth of color, varying, to some extent, in the different varieties. Each of these clusters of dried calyxes contains from two to four of the true seeds, which are quite small, smooth, kidney-shaped, and of a deep reddish-brown color.

These dried clusters, or groups, are usually recognized as the seeds; about fifteen hundred of which will weigh one ounce. They retain their vitality from seven to ten years.

Soil and Fertilisers—The soil best adapted to the beet is a deep, light, well-enriched, sandy loam. When grown on thin, gravelly soil, the roots are generally tough and fibrous; and when cultivated in cold, wet, clayey localities, they are often coarse, watery, and insipid, worthless for the table, and comparatively of little value for agricultural purposes.

A well-digested compost, formed of barnyard manure, loam and salt, makes the best fertilizer. Where this is not to be obtained, guano, superphosphate of lime, or bone-dust, may be employed advantageously as a substitute. Wood-ashes, raked or harrowed in just previous to sowing the seed, make an excellent surface-dressing, as they not only prevent the depredations of insects, but give strength and vigor to the young plants. The application of coarse, undigested, strawy manure, tends to the production of forked and misshapen roots, and should be avoided.

Propagation and Culture—Beets are always raised from seed. For early use, sowings are sometimes made in November; but the general practice is to sow the seed in April, as soon as the frost is out of the ground, or as soon as the soil can be worked. For use in autumn, the seed should be sown about the middle or 20th of May; and, for the winter supply, from the first to the middle of June. Lay out the ground in beds five or six feet in width, and of a length proportionate to the supply required; spade or fork the soil deeply and thoroughly over; rake the surface smooth and even; and draw the drills across the bed, fourteen inches apart, and about an inch and a half in depth. Sow the seeds thickly enough to secure a plant for every two or three inches, and cover to the depth of the drills. Should the weather be warm and wet, the young plants will appear in seven or eight days. When they are two inches in height, they should be thinned to five or six inches apart; extracting the weaker, and filling vacant spaces by transplanting. The surplus plants will be found an excellent substitute for spinach, if cooked and served in like manner. The afterculture consists simply in keeping the plants free from weeds, and the earth in the spaces between the rows loose and open by frequent hoeings.

Mr. Thompson[1] states that:

> "...the drills for the smaller varieties should be about sixteen inches apart, and the plants should be thinned out to nine inches apart in the rows. The large sorts may have eighteen inches between the rows, but still not more than nine inches from plant to plant in the row. When large-sized roots are desired, the rows may be eighteen inches or two feet apart, and the plants twelve or fifteen inches distant from each other in the rows. But large roots are not the best for the table; and it is better to have two medium-sized roots, grown at nine inches apart, than one of perhaps double the size from twice the space. As a square foot of ground should afford plenty of nourishment to produce a root large enough for the table, the area for each plant may, therefore, be limited to that extent. If the rows are sixteen inches apart, and the plants thinned to nine inches in the row, each plant will have a space equal to a square foot. Such, of course, would also be the case if the rows were twelve inches apart, and the plants the same distance from each other in the row. But it is preferable to allow a greater space between the rows than between the plants in the row: for, by this arrangement, the leaves have better scope to grow to each side, and the plants so situated grow better than those which have an equal but rather limited space in all directions; whilst the ground can also be more easily stirred, and kept clean."

Taking the Crop—Roots, from the first sowings, will be ready for use early in July; from which time, until October, the table may be supplied directly from the garden. They should be drawn as fast as they attain a size fit for use; which will allow more time and space for the development of those remaining.

For winter use, the roots must be taken up before the occurrence of heavy frosts, as severe cold not only greatly impairs their quality, but causes them to decay at the crown. Remove the leaves, being careful not to cut or bruise the crown; spread the roots in the sun a few hours to dry; pack them in sand or earth slightly moist; and place in the cellar, out of reach of frost, for the winter.

"The London market-gardeners winter their beets in large sheds, stored in moderately damp mould, and banked up with straw. Mr. Cuthill states that it is a mistake to pack them in dry sand or earth for the winter; and that the same may be said of parsnips, carrots, salsify, scorzonera, and similar roots.

"The object here is, that the moist soil may not draw the natural sap out of the roots so readily as dry sand would do; and hence they retain their fresh, plump appearance, and their tenderness and color are better preserved. In taking up the roots, the greatest care must be exercised that they are neither cut, broken, wounded on the skin, nor any of the fibres removed; and, when the small-leaved varieties are grown, few, if any, of the leaves should be cut off."[2]

If harvested before receiving injury from cold, and properly packed, they will retain, in a good degree, their freshness and sweetness until the new crop is suitable for use.

Seed—To raise seed, select smooth and well-developed roots having the form, size, and color by which the pure variety is distinguished; and, in April, transplant them eighteen inches or two feet apart, sinking the crowns to a level with the surface of the ground. As the stalks increase in height, tie them to stakes for support. The plants will blossom in June and July, and the seeds will ripen in August.

In harvesting, cut off the plants near the ground, and spread them in a light and airy situation till they are sufficiently dried for threshing, or stripping off the seeds; after which the seeds should be exposed, to evaporate any remaining moisture.

An ounce of seed will sow from one hundred to one hundred and fifty feet of drill, according to the size of the variety; and about four pounds will be required for one acre.

Use—

"The roots are the parts generally used, and are boiled, stewed, and also eaten cold, sliced in vinegar and oil. They enter into mixed salads, and are much used for garnishing; and, for all these purposes, the deeper colored they are, the more they are appreciated. Some, however, it ought to be noticed, prefer them of a bright-red color; but all must be of fine quality in fibre, solid, and of uniform color. The roots are also eaten cut into thin slices, and baked in an oven. Dried, roasted, and ground, they are sometimes mixed with coffee, and are also much employed as a pickle. Mixed with dough, they make a wholesome bread; but, for this purpose, the white or yellow rooted sorts are preferred. The roots of all the varieties are better baked than boiled."[3]

The young plants make an excellent substitute for spinach; and the leaves of some of the kinds, boiled when nearly full grown, and served as greens, are tender and well-flavored.

Some of the larger varieties are remarkably productive, and are extensively cultivated for agricultural purposes. From a single acre of land in good condition, thirty or forty tons are frequently harvested; and exceptional crops are recorded of fifty, and even sixty tons. In France, the White Sugar-beet is largely employed for the manufacture of sugar,—the amount produced during one year being estimated to exceed that annually made from the sugar-cane in the State of Louisiana.

For sheep, dairy-stock, and the fattening of cattle, experience has proved the beet to be at once healthful, nutritious, and economical.

Varieties—The varieties are quite numerous, and vary to a considerable extent in size, form, color, and quality. They are obtained by crossing, or by the intermixture of one kind with another. This often occurs naturally when two or more varieties are allowed to run to seed in close proximity, but is sometimes performed artificially by transferring the pollen from the flower of a particular variety to the stigma of the flower of another.

The kinds now in cultivation are as follows:

Bark-Skinned

(Oak Bark-Skinned)

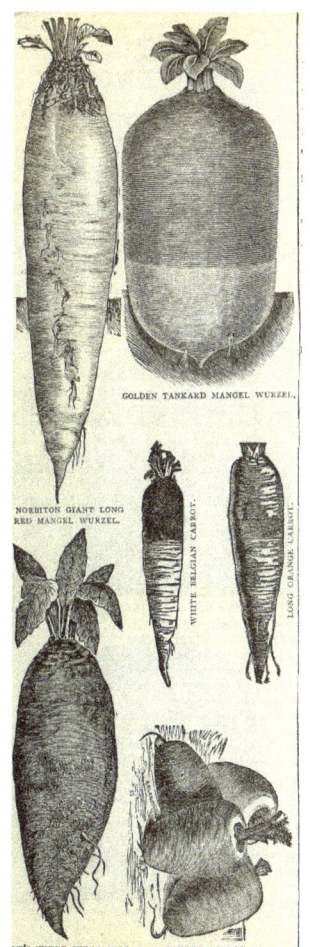

Beet Varieties
Annual catalogue of Price & Reed, 1895

Root produced entirely within the earth, broadest near the crown, and thence tapering regularly to a point; average specimens measuring four inches in their greatest diameter, and about one foot in depth. Skin dark brown, thick, hard, and wrinkled, or striated, sometimes reticulated or netted, much resembling the bark of some descriptions of trees; whence the name. Flesh very deep purplish-red, circled, and rayed with paler red, fine-grained, sugary, and tender. Leaves numerous, spreading, bright green, slightly stained with red; the leaf-stems and nerves bright purplish-red.

An early and comparatively new French variety, of fine flavor, excellent for summer use, and, if sown as late as the second week in June, equally valuable for the table during winter. Not recommended for field culture.

Sow in rows fourteen inches apart, and thin to six inches apart in the rows.

Barrott's New Crimson

Root similar in form to the Castelnaudary, but somewhat larger; smooth and regular, and not apt to fork. Flesh dark crimson, fine-grained and tender. Leaf-stalks yellow.

Bassano.

(Early Flat Bassano; Turnip-Rooted Bassano; Rouge Plate de Bassano)

Bulb flattened; six or seven inches in diameter by three or four inches in depth; not very regular or symmetrical, but often somewhat ribbed, and terminating in a very small, slender tap-root. Skin of fine texture; brown above ground; below the surface, clear rose-red. Flesh white, circled or zoned with bright pink; not very close-grained, but very sugary and well-flavored. Leaves numerous, erect, of a lively green color, forming many separate groups, or tufts, covering the entire top, or crown, of the root. Leaf-stems short, greenish-white, washed or stained with rose.

An Italian variety, generally considered the earliest of garden-beets, being from seven to ten days earlier than the Early Blood Turnip-rooted. The flesh, although much coarser than that of many other sorts, is tender, sweet, and of good quality. Roots from early sowings are, however, not suited for winter use; as, when overgrown, they almost invariably become too tough, coarse, and fibrous for table use. To have them in perfection during winter, the seed should not be sown till near the close of June.

In moist, favorable seasons, it succeeds well in comparatively poor, thin soil.

Cultivate and preserve as directed for the Early Turnip-rooted.

Cattell's Dwarf Blood

Root small, regularly tapering. Flesh deep blood-red. Leaves small, bright red, spreading, or inclined to grow horizontally. Quality good,—similar to that of the Red Castelnaudary; which variety it much resembles in its general character.

On account of its small size, it requires little space, and may be grown in rows twelve inches apart.

Cow-Horn Mangel Wurzel

(Serpent-like Beet; Cow-horn Scarcity)

A sub-variety of the Mangel Wurzel, producing its roots almost entirely above ground; only a small portion growing within the earth. Root long and slender, two feet and a half in length, and nearly three inches in diameter at its broadest part; often grooved or furrowed lengthwise, and almost invariably bent and distorted,—the effect either of the wind, or of the weight of its foliage. Flesh greenish white, circled with red at the centre. Leaves of medium size, green, erect; the leaf-stems and nerves pale red or rose color.

It derives its different names from its various contorted forms; sometimes resembling a horn, and often assuming a shape not unlike that of a serpent.

The variety is much esteemed and extensively cultivated in some parts of Europe, although less productive than the White Sugar or Long Red Mangel Wurzel.

Early Mangel Wurzel.

(Early Scarcity; Disette Hâtive)

Aside from its smaller size, this variety much resembles the Common Red Mangel Wurzel. Root contracted towards the crown, which rises two or three inches above the surface of the soil, and tapering within the earth to a regular cone. Skin purplish rose, deeper colored than that of the last named. Flesh white, circled or zoned with pale red. Leaves spreading, green; the leaf-stems rose-colored.

It is remarkable for the regular and symmetrical form of its roots, which grow rapidly, and, if pulled while young, are tender, very sweet, and well flavored. Planted the last of June, it makes a table-beet of more than average quality for winter use.

When sown early, it attains a comparatively large size, and should have a space of twenty inches between the rows; but, when sown late, fifteen inches between the rows, and six inches between the plants in the rows, will afford ample space for their development.

Early Blood Turnip-Rooted

(Early Turnip Beet)

Early Blood Turnip-Rooted

The roots of this familiar variety are produced almost entirely within the earth, and measure, when of average size, from four inches to four and a half in depth, and about four inches in diameter. Form turbinate, flattened, smooth, and symmetrical. Neck small, tap-root very slender, and regularly tapering. Skin deep purplish-red. Flesh deep blood-red, sometimes circled and rayed with paler red, remarkably sweet and tender. Leaves erect, not very numerous, and of a deep-red color, sometimes inclining to green; but the stems and nerves always of a deep brilliant red.

The Early Blood Turnip Beet succeeds well from Canada to the Gulf of Mexico; and in almost every section of the United States is more esteemed, and more generally cultivated for early use, than any other variety. Among market-gardeners, it is the most popular of the summer beets. It makes a rapid growth, comes early to the table, and, when sown late, keeps well, and is nearly as valuable for use in winter as in summer and autumn.

In common with most of the table sorts, the turnip-rooted beets are much sweeter and more tender if pulled before they are fully grown; and consequently, to have a continued supply in their greatest perfection, sowings should be made from the beginning of April to the last of June, at intervals of two or three weeks.

The roots, especially those intended for seed, should be harvested before severe frosts, as they are liable to decay when frozen at the crown, or even chilled. Sow in drills fourteen inches apart; and, when two inches in height, thin out the plants to six inches apart in the drills. An acre of land in good cultivation will yield from seven to eight hundred bushels.

German Red Mangel Wurzel

(Disette d'Allemagne)

An improved variety of the Long Red Mangel Wurzel, almost regularly cylindrical, and terminating at the lower extremity in an obtuse cone. It grows much out of ground, the neck or crown is comparatively small, it is rarely forked or deformed by small side roots, and is generally much neater and more regular than the Long Red. Size very large; well-developed specimens measuring from eighteen to twenty inches in length, and seven or eight inches in diameter. Flesh white, with red zones or rings; more colored than that of the last named. Leaves erect, green; the stems and nerves washed or stained with rose-red.

For agricultural purposes, this variety is superior to the Long Red, as it is larger, more productive, and more easily harvested.

German Yellow Mangel Wurzel.

(Green Mangel Wurzel; Jaune d'Allemagne)

Root produced half above ground, nearly cylindrical for two-thirds its length, terminating rather bluntly, and often branched or deformed by small side-roots. Size large; when well grown, measuring sixteen or eighteen inches deep, six or seven inches in diameter, and weighing from twelve to fifteen pounds. Skin above ground, greenish-brown; below, yellow. Flesh white, occasionally zoned or marked with yellow. Leaves of medium size, rather numerous, erect, very pale, or yellowish green; the stems and ribs light green.

While young and small, the roots are tender and well-flavored; but this is a field rather than a table beet. In point of productiveness, it differs little from the Common Long Red, and should be cultivated as directed for that variety.

Half Long Blood

(Dwarf Blood; Fine Dwarf Red; Early Half Long Blood; Rouge Nain)

Root produced within the earth, of medium size, or rather small; usually measuring about three inches in thickness near the crown, and tapering regularly to a point; the length being ten or twelve inches. Skin smooth, very deep purplish-red. Flesh deep blood-red, circled and rayed with paler red, remarkably fine grained, of firm texture, and very sugary. Leaves small, bright red, blistered on the surface, and spreading horizontally. Leaf-stems short.

An excellent, half-early, garden variety, sweet, and well flavored, a good keeper, and by many considered very superior to the Common Long Blood. When full grown, it is still tender and fine-grained, and much less stringy and fibrous than the last named, at an equally advanced stage of growth. It may be classed as one of the best table-beets, and is well worthy cultivation.

Improved Long Blood

(Long Smooth Blood)

This is an improved variety of the Common Long Blood, attaining a much larger size, and differing in its form, and manner of growth. When matured in good soil, its length is from eighteen inches to two feet; and its diameter, which is retained for more than half its length, is from four to five inches. It is seldom very symmetrical in its form; for, though it has but few straggling side-roots, it is almost invariably bent and distorted. Skin smooth, very deep or blackish purple. Flesh dark blood-red, sweet, tender, and fine grained, while the root is young and small, but liable to be tough and fibrous when full grown. Leaves small, erect-red, and not very numerous. Leaf-stems blood-red.

This beet, like the Common Long Blood, is a popular winter sort, retaining its color well when boiled. It is of larger size than the last named, grows more above the surface of the ground, and has fewer fibrous and accidental small side-roots. While young, it compares favorably with the old variety; but, when full grown, can hardly be said to be much superior. To have the variety in its greatest perfection for winter use, the seed should not be sown before the 10th of June; as the roots of this, as well as those of nearly all the table-varieties, are much more tender and succulent when very rapidly grown, and of about two-thirds their full size.

Sow in drills fifteen inches apart, and thin to eight inches apart in the drills; or sow on ridges eighteen inches apart.

Long Blood

(Common Long Blood)

The roots of this familiar variety are long, tapering, and comparatively slender; the size varying according to the depth and richness of the soil. Skin dark purple, sometimes purplish-black. Flesh deep blood-red, very fine grained and sugary, retaining its color well after being boiled. Leaves rather numerous, of medium size, erect, deep purplish-red; the leaf-stems blood-red.

One of the most popular of winter beets; but, for late keeping, the seed should not be sown before the middle of June, as the roots, when large, are frequently tough and fibrous.

The Improved Long Blood is a variety of this, and has, to a considerable extent, superseded it in the vegetable garden; rather, it would seem, on account of its greater size, than from any real superiority as respects its quality or keeping properties.

Long Red Mangel Wurzel

(Red Mangel Wurzel; Marbled Field Beet)

Root fusiform, contracted at the crown, which, in the genuine variety, rises six or eight inches above the surface of the ground. Size large, when grown in good soil; often measuring eighteen inches in length, and six or seven inches in diameter. Skin below ground purplish-rose; brownish-red were exposed to the air and light. Leaves green; the stems and nerves washed or stained with rose-red. Flesh white, zoned and clouded with different shades of red.

The Long Red Mangel Wurzel is hardy, keeps well, grows rapidly, is very productive, and in this country is more generally cultivated for agricultural purposes than any other variety. According to Lawson, the marbled or mixed color of its flesh seems particularly liable to vary: in some specimens, it is almost of a uniform red; while, in others, the red is scarcely, and often not at all, perceptible. These variations in color are, however, of no importance as respects the quality of the roots.

The seed may be sown from the middle of April to the last of May. If sown in drills, they should be at least eighteen inches apart, and the plants should be thinned to ten inches in the drills. If sown on ridges, the sowing should be made in double rows; the ridges being three and a half or four feet apart, and the rows fifteen inches apart. The yield varies with the quality of the soil and the state of cultivation; thirty and thirty-five tons being frequently harvested from an acre.

While young, the roots are tender and well-flavored, and are sometimes employed for table use.

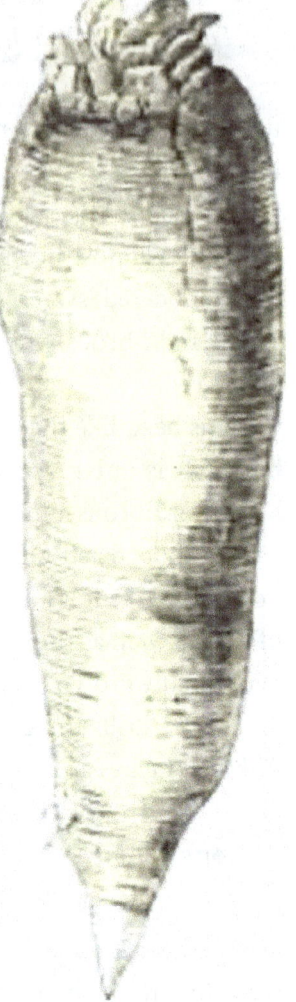

Long Yellow Mangel Wurzel

Long White Green-Top Mangel Wurzel

(Green-top White Sugar; Long White Mangel Wurzel; Disette Blanche à Collet Verte)

An improved variety of the White Sugar Beet. Root produced much above ground, and of very large size; if well grown, measuring nearly six inches in diameter, and eighteen inches in depth,—the diameter often retained for nearly two-thirds the length. Skin green, were exposed to light and air; below ground, white. Flesh white. Leaves green, rather large, and not so numerous as those of the White Sugar.

Very productive, and superior to the last named for agricultural purposes; the quality being equally good, and the yield much greater.

Long Yellow Mangel Wurzel

(Jaune Grosse)

Root somewhat fusiform, contracted towards the crown, which rises six or eight inches above the surface of the ground. Size remarkably large; when grown in deep rich soil, often measuring twenty inches in length, and five or six inches in thickness. Skin yellow, bordering on orange-color. Flesh pale yellow, zoned or circled with white, not close-grained, but sugary. Leaves comparatively large, pale green; the stems and nerves yellow; the nerves paler.

The variety is one of the most productive of the field-beets; but the roots are neither smooth nor symmetrical, a majority being forked or much branched.

In the vicinity of Paris, it is extensively cultivated, and is much esteemed by dairy farmers on account of the rich color which it imparts to milk when fed to dairy-stock. Compared with the German Yellow, the roots of this variety are longer, not so thick, more tapering; and the flesh is of a much deeper color. It has also larger foliage.

Pine-Apple Short-Top

Root of medium size, fusiform. Skin deep purplish-red. Flesh very deep blood-red, fine-grained, as sweet as the Bassano, tender, and of excellent quality for table use. Leaves very short and few in number, reddish-green; leaf-stems and nerves blood-red.

In its foliage, as well as in the color of the root, it strongly resembles some of the Long Blood varieties; but it is not so large, is much finer in texture, and superior in flavor. It is strictly a garden or table beet, and, whether for fall or winter use, is well deserving of cultivation.

Red Castelnaudary

This beet derives its name from a town in the province of Languedoc in France, where the soil is particularly adapted to the growth of these vegetables, and where this variety, which is so much esteemed in France for its nut-like flavor, was originally produced.

The roots grow within the earth. The leaves are thickly clustered around the crown, spreading on the ground. The longest of the leaf-stems do not exceed three inches: these and the veins of the leaves are quite purple, whilst the leaves themselves are green, with only a slight stain of purple. The root is little more than two inches in diameter at the top, tapering gradually to the length of nine inches. The flesh, which is of a deep purple, and exhibits dark rings, preserves its fine color when boiled, is very tender and sweet, and presents a delicate appearance when cut in slices.

Being small in its whole habit, it occupies but little space in the ground, and may be sown closer than other varieties usually are.

Not generally known or much cultivated in this country.

Red Globe Mangel Wurzel

(Betterave Globe Rouge)

Root nearly spherical, but tapering to pear-shaped at the base; nearly one-third produced above ground. Size large; well-grown specimens measuring seven or eight inches in diameter, and nine or ten inches in depth. Skin smooth, and of a rich purplish rose-color below ground; brown above the surface, were exposed to the sun. Flesh white, rarely circled, with rose-red. Leaves pale green, or yellowish green; the stems and ribs or nerves sometimes veined with red.

This variety is productive, keeps well, and, like the Yellow Globe, is well adapted to hard and shallow soils. It is usually cultivated for agricultural purposes, although the yield is comparatively less than that of the last named.

In moist soils, the Yellow Globe succeeds best; and, as its quality is considered superior, it is now more generally cultivated than the Red.

White Globe Mangel Wurzel

A sub-variety of the Yellow and Red Globe, which, in form and manner of growth, it much resembles. Skin above ground, green; below, white. Leaves green. Flesh white and sugary; but, like the foregoing sorts, not fine grained, or suited for table use.

Productive, easily harvested, excellent and profitable for farm purposes, and remarkably well adapted for cultivation in hard, shallow soil.

White Sugar.

(White Silesian; Betterave Blanche)

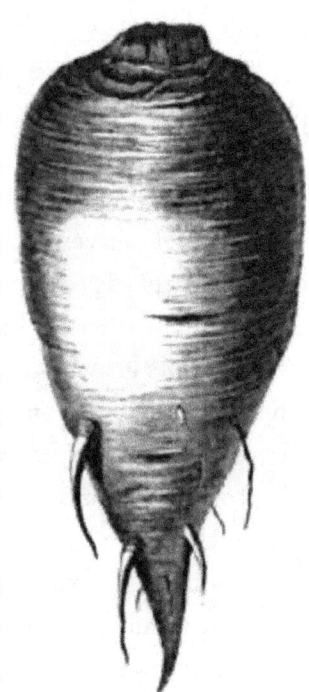

White Sugar Beet

Root fusiform, sixteen inches in length, six or seven inches in its greatest diameter, contracted towards the crown, thickest just below the surface of the soil, but nearly retaining its size for half the depth, and thence tapering regularly to a point. Skin white, washed with green or rose-red at the crown. Flesh white, crisp, and very sugary. Leaves green; the leaf-stems clear green, or green stained with light red, according to the variety.

The White Sugar Beet is quite extensively grown in this country, and is employed almost exclusively as feed for stock; although the young roots are sweet, tender, and well flavored, and in all respects superior for the table to many garden varieties. In France, it is largely cultivated for the manufacture of sugar and for distillation.

Of the two sub-varieties, some cultivators prefer the Green-top; others, the Rose-colored or Red-top. The latter is the larger, more productive, and the better keeper; but the former is the more sugary. It is, however, very difficult to preserve the varieties in a pure state; much of the seed usually sown containing, in some degree, a mixture of both.

It is cultivated in all respects as the Long Red Mangel Wurzel, and the yield per acre varies from twenty to thirty tons.

White Turnip-Rooted.

A variety of the Early Turnip-rooted Blood, with green leaves and white flesh; the size and form of the root, and season of maturity, being nearly the same. Quality tender, sweet, and well flavored; but, on account of its color, not so marketable as the last named.

Wyatt's Dark Crimson

(Whyte's Dark Crimson; Rouge de Whyte)

Root sixteen inches long, five inches in diameter, fusiform, and somewhat angular in consequence of broad and shallow longitudinal furrows or depressions. Crown conical, brownish. Skin smooth, slate-black. Flesh very deep purplish-red, circled and rayed with yet deeper shades of red, very fine-grained, and remarkably sugary. Leaves deep red,[Pg 19]shaded with brownish-red: those of the centre, erect; those of the outside, spreading or horizontal.

The variety is not early, but of fine quality; keeps remarkably well, and is particularly recommended for cultivation for winter and spring use. Much esteemed in England.

Yellow Castelnaudary

Root produced within the earth, broadest at the crown, where its diameter is nearly three inches, and tapering gradually to a point; the length being about eight inches. Skin orange-yellow. Flesh clear yellow, with paler zones or rings. Leaves spreading, those on the outside being on stems about four inches in length; the inner ones are shorter, numerous, of a dark-green color, and rather waved on the edges: the leaf-stems are green, rather than yellow.

An excellent table-beet, being tender, yet firm, and very sweet when boiled, although its color is not so agreeable to the eye.

Yellow Globe Mangel Wurzel

(Betterave Jaune Globe)

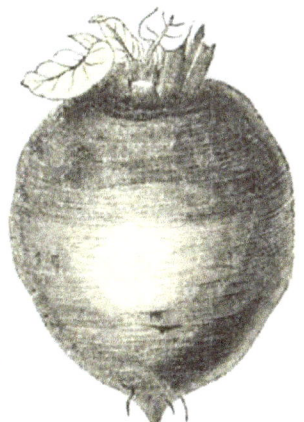

Yellow Globe Mangel Wurzel

This is a globular-formed beet, measuring about ten inches in diameter, and weighing ten or twelve pounds; about one-half of the root growing above ground. Skin yellow, where it is covered by the soil; and yellowish-brown above the surface, where exposed to light and air. Flesh white, zoned or marked with yellow, close-grained and sugary. Leaves not large or numerous, rather erect, green; the stems and ribs paler, and sometimes yellowish.

The Yellow Globe is one of the most productive of all the varieties; and, though not adapted to table use, is particularly excellent for stock of all descriptions, as the roots are not only remarkably sugary, but contain a considerable portion of albumen. It retains its soundness and freshness till the season has far advanced, does not sprout so early in spring as many others, and is especially adapted for cultivation in hard, shallow soil.

The yield varies from thirty to forty tons per acre, according to soil, season, and culture; although crops are recorded of fifty tons and upwards.

Sow from the last of April to the last of May; but early sowings succeed best. If sown in drills, they should be made twenty inches apart, and the plants should be thinned to ten inches apart in the drills; if sown on ridges, sow in double rows, making the ridges three feet and a half, and the rows sixteen inches apart. On account of its globular form, the crop can be harvested with great facility by the use of a common plough.

Yellow Turnip-Rooted

A sub-variety of the Blood Turnip-rooted, differing principally in color, but to some extent also in its form, which is less compressed. Leaves large, yellowish-green; the leaf-stems and nerves yellow. Flesh yellow, comparatively close-grained, sweet and tender.

Not much cultivated on account of its color; the red varieties being preferred for table use.

1. *The Gardener's Assistant.*, by Robert Thompson

2. *The Book of the Garden. By Charles M'Intosh. 2 vols. Edinburgh and London, 1855.*

3. *The Book of the Garden, by Charles M'Intosh. 2 vols. Edinburgh and London, 1855.*

Beet Varieties
Huntington & Page, Seedsmen

2

The Carrot (Daucus carota)

The Carrot, in its cultivated state, is a half-hardy biennial. It is indigenous to some parts of Great Britain, generally growing in chalky or sandy soil, and to some extent has become naturalised in this country; being found in gravelly pastures and mowing fields, and occasionally by roadsides, in loose places, where the surface has been disturbed or removed. In its native state, the root is small, slender, and fibrous, or woody, of no value, and even of questionable properties as an article of food.

Soil, Sowing, and Culture—The Carrot flourishes best in a good, light, well-enriched loam. Where there is a choice of situations, heavy and wet soils should be avoided; and, where extremes are alternatives, preference should be given to the light and dry. If possible, the ground should be stirred to the depth of twelve or fifteen inches, incorporating a liberal application of well-digested compost, and well pulverizing the soil in the operation. The surface should next be levelled, cleared as much as possible of stones and hard lumps of earth, and made mellow and friable; in which state, if the ground contains sufficient moisture to color the surface when it is stirred, it will be ready for the seed. This may be sown from the first of April to the 20th of May; but early sowings succeed best. The drills should be made an inch in depth; and for the smaller, garden varieties, about ten inches apart. The larger sorts are grown in drills about fourteen inches apart; the plants in the rows being thinned to five or six inches asunder.

Harvesting—The roots attain their full size by the autumn of the first year; and, as they are not perfectly hardy, should be dug and housed before the ground is frozen. When large quantities are raised for stock, they are generally placed in bulk in the cellar, without packing; but the finer sorts, when intended for the table, are usually packed in earth or sand, in order to retain their freshness and flavor. With ordinary precaution, they will remain sound and fresh until May or June.

General Seed Catalogue no. 206 (1898), F.C. Heinemann

Seed—To raise seed, select good-sized, smooth, and symmetrical roots; and as early in spring as the frost is out of the ground, and the weather settled, transplant to rows three feet apart, and fifteen inches apart in the rows, sinking the crowns just below a level with the surface of the ground. The seed-stalks are from four to six feet in height, with numerous branches. The flowers appear in June and July; are white; and are produced at the extremities of the branches, in umbels, or flat, circular groups or clusters, from two to five inches in diameter. The seed ripens in August; but, as all the heads do not ripen at once, they should be cut off as they successively mature. The stiff, pointed hairs or bristles with which the seeds are thickly covered, and which cause them to adhere together, should be removed either by threshing or by rubbing between the hands; clearing them more or less perfectly, according to the manner of sowing. If sown by a machine, the seeds should not only be free from broken fragments of the stems of the plant, but the surface should be made as smooth as possible. For hand-sowing, the condition of the seed is less essential; though, when clean, it can be distributed in the drill more evenly and with greater facility.

The seeds of the several varieties differ little in size, form, or color, and are not generally distinguishable from each other. They will keep well two years; and if preserved from dampness, and placed in a cool situation, a large percentage will vegetate when three years old.

In the vegetable garden, an ounce of seed is allowed for one hundred and fifty feet of drill; and, for field culture, about two pounds for an acre.

An ounce contains twenty-four thousand seeds.

Use—Though not relished by all palates, carrots are extensively employed for culinary purposes, and are generally considered healthful and nutritious. They form an important ingredient in soups, stews, and French dishes of various descriptions; and by many are much esteemed, when simply boiled, and served with meats or fish.

> "Carrots may be given to every species of stock, and form in all cases a palatable and nourishing food. They are usually given in their raw state, though they may be steamed or boiled in the same manner as other roots.
>
> "Horses and dairy-cows are the live-stock to which they are most frequently given. They are found in an eminent degree to give color and flavor to butter; and, when this is the end desired, no species of green-feeding is better suited to the dairy. To horses they may be given with cut straw and hay; and, thus given, form a food which will sustain them on hard work. They afford excellent feeding for swine, and quickly fatten them. When boiled, they will be eaten by poultry; and, mixed with any farinaceous substance, form an excellent food for them. They are also used for distillation, affording a good spirit."

The varieties are as follow:

Altringham
(Altringham; Long Red Altringham)

The Altrincham Carrot measures about fourteen inches in length, by two inches in diameter. It retains its thickness for nearly two-thirds its length: but the surface is seldom regular or smooth; the genuine variety being generally characterized by numerous crosswise elevations, and corresponding depressions. Neck small and conical, rising one or two inches above the surface of the soil. Skin nearly bright-red; the root having a semi-transparent appearance. Flesh bright and lively, crisp and breaking in its texture; and the heart, in proportion to the size of the root, is smaller than that of the Long Orange. Leaves long, but not large or very numerous.

According to Lawson, it is easily distinguished from the Long Orange by the roots growing more above ground, by its more convex or rounded shoulders, and by its tapering more irregularly, and terminating more abruptly. It is, however, exceedingly difficult to procure the variety in its purity, as it is remarkably liable to sport, although the roots grown for seed be selected with the greatest care.

It is a good field-carrot, but less productive than the Long Orange and some others; mild and well flavored for the table, and one of the best sorts for cultivation for market.

Thompson[1] states that "it derives its name from a place called Altrincham, in Cheshire, England, where it is supposed to have originated. In seedsmen's lists it is frequently, but erroneously, called the Altringham."

Early Frame
(Early Forcing Horn; Earliest Short Forcing Horn; Early Short Scarlet)

Root grooved or furrowed at the crown, roundish, or somewhat globular; rather more than two inches in diameter, nearly the same in depth, and tapering suddenly to a very slender tap-root. Skin red, or reddish-orange; brown or greenish where it comes to the surface of the ground. Foliage small and finely cut or divided, not so large or luxuriant as that of the Early Horn.

The Early Frame is the earliest of all varieties, and is especially adapted for cultivation under glass, both on account of its earliness, and the shortness and small size of its roots. It is also one of the best sorts for the table, being very delicate, fine-grained, mild, and remarkably well flavored.

Where space is limited, it may be grown in rows six inches apart, thinned to three inches apart in the rows; or sown broadcast, and the young plants thinned to three inches apart in each direction.

Early Frame Carrot

Early Half-Long Scarlet

(Half-Long Red)

Root slender and tapering, measuring seven or eight inches in length, and two inches in its greatest diameter. Crown hollow. Skin red below the surface of the ground, green or brown above. Flesh reddish-orange, fine-grained, mild, and well flavored. Foliage similar to that of the Early Frame, but not abundant.

The variety is remarkably productive; in good soil and favorable seasons, often yielding an amount per acre approaching that of the Long Orange. Season intermediate between the early garden and late field sorts.

Early Horn

(Early Scarlet Horn; Early Short Dutch; Dutch Horn)

Early Horn Carrot

Root six inches in length, two inches and a half in diameter, nearly cylindrical, and tapering abruptly to a very slender tap-root. Skin orange-red, but green or brown where it comes to the surface of the ground. Flesh deep orange-yellow, fine-grained, and of superior flavor and delicacy. The crown of the root is hollow, and the foliage short and small.

The variety is very early, and as a table-carrot much esteemed, both on account of the smallness of its heart and the tenderness of its fibre. As the roots are very short, it is well adapted for shallow soils; and on poor, thin land will often yield a greater product per acre than the Long Orange or the White Belgian, when sown under like circumstances.

Sow in rows one foot apart, and thin to four inches in the rows.

Flander's Large Pale Scarlet

(Flander's Pale Red)

Root produced within the earth, fourteen or fifteen inches long, three or four inches in diameter at the broadest part, fusiform, not very symmetrical, but often quite crooked and angular. The crown is flat, very large, and nearly covered by the insertion of the leaves. Flesh reddish-yellow, and rather coarse-grained. Foliage large and vigorous.

The roots are formed early and with great certainty. It is also very productive, of large size, keeps remarkably well; and, though of coarse texture, one of the best sorts for cultivation for farm-purposes.

It originated in Flanders, and is comparatively an old variety, but is little disseminated, and not grown to any extent, in this country.

Long Orange

Root long, thickest at or near the crown, and tapering regularly to a point. Size very variable, being much affected by soil, season, and cultivation: well-grown specimens measure fifteen inches in length, and three inches in diameter at the crown. Skin smooth, of a reddish-orange color. Flesh comparatively close-grained, succulent, and tender, of a light-reddish vermilion or orange color, the heart lighter, and large in proportion to the size of the root. Foliage not abundant, but healthy and vigorous, and collected into a comparatively small neck. The roots are usually produced entirely within the earth.

If pulled while very young and small, they are mild, fine-grained, and good for table use; but, when full grown, the texture is coarser, and the flavor stronger and less agreeable.

The Long Orange is more cultivated in this country for agricultural purposes than all other varieties. With respect to its value for stock, its great productiveness, and its keeping properties, it is considered the best of all the sorts for field culture. A well-enriched soil will yield from six hundred to eight hundred bushels per acre. The seed is usually sown in drills, about fourteen inches apart, but sometimes on ridges, eighteen or twenty inches apart, formed by turning two furrows together; the ridges yielding the largest roots, and the drills the greatest quantity.

Two pounds of seed are usually allowed to an acre; but, if sown by a well-regulated machine, about one-half this quantity will be sufficient.

Long Red Belgian

(Yellow Belgian; Yellow Green-Top Belgian)

Root very long, fusiform, contracted a little towards the crown, but nearly
of uniform thickness from the top down half the length. Size large; when grown in deep soil, often measuring twenty inches in length, and nearly three inches in diameter. The crown rises four or five inches above the surface of the ground, and is of a green color; below the surface, the skin is reddish-yellow. Flesh orange-red.

This variety, like the White, originated in Belgium. In Europe it is much esteemed by agriculturists, and is preferred to the White Belgian, as it is not only nearly as productive, but has none of its defects.

Long Yellow

(Long Lemon)

Root fusiform, three inches in diameter at the crown, and from, twelve to fourteen inches in depth. Skin pale yellow, or lemon-color, under ground; but greenish on the top, or crown, which rises a little above the surface of the soil. Flesh yellow, the heart paler, and, like that of the Long Orange, of large size. While young, the roots are delicate, mild, and well flavored; but, when full grown, valuable only for stock.

The Long Lemon is easily harvested, and is very productive, yielding nearly the same quantity to the acre as the Long Orange; which variety it much resembles in its general character, and with which it is frequently, to a greater or less extent, intermixed.

Long Surrey

(Long Red; James's Scarlet)

This variety much resembles the Long Orange: the roots, however, are more slender, the heart is smaller, and the color deeper.

It is popular in some parts of England, and is extensively grown over the Continent.

Long White

(Common White)

Root produced entirely below ground, regularly fusiform, fifteen inches long, by about three inches in its largest diameter. Skin white, stained with russet-brown. Flesh white, and generally considered sweeter than that of the colored varieties.

The Common White has been but little cultivated since the introduction of the White Belgian; a variety much more productive, though perhaps not superior either in flavor, or fineness of texture.

New Intermediate

An English variety, comparatively of recent introduction. Root broadest at the crown, and thence tapering very regularly to a point. Size full medium; well-grown specimens measuring nearly three inches in diameter at the broadest part, and about one foot in length. Skin bright orange-red. Flesh orange-yellow, fine-grained, sweet, well flavored, and, while young, excellent for table use.

Very hardy, and also very productive; yielding, according to the best English authority, a greater weight per acre than any other yellow-fleshed variety.

Purple or Blood Red

(Violette)

Root fusiform, and very slender, fourteen inches in length, by two inches and a half in diameter at the top or broadest part. Skin deep purple, varying to some extent in depth of shade, but generally very dark. Flesh purple at the outer part of the root, and yellow at the centre or heart; fine grained, sugary, and comparatively well flavored.

Not much cultivated for the table, on account of the brown color it imparts to soups or other dishes of which it may be an ingredient. It is also inclined to run to seed the year it is sown. It has, however, the reputation of flourishing better in wet, heavy soil, than any other variety.

Short White

(Blanche des Vosges)

Root obtusely conical, seven or eight inches long, by about four inches in diameter at the crown, which is large, flat, greenish, and level with the surface of the ground. Skin white, tinted with amber, smooth and fine. Flesh yellowish-white, remarkably solid, and fine in texture; sweet and well flavored. Foliage rather finely divided, and as vigorous as the Long Orange.

The Short White yields well, retains its qualities during winter, and is well adapted for cultivation in soils that are hard and shallow.

Carrot Diversity

Studley

(Long Red Brunswick)

Root fusiform, very long, and regular; the crown level with the surface of the soil. In good cultivation, the roots attain a length of sixteen inches, and a diameter of nearly two inches. Color bright reddish-orange, like the Altrincham.

An excellent table-carrot, but flourishes well only in deep, mellow soil.

White Belgian

(Green-Top White)

Root very long, fusiform, eighteen to twenty inches in length, and four or five inches in diameter. In the genuine variety, the crown rises five or six inches from the surface of the ground; and, with the exception of a slight contraction towards the top, the full diameter is retained for nearly one-half of the entire length. Skin green above, white below ground. Flesh white, tending to citron-yellow at the centre or heart of the root; somewhat coarse in texture. Foliage rather large and vigorous.

The White Belgian Carrot is remarkable for its productiveness, surpassing in this respect all other varieties, and exceeding that of the Long Orange by nearly one-fourth. It can be harvested with great facility, and gives a good return even on poor soils.

The variety is not considered of any value as a table esculent, and is grown almost exclusively for feeding stock; for which purpose, it is, however, esteemed less valuable than the yellow-fleshed sorts, because less nutritious, and more liable to decay during winter.

Since its introduction, it has somewhat deteriorated; and, as now grown, differs to some extent from the description given above. The roots are smaller, seldom rise more than two or three inches above the soil, and taper directly from the crown to the point. A judicious selection of roots for seed, continued for a few seasons, would undoubtedly restore the variety to its primitive form and dimensions.

The same amount of seed will be required as of the Long Orange: and the general method of culture should be the same; with the exception, that, in thinning out the plants, the White Belgian should have more space.

White Belgian Horn

(Transparent White)

Root seven or eight inches in length, and two inches in its greatest diameter, tapering regularly from the crown to the point. Skin fine, clear white. Flesh very white, and almost transparent, mild, tender, and delicate.

A French variety, remarkable for the peculiar, pure white color of its skin and flesh.

1. The Gardener's Assistant. By Robert Thompson.

3

Parsnip Chervil (Chærophyllum bulbosum)

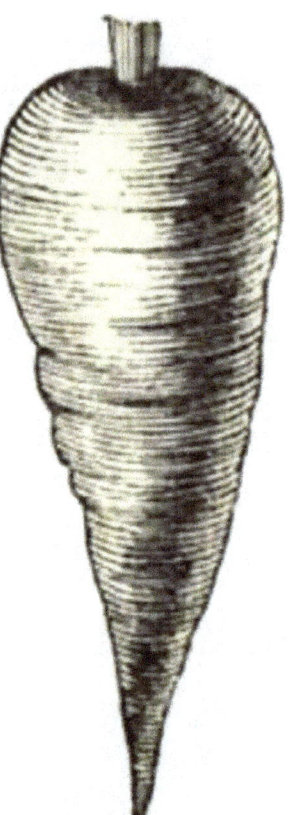

Turnip-Rooted Chervil

(Turnip-Rooted Chervil)

A hardy, biennial plant, from the south of Europe. The root is fusiform, four or five inches long, and nearly an inch and a half in diameter; skin, grayish-black; flesh, white. The leaves are compound, the leaflets very deeply cut, and the divisions of the upper leaves very narrow and slender. The flowers are white, and terminate the top of the plant in umbels, or large, circular, flat, spreading bunches. The seeds are long, pointed, furrowed, concave on one side, of a brownish color, and retain their power of germination but one year. An ounce contains sixty-five hundred seeds.

Soil and Cultivation —The seeds may be sown in drills, in October or April, in the manner of sowing the seeds of the common carrot: preference to be given to rich, mellow soil. The roots will attain their full size by the following August or September, when they should be harvested. With a little care to prevent sprouting, they may be preserved until April.

Seed —The roots intended for seed should be set in the open ground in autumn or in spring. The seeds will ripen in August, and should be sown within a month or two of the time of ripening; or, if kept till spring, should be packed in earth or sand: for, when these precautions are neglected, they will often remain dormant in the ground throughout the year.

FIELD AND GARDEN VEGETABLES OF THE LATE NINETEENTH CENTURY | 29

Top Left: Parsnip Chervil by Franz Eugen Köhler. Top Right: Young parsnip chervil near Gdańsk.
Bottom: Close up of parsnip chervil roots.

4

Chinese Potato/Japanese Yam (Dioscorea batatas)

Stem twelve feet or more in length, of a creeping or climbing habit; leaves heart-shaped, though sometimes halberd-formed; flowers small, in clusters, white.

"The root is of a pale russet color, oblong, regularly rounded, club-shaped, exceedingly tender, easily broken, and differs from nearly all vertical roots in being largest at the lower end."

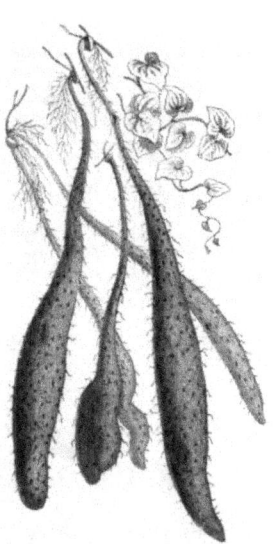

Chinese Yam
Vilmorin-Andrieux & Ci

Propagation and Cultivation —The Chinese Potato requires a very deep, light, rather sandy, and tolerably rich soil; and this should be thoroughly stirred to the depth of at least two feet. No fresh manure should be used, but fine, well-decomposed compost applied, and deeply as well as very thoroughly incorporated with the soil; avoiding however, if possible, its direct contact with the growing roots. It is propagated either by small roots; by the top or neck of the large roots, cut off to the length of five or six inches; or by the small bulbs, or tubers, which the plants produce in considerable numbers on the stem, in the axils of the leaves. These should be planted the last of April, or as soon as the ground is in good working condition. Lay out the land in raised ridges two feet and a half or three feet asunder; and on the summit set the bulbs, or tubers, with the point or shoot upwards, eight or ten inches apart; and cover about an inch deep. Cultivate in the usual manner during the summer; and late in autumn, after the tops are dead, and just before the closing-up of the ground, take up the roots, dry them a short time in the sun, and store them in the cellar for use. The roots are perfectly hardy, and will sustain no injury from the coldest winter, if left unprotected in the open ground. During the second season, the growth of the old root is not continued, but gradually decays as the new roots are formed. A well-grown root will measure about two feet in length, and two inches and a half at its broadest diameter.

Use—The flesh is remarkably white, and very mucilaginous in its crude state. The roots are eaten either boiled or roasted, and require rather more than half the time for cooking that is usually given to the boiling or roasting of the common potato. When cooked, they possess a rice-like taste and consistency, are quite farinaceous, and unquestionably nutritive and valuable for food.

Varieties related to that mentioned above. Left: Japanese mountain yam (Family: Dioscoreaceae; Species: Dioscorea japonica Thunb.) Right: Chinese yam. (Family: Dioscoreaceae; Species: Dioscorea polystachya Turcz)
Seikei Zusetsu vol. 22, page 021

5

Chufa/Earth Almond (Cyperus esculentus)

A perennial plant, from the south of Europe. The roots are long and fibrous, and produce at their extremities numerous small, rounded or oblong, jointed, pale-brown tubers, of the size of a filbert. The flesh of these roots, or tubers, is of a yellowish color, tender, and of a pleasant, sweet, and nut-like flavor. The leaves are rush-like, about eighteen inches high, a little rough, and sharply pointed. The flower-stalks are nearly of the same height as the leaves, three-cornered, hard, and leafless, with the exception of five or six leaflike bracts at the top, from the midst of which are produced the spikelets of flowers, which are of a pale-yellow color.

Propagation and Culture—It is propagated by planting the tubers in April or May, two inches deep, in drills two feet apart, and six inches apart in the drills. They will be ready for harvesting in October. In warm climates, the plant, when once introduced into the garden, spreads with great rapidity, and is exterminated with much difficulty. In the Northern and Middle States, the tubers remaining in the open ground are almost invariably destroyed by the winter.

Use—It is cultivated for its small, almond-like tubers, which, when dried, have somewhat the taste of the almond, and keep a long period. They are eaten either raw or roasted.

"The plant grows spontaneously in the light, humid soils of Spain; and is cultivated in Germany and the south of France. The tubers are chiefly employed for making an orgeat,—a species of drink much used in Spain,[1] Cuba, and other hot climates where it is known. When mashed to a flour,—which is white, sweet, and very agreeable to the taste,—it imparts to water the color and richness of milk."[2]

1. *This drink is a traditional beverage of the Valencian Community, Spain, and is known as 'Horchata'.*
2. *The Horticulturist, and Journal of Art and Rural Taste. Monthly. By P. Barry and J. Jay Smith. Philadelphia.*

FIELD AND GARDEN VEGETABLES OF THE LATE NINETEENTH CENTURY | 33

Chufa sedge — cyperus esculentus — cultivation.
Andrés Marín Jarque

6

German Rampion/Evening Primrose (Œnothera biennis)

The German Rampion, or Evening Primrose, common in this country to gravelly pastures and roadsides, is a hardy biennial plant, and, when in full perfection, measures three or four feet in height, with long, flat, pointed leaves, and large, yellow, fragrant flowers. The seed-pods are oblong, four-sided; the seeds are small, angular, of a brown color, and retain their germinative properties three years.

Sowing and Cultivation—The seeds should be sown annually, in April, in a rich and shady situation; for if grown in a dry, sunny exposure, and sown very early in the season, the plants are inclined to run to flower during the summer: which renders the roots worthless; for they then become hard and fibrous. Sow in drills an inch deep, and fourteen inches apart; thin to six or eight inches in the rows; cultivate in the usual form; and, in September, the roots will be ready for use. For winter use, take up the roots before freezing weather, and pack in sand. For spring use, they may be taken directly from the ground.

To raise Seed—Two or three plants, left in the ground through the winter, will yield an abundant supply of seeds the following summer.

Use—The root is the only part used. This, when full grown, is generally from ten to twelve inches long, fusiform, occasionally with a few strong fibres, whitish on the outside, and white within. The thick, outer covering separates readily, and should be removed when the root is eaten in its crude state. It possesses a nutty flavor; but is inferior to the true Rampion, having a slight pungency. If required as a raw salad, it should be eaten while young. When the roots have attained their full size, they are usually dressed in the manner of Skirret and Scorzonera.

7

Jerusalem Artichoke (Helianthus tuberosus)

The Jerusalem Artichoke is a hardy perennial. In its manner of growth and flowering, it much resembles the common sunflower; of which, as its scientific term suggests, it is really a species. Stem six to eight feet high, very rough, and much branched; leaves alternate, large, rough, heart-shaped at the base, pointed at the ends, and indented on the borders; flowers large, yellow,—produced on the top of the plant, at the extremities of the branches.

Soil, Propagation, and Culture—

"It thrives best in a light, mellow soil, made rich by the application of old, decomposed manure; but the roots will flourish well if planted in any corner of the garden less suited for other descriptions of vegetables. To obtain fine roots, however, the soil should be trenched fifteen or eighteen inches in depth.

"It is propagated by planting the small tubers, or offsets: the large tubers may also be cut or divided into several pieces, each having one eye, as practised with the potato. In April, or early in May, lay out the rows three feet apart, drop the tubers one foot apart in the rows, and cover three inches deep. As the plants come up, hoe the ground between the rows from time to time; and draw a little earth around their stems, to support them, and to afford the roots a thicker covering."

Taking the Crop—The new tubers will be suitable for use in the autumn. In digging, great care should be taken to remove the small as well as the full-grown; for those not taken from the ground will remain fresh and sound during the winter, and send up in the spring new plants, which, in turn, will increase so rapidly, as to encumber the ground, and become troublesome. In localities where the crop has once been cultivated, though no plants be allowed to grow for the production of fresh tubers, yet the young shoots will continue to make their appearance from time to time for many years.

Use—

"The roots, or tubers, are the parts of the plant eaten. These are boiled in water till they become tender; when, after being peeled, and stewed with butter and a little wine, they will be as pleasant as the real Artichoke, which they nearly resemble both in taste and flavor."

M'Intosh[1] says that the tubers may be used in every way as the potato; and are suited to persons in delicate health, when debarred from the use of most other vegetables.

*Varieties—*For a long period, there was but a single variety cultivated, or even known. Recent experiments in the use of seeds as a means of propagation have developed new kinds, varying greatly in their size, form, and color, possessing little of the watery and insipid character of the heretofore grown Jerusalem Artichoke, and nearly or quite equalling the potato in flavor and excellence.

Common White

Tubers large, and often irregular in form; skin and flesh white; quality watery, and somewhat insipid. It is unfit for boiling, but is sometimes served baked or roasted. It makes a very crisp and well-flavoured pickle.

Purple-Skinned

A French variety, produced from seed. Tubers purplish rose-color; flesh dryer when cooked, and finer flavored, than that of the foregoing.

Red-Skinned

Like the Purple-skinned, produced from seed. Skin red. Between this and the last named there are various intermediate sorts, differing in shades of color, as well as in size, form, and quality.

Yellow-Skinned

The tubers of this variety are of a yellowish color, and are generally smaller, and even more irregularly shaped, than those of the Common White. They are, however, superior in quality, and of a more agreeable taste when cooked.

1. The Book of the Garden. By Charles M'Intosh. 2 vols. Edinburgh and London, 1855.

8

Kohl Rabi (Brassica caulo-rapa)

The Kohl Rabi is a vegetable intermediate between the cabbage and the turnip. The stem, just above the surface of the ground, swells into a round, fleshy bulb, in form not unlike a turnip. On the top and about the surface of this bulb are put forth its leaves, which are similar to those of the Swede turnips; being either lobed or entire on the borders, according to the variety. The seeds are produced the second year; after the ripening of which, the bulb perishes.

Sowing and Cultivation—Mr. Thompson's[1] directions are as follows:

"Kohl Rabi may be sown thinly, broadcast, or in drills four inches apart, in April, May, or June. When the young plants are an inch or two in height, they may be transplanted into any good, well-enriched piece of ground, planting them eight inches apart, in rows fifteen inches asunder, and not deeper in the ground than they were in the seed-bed. Water should be given till they take fresh root, and subsequently in dry weather as required; for though the plants suffer little from droughts, yet the tenderness of the produce is greatly impaired by an insufficient supply of moisture. With the exception of stirring the ground and weeding, no further culture is required. The crop will be fit for use when the bulbs are of the size of an early Dutch turnip: when allowed to grow much larger, they are only fit for cattle. Of field varieties, the bulbs sometimes attain an immense size; weighing, in some cases, fourteen pounds."

Seed—Take up a few plants entire in autumn; preserve them during winter in the manner of cabbages or turnips; and transplant to the open ground in April, two feet apart in each direction. The seeds are not distinguishable from those of the Swede or Ruta-baga Turnip, and retain their vitality from five to seven years.

Use—The part chiefly used is the turnip-looking bulb, formed by the swelling of the stem. This is dressed and eaten with sauce or with meat, as turnips usually are. While young, the flesh is tender and delicate, possessing the combined flavor of the cabbage and turnip.

They are said to keep better than any other bulb, and to be sweeter and more nutritious than the cabbage or white turnip.

> "In the north of France, they are extensively grown for feeding cattle,—a purpose for which they seem admirably adapted, as, from having a taste similar to the leaves of others of the species, they are found not to impart any of that peculiar, disagreeable taste to the milk, which it acquires when cows are fed on turnips."

Varieties—These are as follow:

Artichoke-Leaved
(Cut-leaved)

Of German origin, deriving its name from the resemblance of the leaves to those of the Artichoke. Bulb small, and not smooth or symmetrical. The leaves are beautifully cut, and are very ornamental; but the bulb is comparatively of little value. Not much cultivated.

Kohl Rabi
Adolphe Millot

Early Dwarf White

Bulb white, smaller than that of the Common White, and supported close to the ground. The leaves are also smaller, and less numerous.

It is earlier, and finer in texture, than the last named; and, while young, excellent for the table. Transplant in rows fifteen inches apart, and ten inches asunder in the rows.

Early Purple Vienna.

This corresponds with the Early White Vienna, except in color, which, in this variety, is a beautiful purple, with a fine glaucous bloom. The leaf-stems are very slender, and the leaves smooth, and few in number.

These two Vienna sorts are by far the best for table use. When taken young, and properly dressed, they form an excellent substitute for turnips, especially in dry seasons, when a crop of the latter may fail or become of inferior quality.

Early White Vienna

Dwarf, small, early; bulb handsome, firm, glossy, white, or very pale-green. The leaves are few, small, with slender stems, the bases of which are dilated, and thin where they spring from different parts on the surface of the bulb. The flesh is white, tender, and succulent, whilst the bulb is young, or till it attains the size of an early white Dutch turnip; and at or under this size it should be used.

Set the plants in rows fifteen inches apart, and ten inches from plant to plant in the lines.

Green

Similar to, if not identical with, the Common White. The bulbs are pale-green, attain a very large size, and the variety is hardy and productive. Not suited to garden culture, but chiefly grown for farm-purposes.

Purple

This variety differs little from the White, except in color; the bulb being purple, and the leaf-stems and nerves also tinged with purple. Like the White, it attains a large size, and is only adapted for field culture; the flesh being too coarse and strong-flavored for table use.

Green Kohl Rabi

White

Bulb large,—when full grown, measuring seven or eight inches in diameter, and weighing from eight to ten pounds; leaves rather large and numerous; skin very pale, or whitish-green; stem about six inches high. Hardy, very late, and chiefly employed for farm-purposes.

The variety should be cultivated in rows eighteen inches apart, and the plants should stand one foot apart in the rows.

1. The Gardener's Assistant. By Robert Thompson.

1898, The Harnden Seed Co.

9

Oxalis/Oca (Oxalis crenata)

Of the Tuberous-Rooted Oxalis, there are two varieties, as follow:

White-Rooted

(Oca Blanca)

White Oxalis

Stem two feet in length, branching, prostrate or trailing, the ends of the shoots erect; leaves trifoliate, yellowish-green, the leaflets inversely heart-shaped; flowers rather large, yellow,—the petals crenate or notched on the borders, and striped at their base with purple. The seeds are matured only in long and very favorable seasons. In its native state, the plant is perennial; but is cultivated and treated, like the common potato, as an annual.

Cultivation—The tubers should be started in a hot-bed in March, and transplanted to the open ground in May, or as soon as the occurrence of settled warm weather. They thrive best in dry, light, and medium fertile soils, in warm situations; and should be planted in hills two feet and a half apart, or in drills two feet and a half apart, setting the plants or tubers an inch and a half deep, and fifteen or eighteen inches apart in the drills; treating, in all respects, as potatoes.

The tubers form late in the season; are white, roundish, or oblong, pointed at the union with the plant, and vary in size according to soil, locality, and season; seldom, however, exceeding an inch in diameter, or weighing above four ounces. The yield is comparatively small.

Use—The tubers are used as potatoes. When cooked, the flesh is yellow, very dry and mealy, of the flavor of the potato, with a very slight acidity. The tender, succulent stalks and foliage are used as salad.

Oxalis, Red Tuberous-Rooted

(Oca Colorada)

Plant similar in habit to the White Tuberous-Rooted; but the branches, as well as the under surface of the leaves, are more or less stained with red. Tubers larger than those of the last named, roundish, tapering towards the connection with the plant, and furnished with numerous eyes in the manner of the common potato; skin smooth, purplish-red; flesh often three-coloured,—the outer portion of the tuber carmine-red, the central part marbled, and the intermediate portion yellow,—the colours, when the root is divided transversely, appearing in concentric zones, or rings. The flesh contains but little farinaceous matter, and possesses a certain degree of acidity, which, to many palates, is not agreeable.

Propagated, and in all respects cultivated, like the White. Either of the varieties may also be grown from cuttings, which root readily.

According to a statement from the London Horticultural Society's Journal, the acidity may be converted into a sugary flavor by exposing the tubers to the action of the sun for eight or ten days,—a phenomenon which is analogous to what takes place in the ripening of most fruits. When treated in this form, the tubers lose all trace of acidity, and become as floury as the best descriptions of potatoes. If the action of the sun is continued for a long period, the tubers become of the consistence and sweet taste of figs. Mr. Thompson[1] states that the disagreeable acid taste may also be removed by changing the water when they are three-quarters boiled.

The plants are tender, and are generally destroyed early in autumn by frost. The tubers must be taken up before freezing weather, packed in sand, and placed in a dry, warm cellar for the winter.

Pink Oxalis
Frank Vincentz

Deppe's Oxalis

(Oxalis Deppei)

A perennial plant from Mexico, very distinct from the tuberous-rooted species before described. Stalk about one foot in height, smooth and branching; leaves four together, the leaflets wedge-shaped, pale yellowish-green, the upper surface marked by two brownish lines or stains in the form of two sides of a triangle; flowers terminal, of a carmine-rose or pink-red color, stained with green at the base of the petals.

"The roots are fleshy, tapering, white, and semi-transparent, and furnished on the top of the crown with a mass of scaly bulbs, sometimes amounting to fifty in number, by means of which the plant can be easily propagated. When well grown, the roots are about four inches in length, and from one inch to one inch and a half in thickness."[2]

Soil and Culture—

"This Oxalis requires a light, rich soil, mixed with decayed vegetable matter; and it prefers a southern aspect, provided the soil is not too dry.

"It may be raised from seed; but is generally propagated by planting the bulbs, which should be set the last of April or beginning of May, or when all danger of frost is over, six inches apart, in rows one foot asunder. The bulbs should be only just covered with soil; for thus they occupy a position, with regard to the surface, similar to that in which they are produced: and this seems indispensable, if fine roots are to be obtained.

"The stems have been observed to spring up from a considerable depth; but, in this case, tap-roots were not formed. During summer, the soil must be kept moist in dry weather; otherwise, when rain falls abundantly, the sudden accession of water to the roots occasions their splitting. The plants should be allowed to grow as long as there is no danger from frost; but, previous to this occurring, they should either be taken up or protected. If protected from frost by frames or otherwise, the roots will continue to increase in size till near November. When taken up, the roots should be divested of the numerous bulbs formed on their crowns, and then stored up for use in a cool, dry place, but secure from frost. A similar situation will be proper for the small bulbs; or they may be kept in dry sand till the season of planting."[3]

The plant has been cultivated with the most complete success, with no especial preparation of the soil; merely planting the bulbs in shallow drills, the ground being dug and manured as for other kitchen-garden crops.

Use—In a communication to the "Gardener's Chronicle," Prof. Morren gives the uses of the plant as follow:

"The uses of the Oxalis are many. The young leaves are dressed like sorrel in soup, or as a vegetable. They have a fresh and agreeable acid, especially in spring. The flowers are excellent in salad, alone, or mixed with corn salad, endive of both kinds, red cabbage, beet-root, and even with the petals of the dahlia, which are delicious when thus employed. When served at table, the flowers, with their pink corolla, green calyx, yellow stripes, and small stamens, produce a fine effect. The roots are gently boiled with salt and water, after having been washed and slightly peeled. They are then eaten like asparagus in the Flemish fashion, with melted butter and the yolk of eggs. They are also served up like scorzonera and endive, with white sauce; and form, in whatever way they are dressed, a tender, succulent dish, easy to digest, agreeing with the most delicate stomach. The analogy of the root with salep indicates that its effect should be excellent on all constitutions."

"The bright rose-colored flowers being very ornamental, the plant is sometimes employed as an edging for walks."[4]

1,2,3,4. The Gardener's Assistant. By Robert Thompson.

10

The Parsnip (Pastinaca sativa)

The Parsnip is a hardy biennial, indigenous to Great Britain and some parts of the south of Europe, and, to a considerable extent, naturalized in this country. In its native state, the root is small and fibrous, and possesses little of the fineness of texture, and delicacy of flavour, which characterise the Parsnip in its cultivated state.

The roots are fusiform, often much elongated, sometimes turbinate, and attain their full size during the first year. The flowers and seeds are produced the second year; the plant then measuring five or six feet in height, with a grooved or furrowed, hollow, branching stem. The flowers are yellow, in large spreading umbels five or six inches in diameter. The seeds ripen in July and August; are nearly circular; about one-fourth of an inch in diameter; flat, thin, very light, membranous on the borders, and of a pale yellowish-brown or yellowish-green color. They vary but little in size, form, or color, in the different varieties; and retain their vitality but two years. About six thousand seeds are contained in one ounce.

***Propagation, Soil, and Cultivation*—**It is always propagated from seed sown annually.

***Soil*—**The soil should be mellow, deep, and of a rich vegetable texture.

"If in moderate condition by the manuring of the previous crop, it will be better than applying manure at sowing. Should it be necessary to do so, let the manure be in the most thorough state of decomposition; or, if otherwise, incorporate it with the soil, as far from the surface as possible. The Parsnip will grow in a stronger soil than the Carrot; and succeeds comparatively well when grown in sand, or even in peat, if well manured."

Preparation of the Ground, and Sowing—

"The seed should be sown as early in spring as the ground is in good working condition. As most of the varieties have long fusiform roots, ordinary ploughing will not stir the soil to a sufficient depth for their greatest perfection; and, as the amount of the crop mainly depends on the length of the roots, it is of the first importance to provide for this fact by making the ground fine and friable above and below, to the depth of at least fifteen inches: eighteen or twenty would be better. When the soil has thus been thoroughly pulverised, level off the surface, and rake it fine and smooth, and sow the seed in drills fourteen inches apart and an inch and a half deep; allowing half an ounce of seed for one hundred feet of drill, and from five to six pounds to the acre. When the young plants are two or three inches high, thin them out to about six inches in the rows; and, as they transplant readily, any vacant space can be filled by resetting the surplus plants. Keep the earth between the rows loose, and free from weeds, and also the spaces in the rows, until the leaves cover the ground; after which, little further care will be required. The roots will attain a good size by the middle of September, from which time a few may be drawn for present use; but the Parsnip is far best at full maturity, which is indicated by the decay of the leaf in October."

Harvesting—The Parsnip sustains no injury when left in the open ground during winter; and it is a common practice to take up in the fall a certain quantity of roots to meet a limited demand in the winter months, allowing the rest to remain in the ground until spring. The roots thus treated are considered to have a finer flavor; that is to say, are better when recently taken from the ground.

In taking up the crop in autumn, which should be done just previous to the closing-up of the ground, be careful to remove the soil to a sufficient depth, so as not to injure the roots. The thrust of the spade that easily lifts a Carrot without essential injury, will, if applied to the Parsnip, break the roots of nine in ten at scarcely half their length from the surface of the ground. As the roots keep much fresher, and retain their flavor much better, when taken up entire, the best method is to throw out a trench beside the rows, to the depth of the roots, when they can be easily, as well as perfectly, removed. They should be dug in pleasant weather, and laid on the ground exposed to the sun for a few hours to dry; "and when all the earth is rubbed off them, and their leaves cut off to within an inch of their crowns, they may be stowed away in sand, dry earth, or in any dry, light material most convenient." When thus packed, they will keep well in almost any location, either in the cellar or storehouse.

If the roots which have remained in the ground during winter be taken up in spring, and the tops removed as before directed, they may be packed in sand or earth, and will remain fresh and in good condition for use until May or June.

To raise Seed—In April, thin out the roots, that have been in the ground during the winter, to about eighteen inches apart; or, at the same season, select a few good-sized and symmetrical roots from those harvested in the fall, and set them eighteen inches apart, with the crowns just below the surface of the ground. They will send up a stalk to the height and in the manner before described, and the seeds will ripen in August. The central umbel of seeds is always the largest, and is considered much the best.

Use—

> "The Parsnip is considered as a wholesome and nutritious article of food, and is served at table in various styles in connection with salted meats and fish. The roots, aside from this manner of using, form what may be called an excellent side-dish; when, after being boiled, not too soft, they are dipped in thin batter of flour and butter or the white of eggs, and afterwards fried brown."

They contain a considerable portion of sugar, and are considered more nutritive than carrots or turnips. The roots form a common ingredient in soups; and are sometimes used for making bread, and also a kind of wine said to resemble Malmsey of Madeira.

Aside from the value of the Parsnip as a table vegetable, it is one of the most economical roots for cultivation for farm purposes, as it not only produces an abundant and almost certain crop, but furnishes very nourishing food particularly adapted to and relished by dairy-stock.

Varieties—The varieties, which are not numerous, are as follow:

Common, or Dutch
(Swelling Parsnip; Long Smooth Dutch)
The leaves of this kind are strong and numerous; generally about two feet long or high. The roots are from twenty to thirty inches in length, and from three to four inches in diameter at the shoulder, regularly tapering to the end, occasionally producing a few strong fangs. The crown is short and narrow, elevated, and contracting gradually from the shoulder, which is generally below the surface of the ground.

Seeds from America, Holland, and Germany, sown in the garden of the London Horticultural Society, all proved alike; though some were superior to others in the size of their roots, owing, it was thought, both to a careful selection of seed-roots and to the age of the seeds. It was found that new seeds uniformly produced the largest roots.

Early Short-Horn
A recently introduced variety, similar to the Turnip-rooted, but shorter. Very delicate and fine-flavored.

Guernsey

(Panais Long, of the French)

The leaves of this kind grow much stronger and somewhat taller than those of the Common Parsnip. The leaflets are also broader. The only distinguishable difference in the roots is, that those of the Guernsey Parsnip are the larger and more perfect, being sometimes three feet long. Roots produced from seed obtained from Guernsey were evidently much superior to those which were grown from seed raised in other localities: from which it would appear that the Guernsey Parsnip is only an improved variety of the Common, arising from soil and cultivation in that island. Dr. M'Culloch states that, in Guernsey, its roots grow to the length of four feet. In its flavor, it differs little from the Common Dutch Parsnip.

Hollow-Crowned

(Long Jersey; Hollow-Crowned Guernsey; Hollow-Headed)

In this variety, the leaves are shorter and not so numerous as those of the Common Parsnip. The roots are oblong, about eighteen inches in length, and four inches in diameter at the shoulder, more swollen at the top, and not tapering gradually, but ending somewhat abruptly with a small tap-root. The crown is short, and quite sunk into the shoulder, so as to form a hollow ring around the insertion of the stalks of the leaves; and grows mostly below the surface of the ground.

It is a good sort for general cultivation, especially as it does not require so deep a soil as either the Common, or Guernsey. There is little difference in the flavor or general qualities of the three varieties.

Siam, or Yellow

(Panais de Siam)

This is said to be more tender and richer in flavor than any of the other varieties. It is mentioned by Dr. Neill in the "Encyclopædia Britannica," and is described by M. Noisette as being yellowish in color, and in form intermediate between the Guernsey and Turnip-rooted Parsnips. He also states that it is the most esteemed. It does not, however, appear to be known at the present day in this country.

Turnip-Rooted

Turnip-Rooted Parsnip

The leaves of this sort are few, and do not exceed twelve to sixteen inches in length. The roots are from four to six inches in diameter, tunnel-shaped, tapering very abruptly, with a strong tap-root; the whole being from twelve to fifteen inches in length. The rind is rougher than either of the other sorts; the shoulder very broad, growing above the surface of the soil; convex, with a small, short crown. It is much the earliest of the parsnips; and, if left in the ground, is liable to rot in the crown. The leaves also decay much sooner than those of most other sorts.

It is particularly adapted to hard and shallow soils; and, from its coming into use much earlier than any other kind, very desirable. In flavor, it is mild and pleasant, though less sugary than the long-rooted kinds. The flesh, when dressed, is more yellow than that of any other variety.

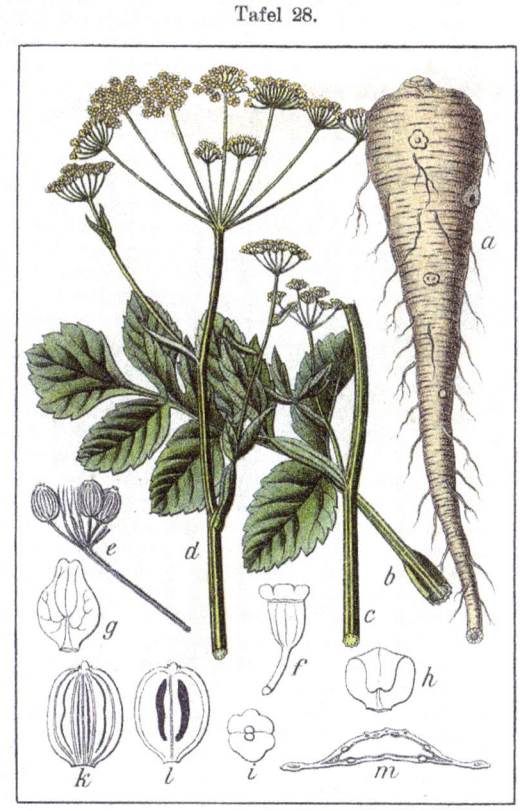

Pastinaca sativa
Jacob Sturm

11

The Potato (Solanum tuberosum)

The Potato is a native of Central or Tropical America. In its wild or natural state, as found growing on the mountains of Mexico or South America, the tubers rarely exceed an inch in diameter, and are comparatively unpalatable. During the last half-century, its cultivation within the United States has greatly increased; and it is now considered the most important of all esculent roots, and next to the cereals in value as an article of human subsistence.

Soil—The soils best suited to the Potato are of the dryer and lighter descriptions; pasture lands, or new land, with the turf freshly turned, producing the most abundant as well as the most certain crops. On land of a stiff, clayey texture, or in wet soils, they are not only extremely liable to disease, but the quality is usually very inferior. "On soils which have been long cropped and heavily manured, they rarely succeed well; and hence garden ground, in most cases, does not produce tubers of so good quality as those obtained from the fields."

Fertilizers—

"In good garden soil, the less manure that is used, the better flavored will be the produce; and it will also be much less affected by the disease. Therefore, whilst the malady prevails, or symptoms of it still remain, it is not advisable to apply much manure.

"Amongst the fertilisers that are employed, may be enumerated, in addition to barnyard and stable manure, leaves, leaf-mould, peat-charcoal, and other carbonaceous substances, lime, gypsum, or plaster, and bone-dust.

"Wood-ashes are useful in supplying potash and other inorganic substances required by the plant; and they may be advantageously applied where the soil contains a large amount of decayed vegetable matter. The same remark will also apply to lime, which is useful in destroying slugs and other vermin, which attack the tubers. Plaster, bone-dust, and superphosphate of lime, are best for humid soils. They induce earliness; and where this is an object, as it must be so long as the disease continues, they may be applied with considerable advantage."[1]

Abridged Catalogue of Burpee's Novelties and Specialties in Seeds (1890)

Propagation—

"This is almost universally from tubers; the seed being seldom sown, except for the production of new varieties. With many it is a doubtful question, whether the tubers cut, or planted whole, yield the greater return. From experiments made in the garden of the London Horticultural Society at Chiswick, it was found, on the mean of two plantations,—one made early in the season, and the other about one month later,—that the produce from cut sets exceeded that from whole tubers by nearly one ton per acre. In the latter planting, the produce from whole tubers was somewhat greater than that from single eyes: but, in the early plantation, the cut sets gave nearly two tons per acre more produce than the whole tubers; the weight of potatoes planted being deducted in every case.

"Another important consideration is, whether small tubers or large ones should be employed for making sets; for if, by using the former, an equally good crop could be obtained, a considerable saving in the expense of sets would be effected. Large tubers, however, are preferable, for the following reasons: In all plants, large buds tend to produce large shoots; and small or weak buds, the reverse. Now, the eyes of potatoes are true buds, and in small tubers they are comparatively weak: they consequently produce weak shoots, and the crop from such is inferior to that obtained from plants originating from larger tubers, furnished with stronger eyes; and this conclusion has been justified by the results of actual experiments.

"The part of the Potato employed for planting is not a matter of indifference. It was found, by an experiment made in the garden of the Horticultural Society, that sets taken from the points of the tubers, and planted early in the season, yielded at the rate of upwards of three tons per acre more produce than was obtained from employing the opposite end of the tubers. In a plantation made a month afterwards, the difference was much less, but still in favor of the point, or top end, of the Potato."[2]

With regard to the quantity of seed per acre, great diversity of opinion exists among cultivators. Much, of course, depends on the variety, as some sorts not only have more numerous eyes, but more luxuriant and stronger plants, than others. Of such varieties, a much less quantity will be required than of those of an opposite character. From a series of experiments carefully made for the purpose of ascertaining the amount of seed most profitable for an acre, it was found that from six to eight bushels, if planted in hills, answered better than more: for, when too much seed was used, there were many small tubers; and where the tubers had been divided into very small parts, or single eyes, the plants were more feeble, and the yield less in number and weight, though usually of larger size.

Methods of Planting and Cultivation—Potatoes are usually planted either in hills or ridges; the former method being the more common in this country. If planted in hills, they should be made from three feet to three and a half apart; the distance to be regulated by the habit of the variety under cultivation. If in ridges or drills, they may be made from two and a half to three feet apart; although some of the earlier and smaller kinds may be successfully grown at eighteen or twenty inches.

> "Of sets formed by the division of an average-sized tuber into four parts, three may be allowed a hill; or, if planted in drills, the sets may be placed from seven to twelve inches asunder,—the distance to be regulated by the habit or size of the plant. On light, warm land, the sets should be covered about four inches in depth; but in wet, cold soil, three inches will be sufficient.
>
> "As soon as the plants are fairly above the surface, hoeing and surface-stirring should be commenced. The earth should gradually be drawn about the hills, or along the ridges, at each successive hoeing, and every encouragement given to the side-roots to extend themselves: for nearly at their extremities the tubers are formed; so that deeply stirring the ground between the hills or ridges tends to their extension. This latter treatment, however, must not be carried beyond a certain stage in the growth of the plant, or after the tubers have reached a considerable size, as the extremities of the roots might be seriously injured. Some varieties of potatoes produce their tubers at a much greater distance from the stem than others. These are chiefly to be found among the later sorts. Most of the early kinds produce theirs close to the stem, or at the extremity of very short runners; seldom more than nine inches from the stalk of the plant."

Forcing—This should be commenced from three to four weeks before the season for planting in the open ground. The earliest varieties should be chosen for the purpose, selecting whole tubers of medium size, and placing them close together, in a single layer, among half-decayed leaves or very light loam, on the surface of a moderate hot-bed.

> "When the shoots have attained the height of two or three inches, and the weather has become sufficiently mild, they should be carefully taken out, and divided into sets; in the process of cutting up the tubers, avoiding as much as possible doing injury to the small fibrous roots, and also to the growing shoots. These sets should then be planted out in hills or drills, in the usual manner and at the usual depth; if possible, leaving the upper portion of the young shoot just above the surface of the ground. Some care is requisite in planting out the sets, particularly in covering; for, if the soil is applied too rudely, the sprouts, which separate very easily from the tubers, are exceedingly liable to be broken off, and the set destroyed for early use. If severe cold or frosty weather occurs, the plants should be protected by straw, or any convenient, light material, placed along the drills or on the hills."

Taking the Crop, and Method of Preservation—

"The early varieties should be dug for use as they attain a suitable size; which, in warm exposure, will be about the beginning of July; and thence till the middle of August, in less favorable places. The practice of partially removing the soil from about the roots, and gathering the largest tubers, leaving the smaller ones, with the expectation that they will attain a larger size, is a mode of proceeding which seldom realizes the hopes of the cultivator; for the Potato, if once disturbed at the roots, seldom recovers the check.

"When no apprehension is felt on account of disease, a week's delay in commencing on the crop will be found of great importance both to the bulk and quality; for just previous to the decay of the tops, if pleasant weather prevails and the ground is sufficiently moist, the tubers increase in size with great rapidity.

"Late varieties usually constitute the great portion of the main crop, and are those which require most care in taking up and storing. So long as the plants continue green, the Potato should be allowed to remain in the ground; as this is quite indicative that the tubers have not arrived at full maturity."

In the preservation of potatoes, it is of the first importance that they be excluded from light. If this is neglected, they become not only injurious, but actually poisonous; and this is especially the fact when they are allowed to become of a green color, which they readily will do on exposure to the light. In a state of complete darkness they should therefore be placed, the day they are taken out of the ground; and it were even better that they were stored in rather a damp state, than that they should be exposed for a day to the light with a view to dry them. Drying has a bad effect on the skin of the Potato; for, if subjected to this, the skin and part of the epidermis are made to part with their natural juices, which ever afterwards renders them incapable of absorbing moisture, even if presented to them. Fermentation is also an important evil to be guarded against, as it changes the whole substance of the Potato, and, so far as seed potatoes are concerned, destroys their vegetative principle. As security against this, they should be stored either in barrels or boxes, or in long, narrow ridges, with partitions of earth between. Potatoes once dried should never be again moistened until just before using.

"Keeping potatoes has the effect of diminishing the quantity of starch contained in them. According to Mr. Johnson, those which in October yielded readily seventeen per cent of starch, gave, in the following April, only fourteen and a half per cent. The effect of frost is also to lessen the quantity of starch. It acts chiefly upon the vascular and albuminous part; but it also converts a portion of the starch into sugar: hence the sweetish taste of frosted potatoes."[3]

Varieties—Messrs. Peter Lawson and Sons[4] describe one hundred and seventy-five varieties: and other foreign authors enumerate upwards of five hundred, describing the habit of the plant; size, form, and color of the tubers; quality and general excellence; and comparative value for cultivation.

They are obtained from seeds; the latter being quite small, flat, and lens-shaped. One hundred and five thousand are contained in an ounce, and they retain their germinative properties three years.

The process is as follows:

> "Select some of the largest and best berries, or balls, when fully ripe, which is denoted by the withering of the stalk; and separate the seeds from the pulp, and dry them thoroughly in the sun. These should be sown in the following spring, and the produce taken up in October. The tubers will then have nearly attained the size of small plums. The best of these should be selected, and the product of each plant carefully and separately preserved. In the month of April following, they should be planted at a distance from one another of from fifteen to eighteen inches; and, when they rise about two inches from the ground, they should be earthed up slightly with the hoe,—an operation which may be repeated during the season. When they have arrived at maturity, they are to be taken up, keeping the product of each stalk by itself; which product is again to be planted the ensuing spring. A judgment of the properties of the varieties will then have been formed, and those are to be reserved for cultivation which are approved of. It will be found, that, whatever had been the character of the parent stock, the seeds will produce numerous varieties, some white, some dark, in color, with tubers of different forms, round, oblong, and kidney-shaped, and varying greatly in the dryness, color, and farinaceous character, of the flesh."[5]

Ash-Leaved Early

Stem nearly two feet in height, erect, with long, smooth, shining, and drooping foliage; flowers very seldom produced; tubers white, roundish, rough-skinned; flesh white, of medium quality. The variety is healthy, and remarkably early; well suited to open culture, but not adapted for growing under glass, on account of its tall habit.

Ash-Leaved Kidney

One of the earliest of the garden varieties, well adapted for forcing under glass or for starting in a hot-bed, and subsequent cultivation in the open ground. The plant is of spreading habit, and about eighteen inches in height; leaves small, recurved; tubers of medium size, kidney-shaped, white; flesh white, dry, and well flavored. Very healthy. Introduced.

Good Seeds at Fair Prices (1898), Northrup, King & Co. Seed Growers

Biscuit

Plant two feet and a half high, spreading; leaves rather rough, large, and of a pale-green color; flowers whitish; tubers rather small, round, smooth, and of a light-brownish color. A very healthy variety, mealy, well flavored, and quite productive. The plants do not decay, nor do the tubers attain full maturity, until nearly the close of the season: the latter are, however, of good quality, and in perfection for the table soon after being harvested.

Black Chenango

(Black Mercer)

Plant vigorous, and generally of healthy habit; tubers nearly of the form of the Lady's Finger, but of larger size; skin very deep purple, or nearly black; flesh purple, both in its crude state and when cooked; quality good, usually dry, and of good flavor.

The Black Chenango is moderately productive, and withstands disease better than almost any other potato; but its dark color is objectionable. Compared with many of the recent varieties, it has little merit, and is not a profitable sort for extensive cultivation.

Buckeye

A Western variety; grown also to a considerable extent in some parts of the Middle States. "It is a handsome, round potato; white throughout, except a little bright pink at the bottom of the eye. It is very early,—ripening as early as the Chenango; attains a good marketable size as soon as the Dykeman; cooks very dry and light; and is fine flavored, particularly when first matured. It throws up a very thick, vigorous, and luxuriant vine; grows compactly in the hill, and to a large size, yielding abundantly."

For planting for early use, it is a promising variety: but for a late or medium crop, upon strong, rich ground, it is said to grow so rapidly, and to so great a size, that many of the tubers are liable to be hollow-hearted; which considerably impairs their value for table use.

Calico

Similar to the Pink-eyed; varying little except in color, which is mostly red, with occasional spots and splashes of white. It is in no respect superior to the last-named variety in quality, and cannot be considered of much value for agricultural purposes or for the table.

California Red

A bright-red potato from California. Tubers variable in form, from long to nearly round, rather smooth; eyes slightly depressed.

It is one of the most productive of all the varieties; but, on account of its extreme liability to disease, cannot be recommended for general cultivation.

Carter

A medium-sized, roundish, flattened, white potato, once esteemed the finest of all varieties, but at present nearly or quite superseded by the Jackson White, of which it is supposed to be the parent. Eyes rather numerous, and deeply sunk; flesh very white, remarkably dry, farinaceous, and well flavored. Originated about thirty years ago, in Berkshire County, Massachusetts, by Mr. John Carter.

Churchill

A variety said to have originated in Maine, and often sold in the market for the "State of Maine;" which it somewhat resembles in size, form, and color. Flesh yellow. Not a desirable sort. It is much inferior to the "State of Maine;" and, in many places, the latter variety has been condemned in consequence of the Churchill having been ignorantly cultivated in its stead.

Cristy

An early sort, of good quality, but rather unproductive. Shape somewhat long, though often nearly round; color white and purple, striped, and blended together. It is of no value as an agricultural variety; and, for table use, cannot be considered superior to many other varieties equally healthy and more prolific.

Cups

Introduced. Plant upright, stocky, surviving till frost; flowers pale purple; tubers pink or reddish, large, oblong, often irregular; flesh dry and farinaceous. Very healthy and productive, but better suited for agricultural purposes than for the table.

Danvers Seedling

(Danvers Red)

Plant healthy and vigorous. The large, full-grown tubers are long; and the smaller, undeveloped ones, nearly round. Color light red, with faint streaks of white; eyes moderately sunk; quality fair.

This variety originated in Danvers, Essex County, Mass.; and, when first introduced, was not only of good size and quality, but remarkably productive. It has, however, much deteriorated; and is now, both as respects quality and yield, scarcely above an average. At one period, it had the reputation of being one of the best varieties for keeping, and of entirely withstanding the attacks of the potato disease.

Davis's Seedling

This variety originated in the town of Sterling, Mass.; and was early disseminated through the influence of the Massachusetts Horticultural Society, at whose exhibitions it attracted much attention on account of its size and beauty. For general cultivation, it is probably one of the most profitable sorts known, as it yields abundantly, even with ordinary attention. Under a high state of cultivation, seven hills have produced a bushel of potatoes.

The tubers are of good size, red, nearly round, though sometimes more or less flattened. Eyes deeply sunk, and not very numerous; flesh nearly white, slightly tinged with pink beneath the skin when cooked; quality good, being dry, farinaceous, and well flavored. It requires the full season for its complete perfection, and resists disease better than most varieties. As a winter potato, or for extensive cultivation for market, it is one of the best of all varieties; and commends itself to the farmer, both as respects quality and yield, as being greatly superior to the Peach-blow, Pink-eye, Vermont White, and many similar varieties, which so abound in city markets.

Dykeman.

Plant of medium strength and vigor, rarely producing seed or blossoms; tubers large, roundish, often oblong; color white, clouded at the stem-end and about the eyes (which are moderately sunk and rather numerous) with purple; flesh white, or yellowish-white, its quality greatly affected by season, and the soil in which the variety may be cultivated. In certain descriptions of rather strong, clayey land, the yield is often remarkably great, and the quality much above medium. In such land, if warm and sheltered, the tubers attain a very large size quite early in the season, and find a ready sale in the market at greatly remunerative prices. Under other conditions, it frequently proves small, waxy, and inferior in quality, and profitless to the cultivator. Notwithstanding these defects, its size, earliness, and productiveness render it worthy of trial.

Early Blue

Tubers of medium size, roundish, of a bright purple or bluish color; eyes moderately deep; flesh, when cooked, white, or yellowish-white, mealy, and well flavored.

This old and familiar variety is one of the earliest of the garden potatoes, of fine quality, and one of the best for forcing for early crops. It retains its freshness and flavor till late in the spring; is of comparatively healthy habit; and, though but moderately productive, is worthy more general cultivation.

Early Cockney

Plant of medium strength and vigor, recumbent, rarely blossoming, and usually ripening and decaying early in the season, or before the occurrence of frost; tubers white, large, roundish, rough; flesh yellowish-white, or nearly white, dry, farinaceous, and of good flavor; hardy, moderately productive, and recommended as a desirable intermediate variety for the garden or for field culture. Introduced.

Early Manly

Plant medium or small, rarely blossoming, and decaying early in the season; tubers of medium size, white, roundish; flesh yellowish-white, dry, mealy, and mild flavored. It yields well, and is a good variety for early garden culture. Introduced.

A.W. Livingston's Sons, True Blue Seeds (1896)

Flour-Ball

Plant reclining, of rather slender habit, rarely blossoming; tubers of medium size, white, round, the skin quite rough or netted; flesh white, dry, farinaceous, and mild flavored. It yields abundantly, and is a good sort for the garden; but would prove less profitable for growing for the market than many other varieties of larger size.

Fluke Kidney

Plant vigorous, with luxuriant, deep-green foliage; continuing its growth till late in the season, or until destroyed by frost. The tuber is remarkable for its singular shape, of a flattened oval, frequently measuring eight or nine inches in length by nearly three inches in width. The peel is thin, and remarkably free from eyes; the surface, very smooth and even; the flesh is very dry, mealy, and farinaceous, exceedingly well flavored, and, in general excellence, surpassed by few, if any, of the late varieties. It is also healthy, hardy, and very productive; but is much better towards spring than when used soon after being harvested.

The variety originated near Manchester, England, about the year 1844; and appears to be a cross or hybrid between the Lapstone Kidney and Pink-eye.

In this country, the variety has never reached the degree of excellence it appears to have attained in England. With us the yield has been small, and it has suffered greatly from disease. The flesh is also yellow when cooked, and quite strong flavored. Not recommended for cultivation.

Forty-Fold

An English variety. Plant healthy, ripening about the middle of September, rarely producing seed or blossoms; tubers white, of medium size, round; skin rough or netted; flesh white, comparatively dry, and well flavored. It yields abundantly; is a good kind for forcing; and, though the plants remain green until frost, the tubers attain a suitable size for use quite early in the season. An English sort, known as Taylor's Forty-fold, is quite distinct; the tubers being oval, much flattened, and of a reddish color.

Garnet Chili

Stem not long or tall, rather erect, sturdy, and branching; flowers abundant, pale purplish-white, and usually abortive; tubers red, or garnet-colored, very large, roundish, and comparatively smooth and regular; flesh white, dry, mealy, and, the size of the tuber considered, remarkably well flavored. The variety is healthy, yields abundantly, is greatly superior to the Peach-blow and kindred sorts for table use, and might be profitably grown for farm-purposes. The plants survive till destroyed by frost.

Gillyflower

Tubers large, oval, or oblong, flattened, white, and comparatively smooth; flesh white, dry, and of fair quality. The plants are healthy, and the variety is very productive: but it is inferior to many others for table use; though its uniform good size, and its fair form, and whiteness, make it attractive and salable in the market. It is similar to, if not identical with, the St. Helena and the Laplander.

Green-Top

Plant strong and vigorous; flowers dull white, generally abortive; tubers quite large, white, roundish, often irregular; eyes deep-set; flesh white, comparatively dry, and well flavored. The variety is productive, and of healthy, hardy habit; not early; the plants continuing green till destroyed by frost. Introduced.

Hill's Early

An old variety, very little, if at all, earlier than the White Chenango. Quality not much above mediocrity; its chief recommendation being its earliness. Skin and flesh yellowish-white; eyes rather deeply sunk; size medium; form roundish; moderately productive. It does not ordinarily cook dry and mealy; and, though desirable as an early potato for a limited space in the garden, cannot be recommended for general cultivation.

Irish Cups

Tubers nearly round, yellowish-white; eyes deep-set; flesh yellow, and strong flavored when cooked. Unfit for table use.

Aside from the difference in form, the variety somewhat resembles the Rohan.

Jackson White

This comparatively new but very excellent variety originated in Maine; and is supposed to be a seedling from the celebrated Carter, which it much resembles. Tubers yellowish-white, varying in size from medium to large; form somewhat irregular, but generally roundish, though sometimes oblong and a little flattened; eyes rather numerous, and deeply sunk; flesh perfectly white when cooked, remarkably dry, mealy, farinaceous, and well flavored.

The variety unquestionably attains its greatest perfection when grown in Maine, or the northern sections of Vermont and New Hampshire; but is nevertheless of good quality when raised in the warmer localities of New England and the Middle States. It is earlier than the Davis Seedling; comparatively free from disease; a good keeper; commands the highest market-price; and, every thing considered, must be classed as one of the best, and recommended for general cultivation.

The plants are very erect, the flowers nearly white; and the balls, or berries, are produced in remarkable abundance.

Rhode-Island Seedling

A variety of comparatively recent introduction. Plant very strong and vigorous; tubers of extraordinary size when grown in strong soils, long and somewhat irregular in form, thickly set on the surface with small knobs, or protuberances, above which the eyes are placed in rather deep basins, or depressions; color red and white intermixed, in some specimens mostly red, while in others white is the prevailing color; flesh yellow when cooked, and quite coarse, but esteemed by many as of good quality for table use.

One of the largest of all the varieties, remarkably productive, quite free from disease, keeps well, and, as an agricultural potato, rivals the Rohan. Requires the full season. It sports more than any potato; being exceedingly variable in size, form, and color.

Lady's Finger

(Ruffort Kidney)

Stem from one foot and a half to two feet high, of straggling habit of growth; leaves smooth, and of a light-green color; blossoms rarely if ever produced; tubers white, smooth, long, and slender, and of nearly the same diameter throughout; eyes very numerous, and slightly depressed.

A very old variety, of pretty appearance, long cultivated, and much esteemed as a baking potato; its peculiar form being remarkably well adapted for the purpose. It is, however, very liable to disease; and as many of the recently introduced seedlings are quite as good for baking, as well as far more hardy and productive, it cannot now be considered as a variety to be recommended for general culture.

Lapstone Kidney

(Nichol's Early)

A variety of English origin. M'Intosh[6] describes it as being "decidedly the best kidney potato grown, and an excellent cropper. Tubers sometimes seven inches in length, and three inches in breadth. It is longer in coming through the ground in spring than most other varieties, and the stems at first appear weakly; but they soon lose this appearance, and grow most vigorously. It is a first-rate potato in August and September; and will keep in excellent condition till May following, without losing either its mealiness or flavor."

Long Red

Form long, often somewhat flattened,—its general appearance being not unlike that of the Jenny Lind, though of smaller size; color red; flesh marbled or clouded with red while crude, but, when cooked, becoming nearly white. The stem-end is often soggy, and unfit for use; and the numerous prongs and knobs which are often put forth on the sides of the tubers greatly impair their value for the table.

A few years since, this variety was exceedingly abundant in the market, and was esteemed one of the best sorts for use late in spring and early in summer. It was also remarkably healthy and very productive, and was considered one of the most valuable kinds for general cultivation. It has somewhat improved in quality by age, although not now to be classed as a potato of first quality. The Jenny Lind and other varieties are now rapidly superseding it in most localities.

Mexican

A very handsome white variety, long and smooth, like the St. Helena, but not quite so large; eyes very slightly depressed. It is of poor quality, quite unproductive, rots badly, and not worthy of cultivation.

Nova-Scotia Blue

This old variety, at one period, was very extensively cultivated, and for many years was considered the most profitable of all the sorts for raising for market or for family use. Form nearly round, the larger specimens often somewhat flattened; color light blue; eyes moderately depressed; flesh white, dry, and good. It yields abundantly; but, in consequence of its great liability to disease, its cultivation is now nearly abandoned.

Old Kidney

Tubers kidney-shaped, white; flesh yellow, rather waxy, and of indifferent flavor.
It is neither very productive, nor very valuable in other respects; and it is now little cultivated.

Peach-Blow

Tubers similar in form to the Davis Seedling, but rather more smooth and regular; color red, the eyes not deeply sunk; flesh yellow when cooked, dry and mealy, but only of medium quality, on account of its comparatively strong flavor.

It is hardy and quite productive; keeps well; and is extensively cultivated for market in the northern parts of New England and the State of New York, as well as in the Canadas. It is common to the markets of most of the large seaport cities; and, during the winter and spring, is shipped in large quantities to the interior and more southern sections of the United States. The Davis Seedling—which is quite as productive, and much superior in quality for table use—might be profitably grown as a substitute.

Pink-Eyed

Tubers nearly round; eyes rather large and deep; color mostly white, with spots and splashes of pink, particularly about the eyes; flesh yellow.

The Pink-eyed is an old but inferior variety, hardly superior in quality to the Vermont White. Though quite productive, it is generally esteemed unworthy of cultivation.

Poggy, or Porgee

(Cow-Horn)

A dark-colored variety, extensively cultivated in the British Provinces, particularly in Nova Scotia; and, during the autumn, imported in considerable quantities into the principal seaports of the United States. It is of excellent quality, and by some preferred to all others, especially for baking; for which purpose, on account of its size and remarkable form, it seems peculiarly adapted.

It is moderately productive, and succeeds well if seed is procured every year or two from the East; but, if otherwise, it soon deteriorates, even under good cultivation.

Size above medium; form long, broadest, and somewhat flattened, at the stem-end, and tapering towards the opposite extremity, which is often more or less sharply pointed. It is also frequently bent, or curved; whence the name "Cow-horn," in some localities. Skin smooth; eyes not depressed; color dark-blue outside, white within when cooked. Not very hardy; requiring a full season for its complete perfection. Unless where well known, its color is objectionable; and it is generally less salable than the white-skinned varieties.

Quarry

A large, white, roundish, English potato, not unlike the variety universally known and cultivated many years since in this country as the Orange Potato. Plant vigorous, and of strong, stocky habit; flowers purple, generally abortive; flesh yellowish-white, of fair quality for table use. A hardy, very productive sort, which might be profitably grown for marketing and for agricultural purposes. The plants survive till frost. Not early.[Pg 70]

Rohan

Tubers very large, in form much resembling the Jenny Lind,—the full-developed specimens being long, and the smaller or immature tubers nearly round; eyes numerous and deep-set; color yellowish-white, with clouds or patches of pink or rose; flesh greenish-white when cooked, yellowish, watery, and strong flavored. The plant is strong and vigorous, and continues its growth till destroyed by frost. The flowers are generally abortive.

Mr. Hyde describes it as a variety famous in history, but infamous as a table potato, and fit only for stock. It formerly gave an immense yield, but now produces only moderate crops; and its cultivation is nearly abandoned.

Shaw's Early

An English variety, much employed for forcing, and extensively cultivated in the vicinity of London for early marketing. It is, for an early sort, a large, beautiful, oblong, white-skinned potato. Its only fault is its hollow eyes. It is very productive.

State of Maine

This variety, as implied by its name, is of Maine origin, and was introduced to general notice six or seven years ago. In form, the tubers are similar to the White Chenango, being long, smooth, and somewhat flattened; though the smaller and undeveloped bulbs are often nearly round. Eyes almost even with the surface, and quite numerous; color white, like the Jackson White. When cooked, the flesh is white, very dry, mealy, and of good flavor.

It is quite early, but more liable to disease than the Davis Seedling and some other varieties. In Maine it is grown in great perfection, nearly equalling the Jackson White and Carter as a table potato. On light soil, it is only moderately productive; but on strong land, in high cultivation, yields abundantly.

St. Helena
(Laplander)

An old and very productive variety. Plant erect, and of a bushy habit, about two feet and a half in height; foliage light green; flowers pale reddish-purple. The tubers are of an oblong form, and remarkably large; specimens having been produced measuring ten inches in length. Eyes numerous, but not deeply set; skin white and smooth; flesh white when cooked, mealy, and of fair quality. It is a very healthy variety, and not easily affected by disease; but belongs to that class of late field potatoes, the foliage of which does not in ordinary seasons decay until injured by frost, and the tubers of which generally require to be kept some time before they are fit for using to the greatest advantage.

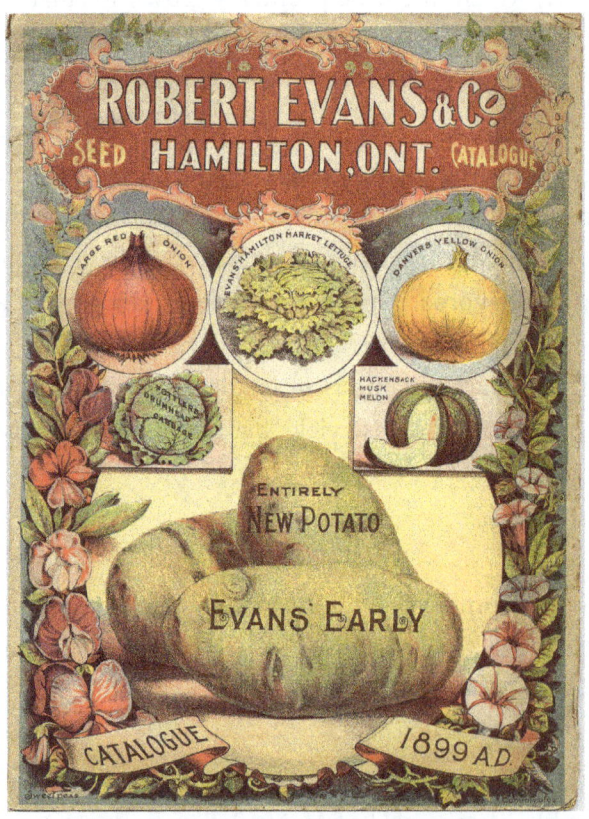

Seed Catalogue (1899), Robert Evans & Co.

Taylor's Forty-Fold
(Forty-Fold)

Plant about one foot and a half high, slender, and spreading in habit; foliage light green; flowers very rarely produced; tubers oval, much flattened, and of medium size; skin rough, and of a dull, reddish color. This variety is very dry and starchy, well flavored, and suffers comparatively little from disease. It is also very productive, and a good early sort for the garden; but not well adapted for field culture, or for cultivation for agricultural purposes.

Tolon

Plant quite low and dwarf, decaying with the season; flowers lilac-purple, large and handsome, generally abortive; tubers of medium size, roundish, of a pink or reddish color; flesh yellow, dry, but not of so mild a flavor as many of the more recent kinds. Moderately productive. Introduced.

Vermont White

A very fair and good-sized but poor variety, grown to a considerable extent in the northern and more interior portions of New England. Color white outside; but the flesh, when cooked, is yellow, soft, not dry, and strong flavored. It is a strong grower, and very productive, but rots badly. It commands only a low price in the market, on account of its very inferior quality; and cannot be recommended for general cultivation.

Veto, or Abington Blue

Tubers long, resembling in form those of the Long Red, and, like that variety, often watery at the stem-end after being cooked; color blue or purplish; flesh white; quality fair as a table potato.

This variety originally was remarkably productive, and at one period was in very general cultivation; but now is rarely planted, as it is extremely liable to disease, and rots badly.

White Chenango

(Chenango; Mercer, of New York)

An old and familiar variety; at one period almost everywhere known, and generally acknowledged as the best of all varieties. As a potato for early planting, whether for family use or for the market, it was a general favorite; but, within a few years past, it has not only greatly deteriorated in quality and productiveness, but has been peculiarly liable to disease and premature decay of the plants. When well grown, the tubers are of good size, rather long, slightly flattened, and comparatively smooth; eyes slightly sunk; color white, with blotches of purple,—before cooking, somewhat purple under the skin; flesh, when cooked, often stained with pale purple; in its crude state, zoned with bright purple. Quality good; dry, mealy, and well flavoured.

The variety is considerably affected by the soil in which it may be cultivated; in some localities, being much more colored than in others. It is now rapidly giving place to new seedling varieties, quite as good in quality, and more healthy and productive.

White Cups

Tubers long and flattened, somewhat irregular; eyes deeply sunk; skin yellowish; flesh white.

It is a very handsome variety, of Maine origin, but is only moderately productive. It is also of ordinary quality, rots easily, and will probably never become popular.

White Mountain

Tubers large, long, white, smooth, uniformly fair and perfect. Appears to be nearly identical with the St. Helena and Laplander. It is very productive, and a good agricultural variety; but, for table use, can be considered only of second quality.

Worcester Seedling

(Dover; Riley)

Tubers of a pinkish-white color, and similar in form to the Jackson White. Eyes deep-set; flesh white, more so than that of the Davis Seedling. It keeps well, and is an excellent variety for cultivation for family use, but less profitable than many others for the market. Stalks upright; blossoms pinkish, but not abundant.

In quality, this comparatively old and well-known variety is nearly or quite equal to the Carter; and, besides, is much more productive. As a garden potato, it deserves general cultivation. Requires the full season.

1,2. The Gardener's Assistant. By Robert Thompson.

3. The Book of the Garden. By Charles M'Intosh. 2 vols. Edinburgh and London, 1855.

4. The Agriculturist's Manual. By Peter Lawson and Son. Edinburgh, 1836.

5. The Elements of Practical Agriculture. By David Low. London, 1843.

6. The Book of the Garden. By Charles M'Intosh. 2 vols. Edinburgh and London, 1855.

12

The Radish (Raphanus sativus)

The Radish is a hardy annual plant, originally from China. The roots vary greatly in form; some being round or ovoid, some turbinate, and others fusiform, or long, slender, and tapering. When in flower, the plant rises from three to four feet in height, with an erect, smooth, and branching stem. The flowers are quite large, and, in the different kinds, vary in color from clear white to various shades of purple. The seed-pods are long, smooth, somewhat vesiculate, and terminate in a short spur, or beak. The seeds are round, often irregularly flattened or compressed: those of the smaller or spring and summer varieties being of a grayish-red color; and those of the winter or larger-rooted sorts, of a yellowish-red. An ounce contains from three thousand three hundred to three thousand six hundred seeds, and they retain their vitality five years.

Soil, Propagation, and Cultivation—All the varieties thrive best in a light, rich, sandy loam; dry for early spring sowings, moister for the summer.

Like all annuals, the Radish is propagated by seeds, which may be sown either broadcast or in drills; but the latter method is preferable, as allowing the roots to be drawn regularly, with less waste. For the spindle-rooted kinds, mark out the drills half an inch deep, and five or six inches apart; for the small, turnip-rooted kinds, three-quarters of an inch deep, and six inches asunder. As the plants advance in growth, thin them so as to leave the spindle-rooted an inch apart, and the larger-growing sorts proportionally farther.

For raising early Radishes without a Hot-bed—Sow in the open ground the last of March or early in April, arch the bed over with hoops or pliant rods, and cover constantly at night and during cold days with garden-matting. In moderate days, turn up the covering at the side next the sun; and, if the weather is very fine and mild, remove it entirely.

Open Culture—Sow in spring as soon as the ground can be worked. If space is limited, radishes may be sown with onions or lettuce. When grown with the former, they are said to be less affected by the maggot. For a succession, a small sowing should be made each fortnight until midsummer, as the early-sown plants are liable to become rank, and unfit for use, as they increase in size.

Radishes usually suffer from the drought and heat incident to the summer; and, when grown at this season, are generally fibrous and very pungent. To secure the requisite shade and moisture, they are sometimes sown in beds of asparagus, that the branching stems may afford shade for the young radishes, and render them more crisp and tender. A good criterion by which to judge of the quality of a Radish is to break it asunder by bending it at right angles. If the parts divide squarely and freely, it is fit for use.

Production and Quantity of Seed—To raise seed of the spring or summer Radishes, the best method is to transplant; which should be done in May, as the roots are then in their greatest perfection. Take them up in moist weather; select plants with the shortest tops and the smoothest and best-formed roots; and set them, apart from all other varieties, in rows two feet and a half distant, inserting each root wholly into the ground, down to the leaves. With proper watering, they will soon strike, and shoot up in branching stalks, producing abundance of seeds, ripening in autumn.

One ounce and a half of seed will sow a bed five feet in width and twelve feet in length. Ten pounds are required for seeding an acre.

The excellence of a Radish consists in its being succulent, mild, crisp, and tender; but, as these qualities are secured only by rapid growth, the plants should be frequently and copiously watered in dry weather. The varieties are divided into two classes; viz., Spring or Summer, and Autumn or Winter, Radishes.

Spring or Summer Radishes

These varieties are all comparatively hardy, and may be sown in the open ground as early in spring as the soil is in good working condition. The earliest spring Radishes are grown as follows:

"In January, February, or March, make a hot-bed three feet and a half wide, and of a length proportionate to the supply required. Put upon the surface of the dung six inches of well-pulverized earth; sow the seeds broadcast, or in drills five inches apart; and cover half an inch deep with fine mould. When the plants have come up, admit the air every day in mild or tolerably good weather by tilting the upper end of the light, or sometimes the front, one, two, or three inches high, that the Radishes may not draw up long, pale, and weak. If they have risen very thick, thin them, while young, to about one inch apart. Be careful to cover the sashes at night with garden mats, woollen carpeting, or like material. Water with tepid water, at noon, on sunny days. If the heat of the bed declines much, apply a moderate lining of warm dung or stable-litter to the sides, which, by gently renewing the heat, will soon forward the Radishes for pulling. Remember, as they advance in growth, to give more copious admissions of air daily, either by lifting the lights in front several inches, or, in fine, mild days, by drawing the lights mostly off; but be careful to draw them on early, before the sun has much declined and the air become cool."

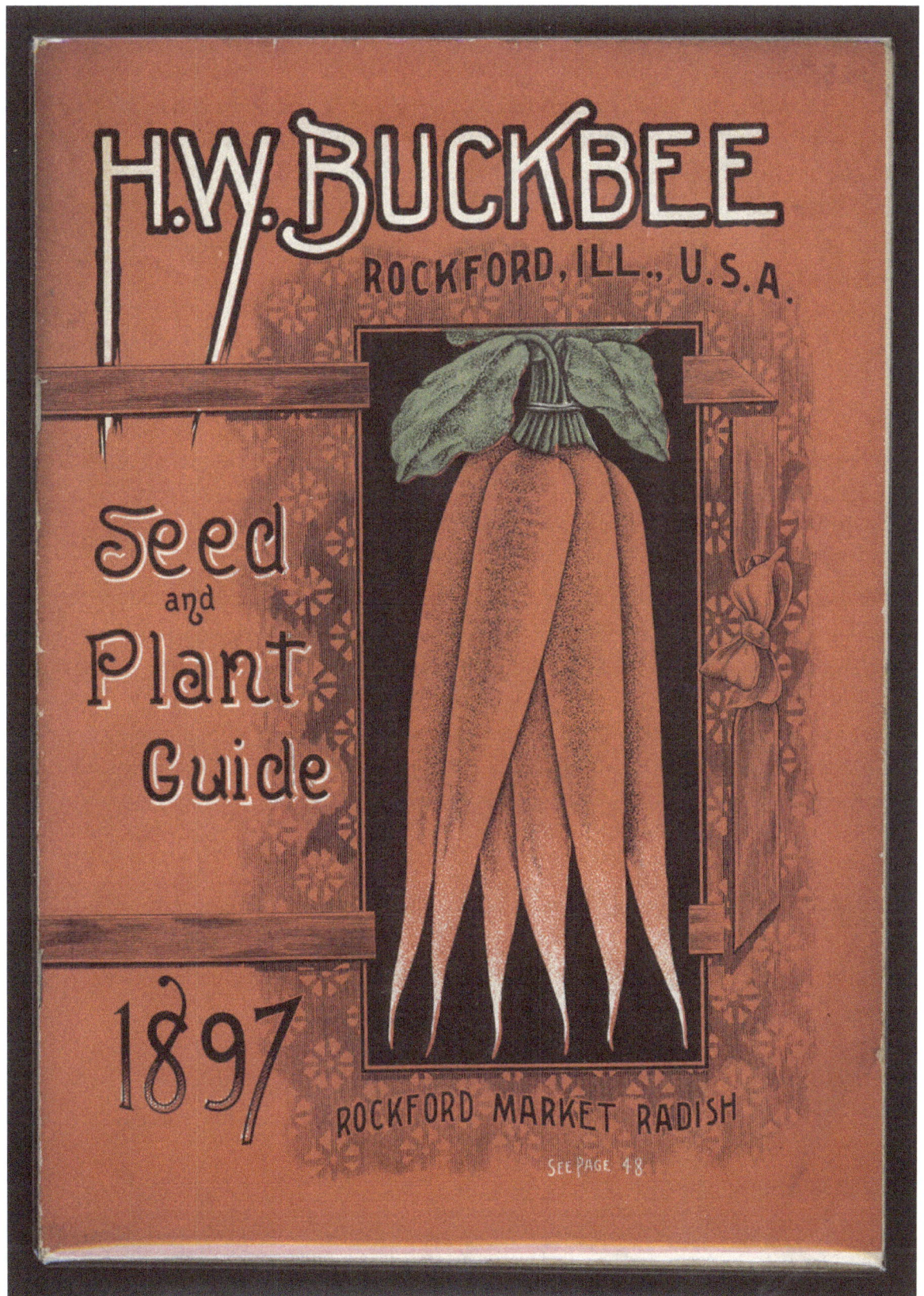

Seed and Plant Guide (1897), H.W. Buckbee

Early Black

(Noir Hatif)

Bulb nearly spherical, slightly elongated or tapering, nearly of the size and form of the Gray Turnip-rooted; skin dull black, rough, and wrinkled; flesh white, solid, crisp, and piquant; leaves of the size of those of the Gray Turnip-rooted. Season intermediate between that of the last named and the Black Spanish.

Long Purple

(Rave Violette Hative)

A sub-variety of the Long Purple, earlier and of smaller size.

Early Purple Turnip-Rooted

A few days earlier than the Scarlet Turnip-rooted. Size, form, and flavor nearly the same.

Early Scarlet Turnip-Rooted

(Rond Rose Hatif)

Bulb spherical, or a little flattened,—often bursting or cracking longitudinally before attaining its full dimensions; skin deep scarlet; flesh rose-colored, crisp, mild, and pleasant; neck small; leaves few in number, and of smaller size than those of the common Scarlet Turnip-rooted. Season quite early,—two or three days in advance of the last named.

As a variety for forcing, it is considered one of the best; but the small size of the leaves renders it inconvenient for bunching, and it is consequently less cultivated for the market than many other sorts.

Extensively grown in the vicinity of Paris.

Early White Turnip-Rooted

(Rond Blanc Hatif)

Skin and flesh white; form similar to that of the Scarlet Turnip-rooted. It is, however, of smaller size, and somewhat earlier. An excellent sort, and much cultivated.

Gray Olive-Shaped

Form similar to the Scarlet Olive-shaped. Skin gray; flesh white, crisp, and well flavoured.

Gray Turnip-Rooted

(Gray Summer; Round Brown)

The form of this variety may be called round, though it is somewhat irregular in shape. It grows large, and often becomes hollow. It should, therefore, be used while young, or when not more than an inch or an inch and a half in diameter. The outside coat is mottled with greenish-brown, wrinkled, and often marked with transverse white lines. The flesh is mild, not so solid as that of many varieties, and of a greenish-white color. The leaves are similar to those of the Yellow Turnip-rooted, growing long and upright, with green footstalks. Half early, and a good variety for summer use.

Long Purple

Root long, a large portion growing above ground; skin deep purple; flesh white, and of good flavor.

The seed-leaves, which are quite large, are used as a small salad. The variety is early, and good for forcing. When the green tops are required for salading, the seeds should be sown in drills, as mustard or cress.

Long Salmon

(Long Scarlet Salmon)

This variety has been considered synonymous with the Long Scarlet; but it is really a distinct sort. The neck of the root rises about an inch above the ground, like that of the Scarlet, but it is of a paler red; and this color gradually becomes lighter towards the middle, where it is a pale-pink or salmon color. From the middle, the color grows paler downwards, and the extremity of the root is almost white. In shape and size, this Radish differs nothing from the Scarlet; nor does it appear to be earlier, or to possess any qualities superior to the Scarlet Radish, the beauty of which, when well grown, exceeds that of any other Long Radish.

Long Scarlet

(Early Scarlet Short-top; Early Frame)

Root long, a considerable portion growing above the surface of the ground,—outside, of a beautiful, deep-pink color, becoming paler towards the lower extremity; flesh white, transparent, crisp, and of good flavor, having less pungency than that of the Scarlet Turnip; leaves small, but larger than those of the last-named variety.

When of suitable size for use, the root measures seven or eight inches in length, and five-eighths or three-fourths of an inch in diameter at its largest part.

The Long Scarlet Radish, with its sub-varieties, is more generally cultivated for market in the Eastern, Middle, and Western States, than any other, or perhaps even more than all other sorts. It is very extensively grown about London, and is everywhere prized, not only for its fine qualities, but for its rich, bright color. It is also one of the hardiest of the Radishes; and is raised readily in any common frame, if planted as early as February.

Olive-Shaped Scarlet

(Oblong Rose-Coloured)

Bulb an inch and a half deep, three-fourths of an inch in diameter, oblong, somewhat in the form of an olive, terminating in a very slim tap-root; skin fine scarlet; neck small; leaves not very numerous, and of small size; flesh rose-colored, tender, and excellent. Early, and well adapted for forcing and for the general crop.

Purple Turnip-Rooted

This is a variety of the Scarlet Turnip-rooted; the size, form, color, and quality being nearly the same. The skin is purple. It is considered a few days earlier than the last named.

Scarlet Turnip-Rooted

(Crimson Turnip-Rooted)

Bulb spherical; when in its greatest perfection, measuring about an inch in diameter; skin fine, deep scarlet; flesh white, sometimes stained with red; leaves rather large and numerous.

The variety is early, and deserves more general cultivation, not only on account of its rich color, but for the crisp and tender properties of its flesh. It is much esteemed in England, and is grown extensively for the London market.

Scarlet Turnip-Rooted Radish

Small, Early, Yellow Turnip-Rooted

Bulb of the size and form of the Scarlet Turnip-rooted; skin smooth, yellow; flesh white, fine-grained, crisp, and rather pungent; foliage similar to that of the scarlet variety; season ten or fifteen days later.

White, Crooked

(Tortillée Du Mans)

Root very long; when suitable for use, measuring twelve inches and upwards in length, and an inch in diameter, nearly cylindrical, often irregular, and sometimes assuming a spiral or cork-screw form; skin white and smooth; flesh white, not so firm as that of most varieties, and considerably pungent; leaves very large.

White Turnip-Rooted

Bulb of the form and size of the Scarlet Turnip-rooted; skin white; flesh white and semi-transparent. It possesses less piquancy than the Scarlet, but is some days later.

Yellow Turnip-Rooted

(Yellow Summer)

Bulb nearly spherical, but tapering slightly towards the tap-root, which is very slender. It grows large,—to full four inches in diameter, when old; but should be eaten young, when about an inch in diameter. The flesh is mild, crisp, solid, and quite white. The skin is of a yellowish-brown color; and the leaves grow long and upright, with green footstalks.

Half early, and well adapted for summer cultivation.

Long White

(White Italian; Naples; White Transparent)

Root long and slender, nearly of the size and form of the Long Scarlet; skin white,—when exposed to the light, tinged with green; flesh white, crisp, and mild.

It is deserving of cultivation, not only on account of its excellent qualities, but as forming an agreeable contrast at table when served with the red varieties.

Long White Purple-Top

A sub-variety of the Long White; the portion of the root exposed to the light being tinged with purple. In size and form, it differs little from the Long Scarlet.

New London Particular

(Wood's Frame)

This is but a sub-variety of the Long Scarlet; the difference between the sorts being immaterial. The color of the New London Particular is more brilliant, and extends farther down the root. It is also said to be somewhat earlier.

Oblong Brown

The Oblong Brown Radish has a pear-shaped bulb, with an elongated tap-root. It does not grow particularly large; and, being hardier than most varieties, is well adapted for use late in the season. The outside is rough and brown, marked with white circles; the flesh is piquant, firm, hard, and white; the leaves are dark green, and rather spread over the ground; the footstalks are stained with purple.

Autumn and Winter Radishes

These varieties may be sown from the 20th of July to the 10th of August; the soil being previously made rich, light, and friable. Thin out the young plants from four to six inches apart; and, in the absence of rain, water freely. During September and October, the table may be supplied directly from the garden. For winter use, the roots should be harvested before freezing weather, and packed in earth or sand, out of danger from frost. Before being used, they should be immersed for a short time in cold water.

To raise Seed—Seeds of the Winter Radishes are raised by allowing the plants to remain where they were sown. As fast as they ripen, cut the stems; or gather the principal branches, and spread them in an open, airy situation, towards the sun, that the pods, which are quite tough in their texture, may become so dry and brittle as to break readily, and give out their seeds freely.

Use—All the kinds are used as salad, and are served in all the forms of the spring and summer radishes.

Varieties—

Black Spanish.

Bulb ovoid, or rather regularly pear-shaped, with a long tap-root. At first the root is slender, and somewhat cylindrical in form: but it swells as it advances in age, and finally attains a large size; measuring eight or ten inches in length, and three or four inches in diameter. The outside is rough, and nearly black; the flesh is pungent, firm, solid, and white; the leaves are long, and inclined to grow horizontally; the leaf-stems are purple. It is one of the latest, as well as one of the hardiest, of the radishes; and is considered an excellent sort for winter use.

Large Purple Winter

(Purple Spanish)

The Large Purple Winter Radish is a beautiful variety, derived, without doubt, from the Black Spanish; and may therefore be properly called the Purple Spanish. In shape and character, it much resembles the Black Spanish: but the outside, when cleaned, is of a beautiful purple, though it appears black when first drawn from the earth; and the coat, when cut through, shows the purple very finely. The footstalks of the leaves have a much deeper tinge of purple than those of the other kinds.

Long Black Winter

A sub-variety of the Black Spanish. Root long and tapering. With the exception of its smaller size, much resembling a Long Orange Carrot.

Long-Leaved White Chinese

Root fusiform, sometimes inversely turbinate, about five inches in length, and an inch in diameter; skin white, and of fine texture; flesh fine-grained, crisp, and though somewhat pungent, yet milder flavored than that of the Black Spanish; leaves large, differing from most other varieties in not being lobed, or in being nearly entire on the borders. Its season is nearly the same as that of the Rose-colored Chinese. The plants produce but few seeds.

Purple Chinese

A sub-variety of the Scarlet, with little variation except in color; the size, quality, and manner of growth, being nearly the same.

Rose-colored Chinese

(Scarlet Chinese Winter)

Bulb rather elongated, somewhat cylindrical, contracted abruptly to a long, slender tap-root; size full medium,—average specimens measuring about five inches in length, and two inches in diameter at the broadest part; skin comparatively fine, and of a bright rose-color; flesh firm, and rather piquant; leaves large,—the leaf-stems washed with rose-red. Season between that of the Gray Summer and that of the Black Spanish.

Winter White Spanish

(Autumn White. Blanc d'Augsbourg)

Root somewhat fusiform, retaining its diameter for two-thirds the length, sharply conical at the base, and, when well grown, measuring seven or eight inches in length by nearly three inches in its fullest diameter; skin white, slightly wrinkled, sometimes tinged with purple where exposed to the sun; flesh white, solid, and pungent, though milder than that of the Black Spanish. It succeeds best, and is of the best quality, when grown in light sandy soil. Season intermediate.

13

Rampion (Campanula rapunculus)

The Rampion is a biennial plant, indigenous to the south of Europe, and occasionally found in a wild state in England. The roots are white, fusiform, fleshy, and, in common with the other parts of the plant, abound in a milky juice; the lower or root leaves are oval, lanceolate, and waved on the borders; the upper leaves are long, narrow, and pointed. Stem eighteen inches or two feet in height, branching; flowers blue, sometimes white, disposed in small, loose clusters about the top of the plant, on the ends of the branches. The seeds are oval, brownish, and exceedingly small; upwards of nine hundred thousand being contained in an ounce. They retain their germinative property five years.

Rampion (Campanula rapunculus)
Wright, Robert Patrick (1857)

The plant flowers in July of the second year, and the seeds ripen in autumn. There is but one variety.

Soil and Cultivation

"Rampion prefers a rich, free, and rather light soil, in a shady situation. It is raised from seed, which should be sown where the plants are to remain, as they do not bear transplanting well. The sowing may be made in April, May, or the beginning of June: but sometimes plants from very early sowings are liable to run up to seed; and, when this is the case, the roots become tough, and unfit for use. The ground should be well dug, and raked as fine as possible. The seed may then be sown either broadcast or in drills, six inches apart, and about one-fourth of an inch deep. As the seeds are very small, it is advisable to mix them with fifteen or twenty times their bulk of fine sand, in order to secure their even distribution in the drills, and to prevent the plants from coming up too closely. The seed should only be very slightly covered with fine earth; and the seed-bed ought to be frequently watered with a fine-rosed watering-pot till the plants come up, which will be in about a fortnight.

"When the young plants are about one inch high, they should be thinned out to four inches apart. After this, no further care is necessary than to water frequently, and to keep the ground free of weeds."[1]

***Taking the Crop**—*The roots will be fit for use from October till April. They may be taken from the ground for immediate use; or a quantity may be taken up in autumn, before the closing-up of the ground, and packed in sand, for use during the winter.

***To raise Seed**—*Leave or transplant some of the best yearling plants, and they will produce an abundance of seed in autumn.

***Use**—*The roots have a pleasant, nut-like flavour; and are generally eaten in their crude state as a salad. The leaves, as well as the roots, are occasionally used in winter salads.

1.The Gardener's Assistant. By Robert Thompson.

Campanula Rapunculus
Flora Batava, Plate Number 0203

14

Ruta-Baga/Swede Turnip (Brassica campestris ruta-baga)

The Ruta-baga, or Swede Turnip, is supposed by De Candolle[1] to be analogous to the Kohl Rabi; the root being developed into a large, fleshy bulb, instead of the stem. In its natural state, the root is small and slender; and the stem smooth and branching,—not much exceeding two feet in height.

The bulbs, or roots, are fully developed during the first year. The plant flowers, and produces its seeds, the second year, and then perishes. Although considered hardy,—not being affected by even severe frosts,—none of the varieties will withstand the winters of the Northern or Middle States in the open ground. The crop should therefore be harvested in October or November, and stored for the winter, out of danger from freezing. Most of the sorts now cultivated retain their freshness and solidity till spring, and some even into the summer; requiring no particular care in their preservation, other than that usually given to the carrot or the potato.

Soil and Cultivation—All the varieties succeed best in a deep, well-enriched, mellow soil; which, previous to planting, should be very deeply ploughed, and thoroughly pulverised by harrowing or otherwise. Some practise ridging, and others sow in simple drills. The ridges are usually formed by turning two furrows against each other; and, being thus made, are about two feet apart. If sown in simple drills, the surface should be raked smooth, and the drills made from sixteen to eighteen inches apart; the distance to be regulated by the strength of the soil.

Seed and Sowing—About one pound of seed is usually allowed to an acre. Where the rows are comparatively close, rather more than this quantity will be required; while three-fourths of a pound will be amply sufficient, if sown on ridges, or where the drills are eighteen inches apart. The sowing may be made from the middle of May to the 25th of July; the latter time being considered sufficiently early for growing for the table, and by some even for stock. Early sowings will unquestionably give the greatest product; while the later-grown bulbs, though of smaller dimensions, will prove of quite as good quality for the table.

To raise Seeds—Select the smoothest and most symmetrical bulbs, and transplant them in April, two feet asunder, sinking the crowns to a level with the surface of the ground.

The seeds are very similar to those of the common garden and field turnip, and will keep from five to eight years.

Varieties—The varieties are as follow:

Ashcroft

Bulb of medium size, ovoid, very smooth and symmetrical; neck very short, or wanting. Above ground, the skin is purple; below the surface, yellow. Flesh yellow, very solid, fine-grained, and of excellent flavor. It forms its bulb quickly and regularly; keeps in fresh and sound condition until May or June; and well deserves cultivation, either for agricultural purposes or for the table.

Common Purple-Top Yellow

An old and long-cultivated sort, from which, in connection with the Green-top, have originated most of the more recent and improved yellow-fleshed varieties. Form regularly egg-shaped, smooth, but usually sending out a few small, straggling roots at its base, near the tap-root; neck short; size rather large,—usually measuring six or seven inches in depth, and four or five inches in its largest diameter; skin purple above ground,—below the surface, yellow; flesh yellow, of close, firm texture, and of good quality. It is very hardy; forms its bulb promptly and uniformly; and in rich, deep soils, yields abundantly. For thin and light soils, some of the other varieties should be selected.

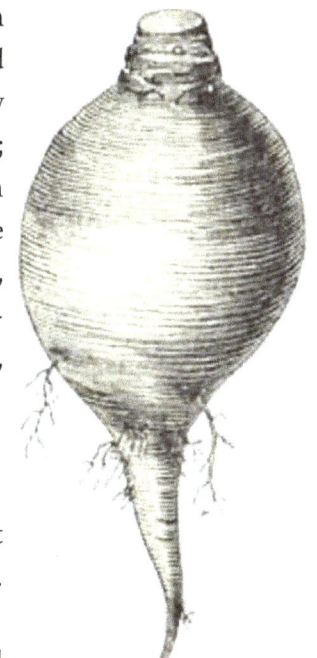

Common Purple-Top Yellow Turnip

Early Stubble

Bulb round, smooth, and regular. The skin, where exposed to light and air, is of a brownish-green; but, where covered by the soil, yellow. The flesh is firm, and well flavored.

The Early Stubble is recommended as forming its bulbs quickly and uniformly, and as being well adapted for late sowing. It yields abundantly; keeps well; is a good sort for the table; and, in some localities, is preferred to the Common Yellow for cultivation for farm purposes.

Green-Top Yellow

In form and foliage, this variety resembles the Common Purple-top; but usually attains a larger size when grown in similar situations. Skin, above the surface of the soil, green; below ground, yellow. The flesh is solid, sweet, and well flavored, but inferior to that of the Purple-top. It keeps well, is of fair quality for the table, and, on account of its great productiveness, one of the best of all varieties for growing for feeding stock.

Green-Top White

Bulb turbinate, smooth, and symmetrical. The skin above ground is of a fine, clear, pea-green; often browned or mellowed where exposed to the direct influence of the sun: below the surface of the ground, it is uniformly white. The flesh is also white, comparatively solid, very sweet, and of fair quality for table use. It differs from the Purple-top White, not only in color, but in size and quality; the bulbs being larger, and the flesh not quite so firm or well flavored.

The Green-top White is productive; continues its growth till the season has far advanced; is little affected by severe weather; and, when sown in good soil, will yield an agricultural crop of twenty-five or thirty tons to an acre.

Laing's Improved Purple-Top

This variety differs from most, if not all, of the varieties of Swedish turnips, in having entire cabbage-like leaves, which, by their horizontal growth, often nearly cover the surface of the ground. In form, hardiness, and quality, it is fully equal to any of the other sorts. Growing late in the autumn, it is not well adapted to a climate where the winter commences early. It has little or no tendency to run to seed in the fall; and even in the spring, when set out for seed, it is a fortnight later in commencing this function than other varieties of Ruta-bagas. It requires good land, in high condition; and, under such circumstances, will yield abundantly, and is worthy of cultivation. The bulb, when well grown, has an almost spherical form; a fine, smooth skin, purple above ground, yellow below, with yellow, solid, and well-flavoured flesh.

Purple-Top White

Bulb oblong, tapering toward the lower extremity, five or six inches in diameter, seven or eight inches in depth, and less smooth and regular than many of the yellow-fleshed varieties. The skin is of a clear rich purple, where it comes to air and light, but, below the ground, pure white; flesh white, very solid and fine-grained, sugary, and well flavored.

The variety is hardy, productive, keeps remarkably well, is good for table use, and may be profitably grown for agricultural purposes. Upwards of twenty-eight tons, or nine hundred and sixty bushels, have been raised from an acre.

River's

Root regularly turbinate, or fusiform, of full medium size, smooth, and with few small or fibrous roots; neck two inches long; skin, above ground, green, washed with purplish-red where most exposed to the sun,—below ground, yellow; flesh yellow, firm, sweet, and well flavored. Esteemed one of the best, either for stock or the table. Keeps fresh till May or June.

Skirving's Purple-Top

(Skirving's Improved Purple-top; Skirving's Liverpool; Southold Turnip, of some localities)

Bulb ovoid, or regularly turbinate, and rather deeper in proportion to its diameter than the common Purple-top Yellow; surface remarkably smooth and even, with few fibrous roots, and seldom deformed by larger accidental roots, although, in unfavorable soils or seasons, a few coarse roots are put forth in the vicinity of the tap-root; size full medium,—five to seven inches in length, and four or five inches in diameter. Sometimes, when sown early in good soil, and harvested late, the average will considerably exceed these dimensions. Neck short, but, when grown in poor soil, comparatively long; skin, above ground, fine, deep purple,—below ground, yellow,—the colors often richly blending together at the surface; flesh yellow, of solid texture, sweet, and well flavored.

This variety was originated by Mr. William Skirving, of Liverpool, Eng. In this country it has been widely disseminated, and is now more generally cultivated for table use and for stock than any other of the Swede varieties. The plants seldom fail to form good-sized bulbs. It is a good keeper; is of more than average quality for the table; and long experience has proved it one of the best sorts for cultivation on land that is naturally shallow and in poor condition. On soils in a high state of cultivation, upwards of nine hundred bushels have been obtained from an acre.

In sowing, allow twenty inches between the rows, and thin to ten or twelve inches in the rows.

Sweet German

Bulb four or five inches in diameter, six or seven inches in depth, turbinate, sometimes nearly fusiform. In good soil and favorable seasons, it is comparatively smooth and regular; but, under opposite conditions, often branched and uneven. Neck two or three inches in length; skin greenish-brown above ground, white beneath; flesh pure white, of extraordinary solidity, very sweet, mild, and well flavored.

It retains its solidity and freshness till spring, and often at midsummer has no appearance of sponginess or decay. As a table variety, it must be classed as one of the best, and is recommended for general cultivation.

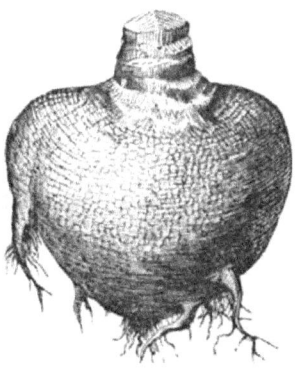

Sweet German Turnip

White French

(Long White French)

The roots of this variety are produced entirely within the earth. They are invariably fusiform; and, if well grown, measure four or five inches in diameter, and from eight to ten inches in length. Foliage not abundant, spreading; skin white; flesh white, solid, mild, sweet, and delicate. It is not so productive as some other varieties, and is therefore not so well adapted to field culture; but for table use it is surpassed by few, if any, of its class.

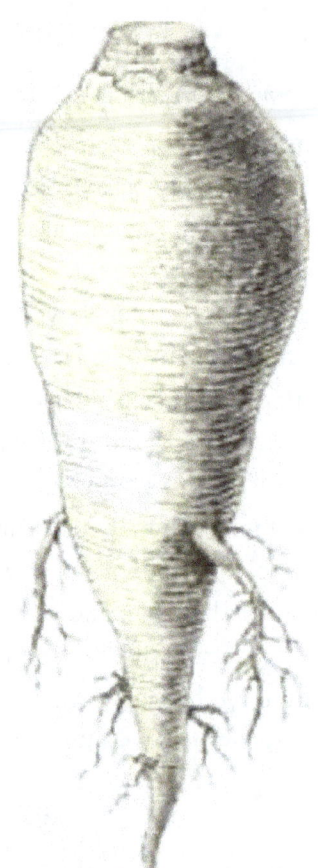

White French Turnip

A rough-leaved, fusiform-rooted variety of the common garden-turnip: is known by the name of "White French" in many localities; but, according to the most reliable authority, that name has not only long been used in connection with, but properly belongs to, the white turnip above described.

1.The Candolle's Systema Naturale. By Prof. De Candolle. 2 vols. 8vo. Paris, 1818, 1821.

15

Salsify/Oyster Plant (Tragopogon porrifolius)

(Leek-leaved Salsify; Vegetable Oyster; Purple Goat's Beard)

The Salsify is a hardy biennial plant, and is principally cultivated for its roots, the flavor of which resembles that of the oyster; whence the popular name.

The leaves are long and grass-like, or leek-like; the roots are long and tapering, white within and without, and, when grown in good soil, measure twelve or fourteen inches in length, and rather more than an inch in diameter at the crown.

Soil and Cultivation—The Oyster-plant succeeds best in a light, well-enriched, mellow soil; which, previous to sowing the seeds, should be stirred to the depth of twelve or fifteen inches. The seeds should be sown annually, in the same manner and at the same time as the seeds of the carrot and parsnip. Make the drills fourteen inches apart; cover the seeds an inch and a half in depth; and thin, while the plants are young, to four or five inches asunder.

Tragopogon Porrifolium
Flora Batava

Early sowings succeed best; as the seeds, which are generally more or less imperfect, vegetate much better when the earth is moist than when dry and parched, as it is liable to become when the season is more advanced. Cultivate in the usual manner during the summer; and, by the last of September or beginning of October, the roots will have attained their full growth, and be ready for use. The plants will sustain no injury during the winter, though left entirely unprotected in the open ground; and the table may be supplied directly from the garden, whenever the frost will admit of their removal. A portion of the crop should, however, be taken up in autumn, and stored in the cellar, like other roots; or, which is perhaps preferable, packed in earth or sand. Roots remaining in the ground may be drawn for use till April, or until the plants have begun to send up their stalks for flowering.

Seeds: Production and Quantity—For the production of seeds, allow a few plants to remain during the winter in the open ground where they were sown. They will blossom in June and July. When fully developed, the stem is about three feet in height, cylindrical, and branching. The flowers are large, of a very rich violet-purple, and expand only by day and in comparatively sunny weather. As the flowers are put forth in gradual succession, so the heads of seeds are ripened at intervals, and should be cut as they assume a brownish color.

The seeds are brownish,—lighter or darker as they are less or more perfectly matured,—long and slender, furrowed and rough on the sides, tapering to a long, smooth point at the top, often somewhat bent or curved, and measure about five-eighths of an inch in length. They will keep four years.

An ounce contains three thousand two hundred seeds, and will sow a row eighty feet in length. Some cultivators put this amount of seed into a drill of sixty feet; but if the seed is of average quality, and the season ordinarily favorable, one ounce of seed will produce an abundance of plants for eighty or a hundred feet.

Use—The roots are prepared in various forms; but, when simply boiled in the manner of beets and carrots, the flavor is sweet and delicate. The young flower-stalks, if cut in the spring of the second year and dressed like asparagus, resemble it in taste, and make an excellent dish.

The roots are sometimes thinly sliced, and, with the addition of vinegar, salt, and pepper, served as a salad. They are also recommended as being remedial or alleviating in cases of consumptive tendency.

There is but one species or variety now cultivated.

FIELD AND GARDEN VEGETABLES OF THE LATE NINETEENTH CENTURY | 87

"Oyster-plant", and Scorzonera: "Black Salsify."
Encyclopedia of Food by Artemas Ward

16

Scolymus (Scolymus hispanicus)

(Spanish Scolymus; Spanish Oyster-Plant)

In its natural state, this is a perennial plant; but, when cultivated, it is generally treated as an annual or as a biennial. The roots are nearly white, fleshy, long, and tapering in their general form, and, if well grown, measure twelve or fifteen inches in length, and an inch in diameter at the crown. When cut or bruised, or where the fibrous roots are broken or rubbed off, there exudes a thick, somewhat viscous fluid, nearly flavorless, and of a milk-white color. The leaf is large, often measuring a foot or more in length, and three inches in diameter, somewhat variegated with green and white, deeply lobed; the lobes or divisions toothed, and the teeth terminating in sharp spines, in the manner of the leaves of many species of thistles. When in flower, the plant is about three feet in height. The flowers, which are put forth singly, are of an orange-yellow, and measure an inch and a half in diameter. The seeds are flat, and very thin, membranous on the borders, of a yellowish color, and retain their vitality three years. An ounce contains nearly four thousand seeds.

Scolyme d'Espagne
Vilmorin-Andrieux 1883

Soil and Cultivation—Any good garden loam is adapted to the growth of the Scolymus. It should be well and deeply stirred as for other deep-growing root crops. The seeds should be sown from the middle of April to the 10th of May, in drills an inch deep, and fourteen inches asunder. Thin the young plants to five inches distant in the rows; and, during the summer, treat the growing crop as parsnips or carrots.

Use—It is cultivated exclusively for its roots, which are usually taken up in September or October, and served at table, and preserved during the winter, in the same manner as the Salsify, or Oyster-plant. They have a pleasant, delicate flavor; and are considered to be not only healthful, but remarkably nutritious.

FIELD AND GARDEN VEGETABLES OF THE LATE NINETEENTH CENTURY | 89

Scolymus hispanicus
Flore coloriée de poche du littoral méditerranéen de Gênes à Barcelone y compris la Corse

17

Black Salsify (Scorzonera hispanica)

(Black Oyster-plant; Black Salsify)

This is a hardy perennial plant, introduced from the south of Europe, where it is indigenous. The root is tapering, and comparatively slender,—when well developed, measuring about a foot in length, and an inch in diameter near the crown, or at the broadest part; skin grayish-black, coarse, somewhat reticulated, resembling the roots of some species of trees; flesh white; leaves long, ovate, broadest near the end, and tapering sharply to the stem. They are also more or less distinctly ribbed, and have a few remote teeth, or serratures, at the extremities. When in flower, the plant measures about four feet in height; the stalk being nearly cylindrical, slightly grooved or furrowed, smooth, and branched towards the top. The flowers are large, terminal, yellow; the seeds are whitish, longer than broad, taper towards the top, and retain their vitality two years. An ounce contains about two thousand five hundred seeds.

Scorzonera hispanica var. glastifolia
Flora Batava, Volume 16

Soil and Culture—Though a perennial, it is generally cultivated as an annual or biennial, in the manner of the carrot or parsnip. Thompson[1] says, "It succeeds best in a light, deep, free soil and an open situation. It is raised from seed, which may be sown in drills one foot apart, covering with soil to the depth of half an inch. As it is apt to run to seed the same year in which it is sown, and consequently to become tough and woody," the planting should not be made too early, particularly in the warmer sections of the country. A second sowing may be made about four weeks from the first:

> "...as a precautionary measure, in case the plants of the first sowing should run. The young plants, when three or four inches high, should be thinned out to eight inches asunder in the rows. Towards the middle or last of September, the roots will have attained sufficient size to be drawn for immediate use: others will come in for use in October and November. In the latter month, they will be in perfection; and, before the closing-up of the ground, a quantity may be taken up, and stored in sand for the winter. When the ground is open, the roots may be drawn from time to time, as required for immediate use. About the middle of April, the roots remaining in the ground will begin to run to flower; after which they soon become hard, woody, and unfit for the table. Before this takes place, however, they may be taken up, and stored in sand, where they may be kept for use till May or June."[2]

To raise Seed—Allow a few well-grown plants to remain in the ground during winter; or select a few good-sized roots from those harvested in autumn, and reset them in April, about eighteen inches apart, covering them to the crowns. The seed will ripen at the close of the summer or early in autumn. Seed saved from plants of the growth of two seasons is considered best; that produced from yearling plants being greatly inferior.

Use—It is cultivated exclusively for its roots; no other portion of the plant being employed in domestic economy. The flesh of these is white, tender, sugary, and well flavored. They are boiled in the manner of the parsnip, and served plain at the table; or they may be cooked in all the forms of salsify or scolymus. Before cooking, the outer, coarse rind should be scraped off, and the roots soaked for a few hours in cold water for the purpose of extracting their bitter flavor.

1,2. The Gardener's Assistant. By Robert Thompson.

18

Skirrret (Sium sisarum)

(Crummock, of the Scotch)

Skirret is a hardy perennial, and is cultivated for its roots, which are produced in groups, or bunches, joined together at the crown or neck of the plant. They are oblong, fleshy, of a russet-brown color without, white within, very sugary, and, when well grown, measure six or eight inches in length, and nearly an inch in diameter.

The leaves of the first year are pinnate, with seven or nine oblong, finely toothed leaflets. When fully developed, the plant measures from three to five feet in height; the stem being marked with fine, parallel, longitudinal grooves, or lines. The flowers are small, white, and are produced in umbels at the extremities of the branches. The seeds, eight thousand of which are contained in an ounce, are oblong, of a greenish-gray color, and closely resemble those of the common caraway. They will keep but two years; and, even when newly grown, sometimes remain in the ground four or five weeks before vegetating.

Soil and Culture—Skirret succeeds best in light, mellow soil, and is propagated by suckers, or seeds. The best method is to sow the seeds annually, as, when grown from slips, or suckers, the roots are liable to be dry and woody; the seeds, on the contrary, producing roots more tender, and in greater perfection.

Sow the seeds in April, in drills one foot apart, and about an inch in depth; thin to five or six inches; and, in September, some of the roots will be sufficiently grown for use. Those required for winter should be drawn before the closing-up of the ground, and packed in sand.

To propagate by Slips, or Suckers—In the spring, remove the required number of young shoots, or sprouts, from the side of the roots that have remained in the ground during winter, not taking any portion of the old root in connection with the slips; and set them in rows ten inches asunder, and six inches apart in the rows. They will soon strike, and produce roots of suitable size for use in August or September.

To raise Seeds — The plants that have remained in the ground during the winter, if not disturbed, will send up stalks as before described, and ripen their seeds at the close of the summer. Two or three plants will yield all the seeds ordinarily required for a single garden.

Use — The roots were formerly much esteemed, but are now neglected for those greatly inferior. When cooked and served as salsify or scorzonera, they are the whitest and sweetest of esculent roots, and afford a considerable portion of nourishment.

There are no varieties.

Sium sisarum

19

Sweet Potato (Ipomœa batatas)

(Spanish Potato; Carolina Potato; Convolvulus batatus)

The Sweet Potato is indigenous to both the East and West Indies. Where its growth is natural, the plant is perennial; but, in cultivation, it is always treated as an annual. The stem is running or climbing, round and slender; the leaves are heart-shaped and smooth, with irregular, angular lobes; the flowers, which are produced in small groups of three or four, are large, bell-shaped, and of a violet or purple color; the seeds are black, triangular, and retain their vitality two or three years,—twenty-three hundred are contained in an ounce.

The plants rarely blossom in the Northern or Middle States, and the perfect ripening of the seeds is of still more rare occurrence. The latter are, however, never employed in ordinary culture; and are sown only for the production of new varieties, as is sometimes practised with the common potato.

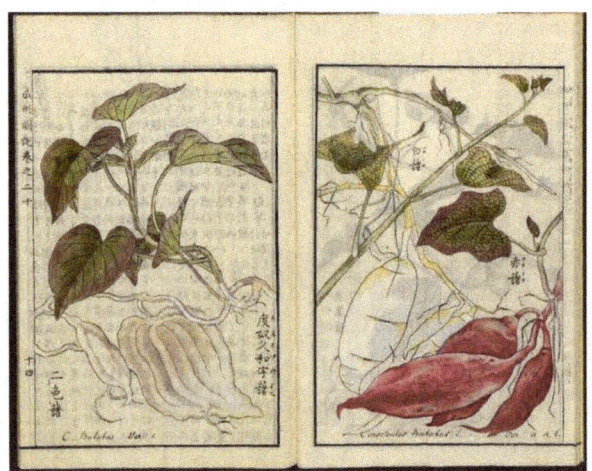

Sweet Potato
Seikei Zusetsu vol. 20, page 14

Soil, Planting, and Cultivation—In warm climates, the Sweet Potato is cultivated in much the same manner as the common potato is treated at the North. It succeeds best in light, warm, mellow soil, which should be deeply stirred and well enriched. The slips, or sprouts, may be set on ridges four feet apart, and fifteen inches from plant to plant; or in hills four or five feet apart in each direction, three plants being allowed to a hill. During the summer, give the vines ordinary culture; and late in September, or early in October, the tubers will have attained their growth, and be ready for harvesting. The slips, or sprouts, are generally obtained by setting the tubers in a hot-bed in March or April, and breaking off or separating the sprouts from the tubers as fast as they reach four or five inches in height or attain a suitable size for transplanting. In favorable seasons, the plucking may be repeated three or four times. In setting out the slips, the lower part should be sunk from one-third to one-half the entire length; and, if very dry weather occurs, water should be moderately applied.

Keeping—The essentials for the preservation of Sweet Potatoes are dryness and a warm and even temperature. Where these conditions are not supplied, the tubers speedily decay. By packing in dry sand, and storing in a warm, dry room, they are sometimes preserved in the Northern States until the time of starting the plants in spring.

Varieties—Though numerous other varieties, less marked and distinctive, are described by different authors, and are catalogued by gardeners and seedsmen, the principal are as follow:

Kentucky Early Red

(Red Nansemond)

Tubers red, or purplish-red, of medium size; flesh yellow, dry, sweet, and of good quality. A very prolific, hardy variety; recommended as the best red Sweet Potato for Northern culture.

Large White

(Patate-blanche of the French)

Tubers from six to ten inches in length,—thickest at the middle, where they measure from two to nearly three inches in diameter; weight from six ounces to a pound and upwards; skin dusky white; flesh nearly white, but with a shade of yellow. Not so fine-grained or so sweet as the Yellow or Purple, but quite farinaceous and well flavored.

It requires a long season in order to its full development; but, being remarkably hardy, it will succeed well in any of the Middle States, and attain a fair size in the warmer sections of New England.

Nansemond

(Yellow Nansemond)

A variety said to have originated in Nansemond County, Virginia; whence the name. Tubers large, yellow, swollen at the middle, and tapering to the ends; flesh yellow, dry, unctuous, sweet, and well flavored.

It is early fit for the table; matures in short seasons; is very productive; succeeds well in almost any tillable soil; and, having been long acclimated, is one of the best sorts for cultivation at the North,—very good crops having been obtained in Maine and the Canadas.

Purple-Skinned

(New-Orleans Purple; Patate violette)

Tubers swollen at the middle, and tapering in each direction to a point,—measuring, when well grown, from seven to nine inches in length, and from two to three inches in diameter; skin smooth, reddish-purple; flesh fine-grained, sugary, and of excellent quality. The plants attain a remarkable length, and the tubers are rarely united about the neck as in most other varieties.

The Purple-skinned is early and productive, but keeps badly. It would probably succeed much better in cool climates than either the White or the Yellow. It is much grown in the vicinity of Paris.

Red-Skinned, or American Red

Tubers fusiform, long, and comparatively slender,—the length often exceeding twelve inches, and the diameter rarely above two inches; weight from three to ten ounces; skin purplish-red, smooth and shining; flesh yellow, very fine-grained, unctuous, sugary, and farinaceous; plant long and slender.

This variety is early, quite hardy, very productive, and excellent, but does not keep so well as the yellow or white sorts. It is well adapted for cultivation in the cooler sections of the United States; where, in favorable seasons, the crop has proved as certain, and the yield nearly as abundant, as that of the common potato.

Rose-Coloured

Tubers somewhat ovoid, or egg-shaped, often grooved, or furrowed, and of extraordinary size. Well-grown specimens will measure eight or nine inches in length, and four inches or more in diameter; frequently weighing two and a half, and sometimes greatly exceeding three pounds. Skin rose-colored, shaded or variegated with yellow; flesh sweet, of a pleasant, nut-like flavor, but less soft or unctuous than that of the other varieties.

It is hardy, remarkably productive, and, its excellent keeping properties considered, one of the best sorts for cultivation.

Yellow-Skinned

(Yellow Carolina)

Tubers from six to ten inches in length, thickest at the middle, where they measure from two to three inches in diameter, and pointed at the extremities; weight varying from four to twelve ounces and upwards; skin smooth, yellow; flesh yellow, fine-grained, unctuous, and remarkably sugary,—surpassing, in this last respect, nearly all other varieties. Not so early as the Red-skinned or the Purple.

When grown in the Southern States, it yields well; perfectly matures its crop; and, in color and flavor, the tubers will accord with the description above given. When grown in the Middle States, or in the warmer parts of New England, it decreases in size; the tubers become longer and more slender; the color, externally and internally, becomes much paler, or nearly white; and the flesh, to a great extent, loses the fine, dry, and sugary qualities which it possesses when grown in warm climates.

20

Tuberous-Rooted Chickling Vetch (Lathyrus tuberosus)

(Tuberous-rooted Pea; Eatable-rooted Pea)

Perennial; stem about six feet high,—climbing, slender, four-sided, smooth, and of a clear green color; flowers rather large, in bunches, of a fine carmine rose-color, and somewhat fragrant; pod smooth; seeds rather large, oblong, a little angular, of a brown color, spotted with black; root spreading, furnished with numerous blackish, irregularly shaped tubers, which are generally from an ounce to three ounces in weight.

The roots are very farinaceous, and, when cooked, are highly esteemed. In taste, they somewhat resemble roasted chestnuts. Where the roots are uninjured by the winter, the plant increases rapidly, and is liable to become a troublesome inmate of the garden.

Lathyrus Tuberosus Root

Lathyrus tuberosus
Flora Batava, Volume 3

21

Tuberous-Rooted Tropaeolu/Ysano (Tropæolum tuberosum)

This is a perennial plant from Peru, and deserves mention as a recently introduced esculent. It produces an abundance of handsome yellow and red tubers, about the size of small pears; the taste of which is not, however, very agreeable. On this account, a particular mode of treatment has been adopted in Bolivia, where, according to M. Decaisne, they are treated in the following manner:

The tubers designated "Ysano," at La Paz, require to be prepared before they are edible. Indeed, when prepared like potatoes, and immediately after being taken up, their taste is very disagreeable. But a mode of making them palatable was discovered in Bolivia; and the Ysano has there become, if not a common vegetable, at least one which is quite edible. The means of making them so consists in freezing them after they have been cooked, and they are eaten when frozen. In this state it is said that they constitute an agreeable dish, and that scarcely a day passes at La Paz without two lines of dealers being engaged in selling the Ysano, which they protect from the action of the sun by enveloping it in a woollen cloth, and straw. Large quantities are eaten sopped in treacle, and taken as refreshment during the heat of the day.

Propagation and Culture—The plant may be propagated by pieces of the tubers, in the same manner as potatoes; an eye being preserved on each piece. The sets should be planted in April or May, according to the season, about four feet apart, in light, rich soil. The stems may be allowed to trail along the ground, or pea-sticks may be placed for their support. In dry soils and seasons, the former method should be adopted; in those which are moist, the latter. The tubers are taken up in October, when the leaves begin to decay, and stored in sand.

Tropæolum Tuberosum, Leaves and Flowers
Paxton's Magazine of Botany and Register of Flowering Plants

22

The Turnip (Brassica rapa)

The common Turnip is a hardy, biennial plant, indigenous to Great Britain, France, and other parts of Europe. The roots of all the varieties attain their full size during the first year. The radical leaves are hairy and rough, and are usually lobed, or lyrate; but, in some of the sorts, nearly spatulate, with the borders almost entire. The flowers are produced in May and June of the second year, and the seeds ripen in July; the flower-stalk rises three feet or more in height, with numerous branches; the leaves are clasping, and much smoother and more glaucous than the radical leaves of the growth of the previous year; the flowers are yellow, and are produced in long, loose, upright, terminal spikes; the seeds are small, round, black, or reddish-brown, and are very similar, in size, form, and color, in the different varieties,—ten thousand are contained in an ounce, and they retain their vitality from five to seven years.

Propagation and Culture—All the sorts are propagated by seeds; which should be sown where the plants are to remain, as they do not generally succeed well when transplanted. Sowings for early use may be made the last of April, or beginning of May; but as the bulbs are seldom produced in perfection in the early part of the season, or under the influence of extreme heat, the sowing should be confined to a limited space in the garden. The seeds may be sown broadcast or in drills: if sown in drills, they should be made about fourteen inches apart, and half an inch in depth. The young plants should be thinned to five or six inches asunder. For a succession, a few seeds may be sown, at intervals of a fortnight, until the last week in July; from which time, until the 10th of August, the principal sowing is usually made for the winter's supply. In the Middle States, and the warmer portion of New England, if the season is favorable, a good crop will be obtained from seed sown as late as the last week in August.

Harvesting—Turnips for the table may be drawn directly from the garden or field until November, but must be harvested before severe freezing weather; for, though comparatively very hardy, few of the varieties will survive the winters of the Northern States in the open ground.

Seed—As the various kinds readily hybridize, or intermix, only one variety should be cultivated in the same neighborhood for seed. Select the best-formed bulbs, and transplant them out in April, in rows two feet apart, and one foot apart in the rows, just covering the crowns with earth, or leaving the young shoots level with the surface of the ground.

An ounce of seed will sow eight rods of land, and a pound will be sufficient for an acre.

Varieties—The varieties are numerous, as follow:

Altrincham
(Yellow Altrincham; Altringham)
This is a yellow-fleshed, field variety, of rather less than average size. The bulb, however, is of a fine, globular shape, with a light-green top, very small neck and tap-root, and possessed of considerable solidity.

Border Imperial
(Border Imperial Purple-Top Yellow)
Bulb five or six inches in diameter, nearly spherical, sometimes flattened, and usually very smooth and symmetrical; skin yellow, the upper surface of a bright purple; flesh yellow, firm, and sugary; leaves large.

The variety is of English origin, and is recommended for its earliness and great productiveness.

Chivas's Orange Jelly
Bulb of a handsome, round form, with a small top; the skin is pale orange; and the flesh yellow, juicy, sweet, and tender. It has very little fibre; so that, when boiled, it almost acquires the consistence of a jelly. It originated in Cheshire, Eng.

Cow-Horn.
(Long Early White Vertus)
Root produced much above ground, nearly cylindrical, rounded at the end, ten or twelve inches in length, nearly three inches in diameter, and weighing from one and a half to two pounds. The skin is smooth and shining,—white below the surface of the ground, and green at the top; the flesh is white, tender, and sugary. Early, very productive, and remarkable for its regular form and good quality. As a field-turnip, it is one of the best; and, when pulled young, good for table use. During winter, the roots often become dry and spongy.

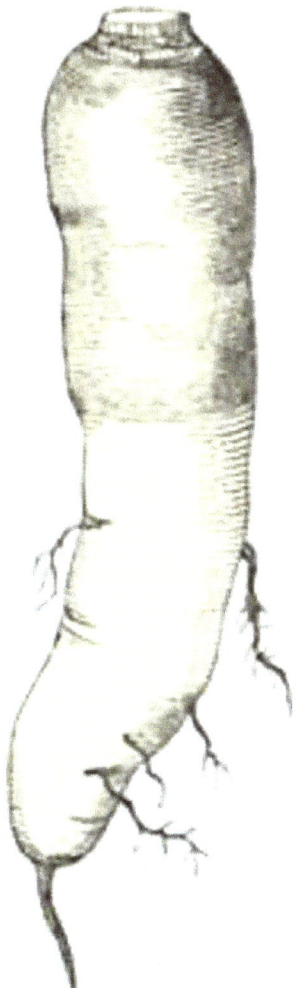

Cow-Horn Turnip

Dales's Hybrid

This variety is of English origin, and is said to be a hybrid from the Green-top Swede and the common White Globe. Its prevailing traits are, however, those of the White Globe; inasmuch as its roots are similar in form and texture.

Foliage strong and luxuriant; root large, oblong, pale yellow; the upper surface light green; neck and tap-root small. The form of the bulb, though generally oblong, is sometimes nearly globular; but its more material characteristics, large size, and luxurance of growth, are uniformly the same. Its reputation as a turnip of very superior quality has not been sustained in this country.

Early Flat Dutch.

(Early White Dutch; White Dutch)

An old and well-known early garden variety; bulb round, very much flattened, and produced mostly within the earth; skin white, somewhat washed with green at the insertion of the leaves, which are of medium size. Before the bulb has attained its full dimensions, the flesh is fine-grained, tender, and sweet; but when ripe, especially in dry seasons, it often becomes spongy and juiceless: in which condition, it is of no value for the table; and, even for stock, is comparatively worthless. Average specimens measure about four inches in diameter, and two inches and a half in depth.

Early Yellow Dutch

(Yellow Dutch)

This variety has a small, globular root, of a pale-yellow color throughout. It somewhat resembles the Yellow Malta, and is a good garden variety. The portion of the bulb above ground, and exposed to the sun, is washed with green. It is of medium size, early, tender, rather close-grained, and sugary; better suited for use in summer and autumn than for winter. By some, the variety is esteemed the best of the yellow garden turnips.

Finland

(Yellow Finland)

This is a beautiful, medium-sized turnip, of a bright yellow throughout, even to the neck; somewhat similar to a firm Yellow Malta, but of finer color. The under part of the bulb is singularly depressed: from this depression issues a small, mousetail-like root. It is somewhat earlier, and also hardier, than the Yellow Malta.

The flesh is tender, close-grained, and of a sweet, sugary flavor; the leaves are small, and few in number; bulb about two inches in thickness by four inches in diameter, weighing eight or ten ounces. An excellent garden variety.

Finland Turnip

Freneuse

Root produced within the earth,—long, tapering, and rather symmetrical; size small,—average specimens measuring five or six inches in length, an inch and a half in diameter at the crown, and weighing eight or ten ounces; skin white, or yellowish-white; flesh white, dry, very firm, and sugary; leaves small, deep green, spreading. Half early, and one of the best of the dry-fleshed varieties.

Golden Ball.

(Yellow Globe)

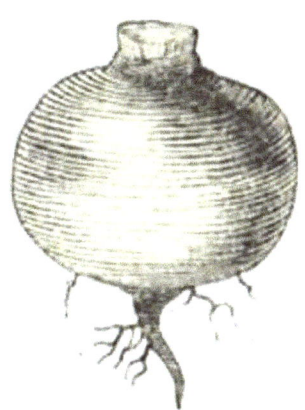

Golden Ball Turnip

Bulb produced mostly within the earth, nearly globular, and very smooth and symmetrical; skin bright yellow below ground, greenish above; leaves comparatively small, spreading; flesh pale yellow, sweet, and well flavored, but not so fine-grained as that of many other varieties. It is a good table turnip; and with the Robertson's Golden Stone, which it greatly resembles, the most valuable for cultivation, where large-sized garden turnips are required. Its size is about that of the last named. Average specimens measure four inches in diameter, nearly the same in depth, and weigh from twelve to fourteen ounces.

Green Globe

(Green-Top White Globe)

Roots of a fine, globular shape, with a small neck and tap-root; very white below, and green above, the surface of the ground; of medium size, hardy, and firm in texture, but scarcely so much so as the Green Round; than which it arrives at maturity rather earlier. It is somewhat larger than the White Norfolk; has large, deep-green foliage; grows strongly; and produces extraordinary crops: but it soon becomes spongy, and often decays in autumn or early in winter.

A sub-variety, of larger size and with softer flesh, is known by the name of Hungarian Green-Top Globe.

Green Norfolk

(Green-top Norfolk; Green Round)

A sub-variety of the White Norfolk, of nearly the same form and size; the bulb differing principally in the color of the top, which is green.

The Norfolk turnips are all of a peculiar flattish form; rather hollowed towards their neck, as also on their under side. When grown to a large size, they become more or less irregular, or somewhat angular. The Green-top variety possesses these characters in a less degree than the White-top; and is generally round, flattened, but not much hollowed, on the upper or under surface. It is hardier than the White or Red varieties.

Green Tankard

Roots more than half above ground; oblong, or tankard-shaped; of a greenish color, except on the under surface, which is white; flesh white and sweet, but of coarse texture.

The term "Tankard" is applied to such common field turnips as are of an oblong shape, and the roots of which, in general, grow much above the surface of the ground. Such oblong varieties, however, as approach nearest to a round or globular form, are sometimes termed "Decanter," or "Decanter-shaped turnips."

In good soils, the Green Tankard sometimes attains a weight of eight or ten pounds. As a garden variety, it is of little value.

Green-Top Flat

Similar in size, form, and quality to the common Purple-top Flat; skin, above ground, green.

Long grown in New England for feeding stock; and, in its young state, often used as a table turnip. Now very little cultivated.

Green-Top Yellow Aberdeen

(Green-Top Yellow Bullock)

An old and esteemed variety, similar in size and form to the Purple-top Yellow Aberdeen: the color of the top is bright green.

Lincolnshire Red Globe

This variety is remarkable for its large, deep-green, luxuriant foliage. Bulb very large, roundish; skin, below ground, white,—above the surface, purple; flesh white, firm, and, when young, well flavored, and adapted to table use. It yields abundantly; is uniformly fair, and free from small roots; an average keeper; and deserving of cultivation, especially for agricultural purposes.

Long Black

Except in the form of its roots, this variety much resembles the Round Black. It possesses the same peculiar, piquant, radish-like flavor; and is served at table in the same manner.

Long White Maltese

(Long White Clairfontaine)

Roots eight or nine inches in length, an inch and a half in diameter, somewhat fusiform, and very smooth and symmetrical. The crown rises two or three inches above the surface of the ground, and is of a green color, except where exposed to the sun, when it often becomes purple or reddish-brown. Below the surface of the soil, the skin is of a dull or dirty white. Flesh white, moderately fine, tender, and of a sugary flavor. Half early.

The variety has some resemblance to the Cow-horn; but is smaller, and the flesh not so white.

Petrosowoodsks

Bulb of medium size, flattened,—comparatively smooth and regular; tap-root very slender, issuing from a basin; skin blackish-purple above and below ground, sometimes changing to yellow about the tap-root of large or overgrown bulbs; flesh yellow, fine-grained, and tender, if grown in cool weather, but liable to be fibrous and strong-flavored when grown during the summer months. The variety is early, and must be classed as a garden rather than as a field turnip.

Pomeranian Globe

Bulb globular, remarkably smooth and regular; the neck is small, and the skin white, smooth, and glossy; the flesh is white, close-grained, tender, and sweet; the leaves are large, and of a dark-green color, with paler or whitish nerves. Half early.

When in perfection, the bulbs measure three and a half or four inches in diameter, about the same in depth, and weigh from fourteen to eighteen ounces. If sown early in good soil, and allowed the full season for development, the roots sometimes attain a weight of eight or ten pounds. It is generally cultivated as a field turnip, but is also sown as a garden variety; the roots being of good quality for the table, if pulled when about half grown.

Preston, or Liverpool Yellow

An early sort, somewhat resembling the Yellow Malta: the bulbs attain a larger size, the foliage is stronger, and the basin, or depression, about the tap-root less deeply sunk.

Purple-Top Flat

(Red-Top Flat)

Bulb round, flattened, nearly one-half growing above ground; neck and tap-root small; skin reddish-purple where exposed to light and air, and white below the surface of the soil; flesh very white, close-grained while young, and of a sugary but often bitter taste. During winter, it usually becomes dry and spongy. Average specimens measure two and a half inches in depth, four or five inches in diameter, and weigh from sixteen to twenty ounces.

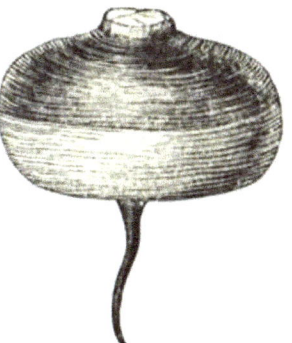

Purple-Top Strap Leaved Turnip

This old and well-known variety, at one period, was the principal field as well as garden turnip of the Northern and Middle States. It is now, however, very little cultivated; being superseded by the Strap-leaved and other more desirable sorts.

Purple-Top Strap-Leaved

Bulb very flat, smooth, and regular in form, produced almost entirely above ground; tap-root slender; leaves few, upright, broad, rounded at the ends, and tapering to the neck, which is very small; skin above, clear, bright purple,—below, pure white, often finely clouded or shaded at the union of the colors; flesh clear white, firm, solid, sugary, mild, and remarkably well-flavored; size medium,—measuring about two inches and a half in depth by four or five inches in diameter, and weighing from ten to twelve ounces. Field-grown roots, with the benefit of a long season and rich soil, attain much greater dimensions; often, however, greatly deteriorating in quality as they increase over the average size.

This variety is unquestionably one of the best of the flat turnips, either for the garden or field. It is early, hardy, very prolific, will thrive in almost any description of soil, is of excellent quality, and rarely fails to yield a good crop. It is the best of all the flat turnips for sowing among corn or potatoes, or upon small patches of the garden from which early pease or beans have been harvested.

Purple-Top Yellow Aberdeen

(Purple-Top Aberdeen; Purple-Top Yellow Bullock)

Bulb globular, reddish-purple above, and deep yellow below; tap-root very small; leaves deep green, comparatively short, and inclined to grow horizontally.

In rich soil and long seasons, the bulbs sometimes attain a weight of eight or ten pounds; but specimens of average size measure about four inches in depth, nearly five inches in diameter, and weigh from sixteen to twenty ounces. The flesh is pale yellow, tender, sugary, and nearly equal to that of the Swedes in solidity. The variety is very hardy, and, although generally grown for farm purposes, is really superior to many sorts cultivated exclusively for table use.

Red Globe.

An old, medium-sized, globular turnip, well suited for cultivation in light soil and on exposed or elevated situations. Skin red, where exposed to the sun,—below ground, white; flesh white, and finer in texture than that of the White Globe. It is not suited for table use; and is generally field-grown, and fed to stock.

Red Norfolk

(Red-Top Norfolk; Red Round)

This is a sub-variety of the White Norfolk, the size and form being nearly the same. Skin washed, or clouded with red were exposed to the light. It is firmer in texture, and more regular in its form, than the last named; and, if there be any difference in size, this is the smaller variety.

Red Tankard

Bulb produced partially above ground, pyriform, eight or nine inches in depth, four or five inches in diameter, and weighing about three pounds; below ground, the skin is white,—above, purple or violet; flesh white, rather firm, sugary, and well flavored; foliage large.

It is recommended for its earliness and productiveness, but must be considered a field rather than a table variety.

Robertson's Golden Stone

An excellent, half-early variety; form nearly globular; color deep orange throughout, sometimes tinged with green on the top; size above medium,—average specimens measuring nearly four inches in depth, four inches in diameter, and weighing from sixteen to eighteen ounces; flesh firm, and well flavored.

The Robertson's Golden Stone is remarkably hardy, keeps well, and is one of the best of the Yellows for autumn or winter use.

Round Black

Leaves few, small, and comparatively smooth; bulb produced almost or altogether under ground, of an irregular, roundish form, often divided, or terminating in thick branches at its lower extremity; skin black, and very tough; flesh white.

The variety is extensively cultivated in some parts of Europe, and is much esteemed for its peculiar, piquant, somewhat radish-like flavor. It is sometimes served in its crude state as a salad.

Six Weeks

(Autumn Stubble; Early Dwarf.)

Bulb produced much above ground, rather large, and of an irregular, globular form. It soon arrives at maturity; but, on account of its natural softness of texture, should always be sown late, and used before severe frosts. As descriptive of its earliness, it has received the above names; being suited for very late sowing, after the removal of early crops; or for making up blanks in turnip-fields, where the first sowing may have partially failed.

It is well flavored, but soon becomes dry and spongy, and is unsuitable for use during winter. Skin white below the surface of the ground, greenish above. Field-grown specimens sometimes weigh three pounds and upwards.

Small Long Yellow

Leaves very small, and spreading; root generally entirely under ground, small, and of an oblong or carrot shape, terminating abruptly at the point; skin pale yellow; flesh yellow, firm, dry, and sugary, with some degree of piquancy. It is a good variety for the table, and also a good keeper.

Snow-Ball

(Navet Boule de Neige)

The bulb of this variety is nearly spherical, very smooth and regular; size medium,—the average dimensions being four inches in diameter, four and a half in depth, and the weight about a pound. The neck is small, and the skin white. The flesh of the young bulbs is white, fine-grained, tender, and sugary; but, if overgrown or long kept, it is liable to become dry and spongy.

The variety is early, and, though classed by seedsmen as a garden turnip, is well adapted for field culture; as it not only yields abundantly, but succeeds well when sown late in the season on land from which early crops have been harvested.

Stone Globe

Bulb globular, and regularly formed, growing mostly beneath the surface of the ground. It belongs to the White-Globe varieties, and is considered the hardiest and the best suited for winter use of any of its class. The leaves are larger, stronger, and deeper colored, than any of the White-globe sorts.

Skin and flesh white; texture moderately close; flavor sweet, and its keeping properties good; size rather large.

Teltow, or Small Berlin

(Teltau)

This is said to be the smallest of turnips; its leaves not exceeding in number those of the radish. The root is fusiform or spindle-shaped, not very regular, and produced entirely under ground; skin dusky white; flesh dry, dull white, very fine-grained, piquant, and sugary; leaves erect, yellowish-green. Early. The roots measure three inches long by about an inch and three-fourths at their largest diameter, and weigh from three to four ounces.

The Teltow Turnip is much esteemed on account of its excellent qualities, and is one of the best early garden varieties.

According to Loudon, it is in high repute in France, Germany, and Holland; and is grown in the sandy fields around Berlin, and also near Altona, whence it is imported to the London market. It is, or was, grown in immense quantities in the neighborhood of Moscow.

The peculiar flavor is in the outer rind. When used, it should not be peeled. It bears transplanting well; and may be set in rows one foot apart, and nine inches apart in the rows.

Waite's Hybrid Eclipse

A recent variety, of English origin, introduced by Mr. John G. Waite, a seed-merchant of London. As figured and described, it is of large size, very richly colored, and remarkably smooth and symmetrical. At the crown, it is broad and round-shouldered, and measures about six inches in diameter; which size is nearly retained to a depth of eight or nine inches, when it contracts in a conical form to a tap-root. Color of upper portion, clear purple, richly clouded, and contrasting finely with the yellow on the lower part. It is represented as a turnip of excellent quality, and as being very productive.

When cultivated in this country, it has generally fallen short of the excellence it is represented as attaining in England. It is apparently not adapted to the dry and warm summers of the United States.

White Globe

(Common Field Globe)

Root globular; skin smooth, perfectly white; flesh also white; neck and tap-root small. Although this description embraces the principal characters of the White Globe, there is considerable variety in the turnips to which this name is applied, arising from the degree of care and attention bestowed by growers in selecting their seed-roots; and the shape is often not a little affected by the soil in which they are grown. Thus Globes of any kind, and particularly those of this variety, when grown on a very superior, rich soil, may be said to be forced beyond their natural size, and thereby acquire somewhat of a monstrous or overgrown appearance; losing, in a great measure, their natural symmetry.

This variety is better adapted to field culture than to the garden, as it is altogether too coarse in texture for table use. It is a poor keeper, and, in unfavorable seasons, sometimes decays before the time of harvesting. Specimens have been grown weighing fifteen and even eighteen pounds.

White Norfolk.

(White Round)

A large English variety, somewhat irregular in form, but usually more or less compressed, and sometimes pyriform; the upper portion of the root being produced four or five inches above ground. Specimens sometimes measure ten or twelve inches in diameter. The leaves are large, and rather numerous; the skin white below the surface, and often white above, but sometimes washed with green; flesh white and coarse-grained, but sweet. Very late.

It is but a sub-variety of the Common Flat Turnip, and oftentimes attains a most extraordinary size. For the garden, it possesses no value. It is grown exclusively as an agricultural or field turnip; but is very liable to rot; soon becomes spongy; and can only be classed as third-rate, even for feeding stock.

White Stone

(Early Stone; White Garden Stone)

This common and well-known garden turnip somewhat resembles the White Dutch; but has stronger foliage, is rounder in form, and finer in texture. A carefully selected and improved variety of this is known by the name of Mouse-tail Turnip; and, in addition, some catalogues contain varieties under the name of Red-topped Mouse-tail, &c.

Skin and flesh white; size full medium, measuring three and a half to four inches in depth by four and a half or five inches in diameter.

White Tankard

(Navet Gras d'Alsace)

Bulb pyriform, cylindrical at the crown, which, like that of the Red Tankard, rises two or three inches from the ground; skin white in the earth, green above; flesh white, tender, sweet, rather firm, and close-grained. Early.

Vilmorin mentions two varieties; one having entire leaves, the other with lyrate or lobed leaves; giving preference, however, to the one with entire leaves.

Like most of the Tankards, the variety seems better adapted to agricultural than to horticultural purposes.

White-Top Flat

Bulb similar in size and form to the Green-top Flat; leaves few and small; skin uniformly white; flesh white, firm, sugary, and well flavored. As a table variety, it is superior to the Purple-top Flat or the Green-top.

White-Top Strap-Leaved

This is a sub-variety of the Purple-top Strap-leaved; differing little, except in color. The leaves are erect, few and small, somewhat lanceolate, and nearly entire on the borders; the bulb is of medium size, much flattened, green above ground, white below, and remarkably smooth and regular in form; tap-root very small; the flesh is white, very fine-grained, saccharine, mild, and excellent.

Early, productive, and recommended as one of the best varieties for field or garden culture.

The Strap-leaved Turnips appear to be peculiarly adapted to the climate of the Northern States, and are greatly superior in all respects to the Common White and Purple-top Flat varieties. Though of comparatively recent introduction, they have been widely disseminated; and, wherever grown, are highly esteemed.

Yellow Malta

(Maltese; Golden Maltese)

A beautiful, very symmetrical, small-bulbed, early variety, slightly flattened above, somewhat concave about, the tap-root, which, as well as the neck, is remarkably small; skin very smooth, bright orange-yellow; foliage small, and not abundant,—on which account the plants may be grown quite close to each other; flesh pale-yellow, fine-grained, and well flavored. It is a good garden variety, and one of the best of the Yellows for summer use. Average bulbs measure two inches in depth, four inches in diameter, and weigh about ten ounces.

Yellow Scarisbrick

Bulb flattened, smooth, and regular; neck small; skin pale yellow,—above ground, green; flesh yellowish-white, tender, and sweet; leaves of medium size, very pale-green. Season late. Well-grown specimens measure four inches in diameter, and about three inches in depth.

Yellow Stone

Very similar to the Golden Ball or Yellow Globe. Compared with these varieties, the bulb of the Yellow Stone is produced more above ground, and the upper surface is more colored with green. One of the best of garden turnips.

Yellow Tankard

Root somewhat fusiform, or of a long, irregular, tankard shape; the crown rising just above the ground. Average specimens measure seven or eight inches in length, three inches and a half in diameter, and weigh about twenty-four ounces. Skin yellowish-white below ground, green above; flesh pale yellow, firm, and sugary; leaves large. It is esteemed for the solidity of its flesh, and for its earliness and productiveness. A good variety for either field or garden.

Some of Gartons Limited's Turnips, Swedes, Kales, and Kohl Rabi's

Part Two: Alliaceous Plants

Chives; Garlic; Leek; Onion; Rocambole; Shallot; Welsh Onion.

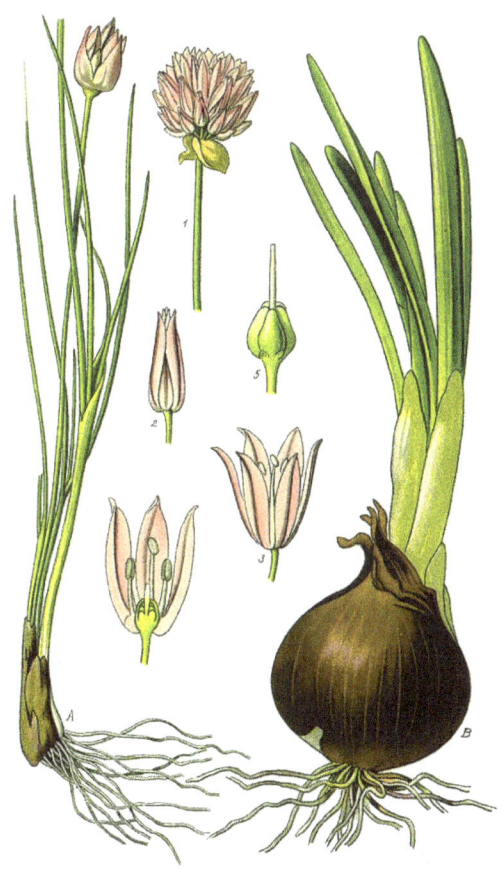

Illustration Allium schoenoprasum and Allium cepa
Prof. Dr. Otto Wilhelm Thomé

23

The Chive (Allium schœnoprasum)

The Cive is a hardy, bulbous-rooted, perennial plant, indigenous to France and Great Britain. The leaves, which are produced in tufts, are seven or eight inches in length, erect and cylindrical, or awl-shaped. The bulbs are white, oval, and of small size; usually measuring about half an inch in diameter. The flower-stalk rises to the height of the leaves, and produces, at its extremity, a globular group of purplish, barren flowers.

Propagation and Culture—As the plant seldom, if ever, produces seeds, it is always propagated by a division of the roots, or bulbs. These are produced in compact groups, or bunches, seven or eight inches in diameter.

> "One of these groups may be divided into a dozen or more parts, each of which will, in a short time, form a cluster equal in size to the original. They should be planted in spring or autumn, in rows eighteen inches apart, and twelve or fifteen inches asunder in the rows. All the cultivation they require is to be kept free from weeds; and they will thrive in any common garden soil. A planting will last many years; but it is well to renew it every third or fourth year."

Use—The young leaves are the parts of the plant used; but, whether used or not, to keep them in a fresh and tender condition, the plants should be frequently shorn to the ground. They possess the flavor peculiar to the Onion family; and are principally used in flavoring soups, and as an ingredient in spring salads. The leaves and bulbs are sometimes taken together, and eaten crude, as a substitute for young onions. In omelets, the Cive is considered almost indispensable.

There are no varieties.

24

Common Garlic (Allium sativum)

This is a perennial plant, from the south of Europe. The root is composed of from ten to fifteen small bulbs, called "cloves," which are enclosed in a thin, white, semi-transparent skin, or pellicle. The leaves are long and narrow. The flower-stem is cylindrical, about eighteen inches in height, and terminates in an umbel, or group, of pale-pink flowers, intermixed with small bulbs. The seeds are black, and, in form, irregular; but are seldom employed for propagation; the cloves, or small bulbs, succeeding better.

Planting and Cultivation—Garlic thrives best in a light, well-enriched soil; and the bulbs should be planted in April or May, an inch deep, in rows or on ridges, fourteen inches apart, and five or six inches apart in the rows.

"All the culture necessary is confined to keeping the ground free from weeds. When the leaves turn yellow, the plants may be taken up; and, having been dried in the sun, they should be tied up in bunches by the stalks, and suspended in a dry, airy room, for use."[1]

Use—It is cultivated for its bulbs, or cloves, which possess more of the flavor of the onion than any other alliaceous plant. These are sometimes employed in soups, stews, and other dishes; and, in some parts of Europe, are eaten in a crude state with bread.

"It is not cultivated to any considerable extent in this country; its strong flavor, and the offensive odor it communicates to the breath, causing it to be sparingly used in our cookery.
"Where attention is paid to culture, the Common Garlic will attain a size of seven and a half inches in circumference, each bulb; whereas, when grown negligently and unskilfully, it does not attain half that size. Twenty ordinary bulbs weigh one pound."[2]

Early Rose Garlic

(Early Pink)

This is a sub-variety of the Common Garlic. The pellicle in which the small bulbs are enclosed is rose-colored; and this is its principal distinguishing characteristic. It is, however, nearly a fortnight earlier.

For culinary purposes, it is not considered superior to the Common Garlic. Propagation and cultivation the same; though, in warm climates, the bulbs are sometimes planted in autumn.

Great-Headed Garlic

(Allium ampeloprasum)

This species is a hardy perennial, and is remarkable for the size of its bulbs; which, as in the foregoing species and variety, separate into smaller bulbs, or cloves. The leaves and stem somewhat resemble those of the leek; the flowers are rose-colored, and are produced at the extremity of the stalk, in large, regular, globular heads, or umbels; the seeds are similar to those of the Common Garlic, but are seldom used for reproduction; the cloves, or small bulbs, being generally employed for this purpose. It is used and cultivated as the Common Garlic.

1. The Gardener's Assistant. By Robert Thompson.

2. The Book of the Garden. By Charles M'Intosh. 2 vols. Edinburgh and London, 1855.

Garlic. Danyang, South Korea
Douglas Perkins

25

The Leek (Allium porrum)

The Leek is a hardy biennial, and produces an oblong, tunicated bulb; from the base of which, rootlets are put forth in great numbers. The plant, when full grown, much resembles what are commonly known as "Scallions;" the lower, blanched portion being the part eaten. This varies in length from four to eight inches, and in diameter from less than an inch to more than three inches. The leaves are long, narrow, smooth, and pointed; and spread in opposite directions, somewhat in the form of a fan. The flower-stem proceeds from the centre of this collection of leaves, and is about four feet in height. The flowers are white, with a stripe of red, and are produced in terminal, globular groups, or umbels; the seeds are black, irregular, but somewhat triangular in form, and, with the exception of their smaller size, are similar to those of the onion. About twelve thousand seeds are contained in an ounce; and they retain their vitality two years.

Soil, Sowing, and Cultivation—The Leek is very hardy, and easily cultivated. It succeeds best in a light but well-enriched soil. When fine leeks are desired, it can hardly be made too rich. It should also be thoroughly spaded over, and well pulverized to the depth of at least twelve inches. The seed should be sown in April, at the bottom of drills made six or eight inches deep, and eighteen inches asunder. Sow the seeds thinly, cover half an inch deep, and thin the young plants to nine inches distant in the drills. As the plants increase in size, draw the earth gradually into the drills, and around the stems of the leeks, until the drills are filled. By this process, the bulbs are blanched, and rendered tender and mild flavored. The seeds are sometimes sown broadcast, and in July transplanted to trenches, and subsequently cultivated, as before directed. The plants are also sometimes set on the surface, and afterwards earthed up to the height of six or eight inches in the process of cultivation. In October, the leeks will be suitable for use; and, until the closing-up of the ground, may be drawn from time to time as required for the table. For winter use, they should be preserved in earth or sand.

Early leeks may be obtained by sowing the seeds in a hot-bed in February or March, and transplanting to the open ground in June or July.

Seed—To obtain seed, some of the finest plants of the growth of the previous year should be set out in April, fifteen inches apart, and the stems sunk to the depth of three or four inches.

> "The seed ripens in autumn, and its maturity is known by the heads changing to a brown color. It is best preserved in the heads; and these should be cut off with a portion of the stalk a foot in length, tied in bunches, and hung in a dry, airy situation. In this manner, the seed will retain its vegetative powers for two or three years: after that time, it is not to be depended on."[1]

Use—The whole plant, except the roots, is used in soups and stews. The white stems, which are blanched by being planted deep for the purpose, are boiled, served with toasted bread and white sauce, and eaten like asparagus. It has the flavor, and possesses the general properties, of the onion.

Varieties—

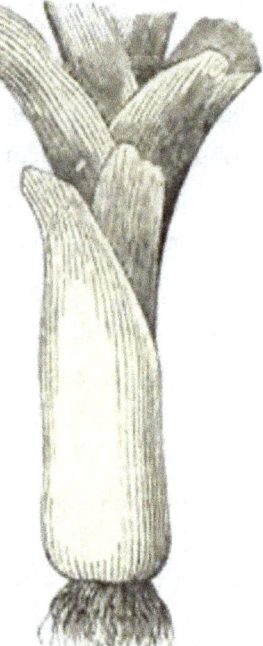

Rouen Leek

Common Flag
(Long Flag)

The stem, or blanched portion, of this variety is about six inches in depth, and an inch in diameter. The leaves are put forth in opposite directions, are comparatively erect, and of a glaucous-green color.

The variety is remarkably hardy, and well suited for open culture.

Large Rouen
(Gros de Rouen)

Leaves very dark-green, broad, and of thick substance; stem rather short, but remarkably thick, sometimes measuring nearly four inches in diameter. It is now the variety most cultivated near Paris; and, since its general dissemination, has been much approved by all who have grown it. It is found to be the best kind for forcing, as it acquires a sufficient thickness of stem sooner than any other. In England, it is pronounced one of the best, if not the best, of all varieties.

Little Montagne

Stem very short and slender; foliage deeper green than that of the Common Flag. It is the smallest of the leeks. Not much cultivated.

London Flag
(Large Flag; Broad Flag; English Flag; Gros Court)

Stem about four inches in length, and nearly an inch and a half in diameter. The leaves are larger, of a paler color, and softer in their texture, than those of the Common Flag.

The London-flag Leek is hardy, and of good quality. It is more generally cultivated in this country than any other variety.

Musselburgh

(Scotch Flag; Edinburgh Improved)

Stem somewhat shorter than that of the London Flag, but of equal thickness. The swelling at the base has the same form. The leaves are broad and tall, and spread regularly in a fan-like manner. Their color is deeper than that of the Long Flag or the Large Rouen, but paler than the London Flag. Hardy, and of excellent quality. It originated in England.

Proliferous Leek

This is a viviparous variety of the common leek, producing young plants on its flower-stalk instead of flowers. The leaves are similar to those of the London Flag; and the plant, in its young state, before it runs to flower, exactly resembles it. The flower-scape is from two to three feet high, and supports a compact, irregular, globose umbel, composed of numerous small bulbs, intermixed with flowers. Some of these bulbs occasionally produce a second umbel, on scapes of from six to eight inches in length, but of much smaller dimensions than the principal one.

The variety is cultivated in rows, like other leeks; and the bulbs will remain sound several months after they have ripened.

Small Early Netherland

(Small Summer Brabant)

Leaves long, narrow, dark-green; stem small. On this account, it is not so valuable as many others for a main crop: besides, if sown at the same time, it is liable to run to seed before winter. A small sowing, however, may be made with advantage for early use.

Yellow Poitou

(Jaune du Poitou)

A remarkably large variety; the leaves having sometimes measured five feet in length, and six inches in breadth. They are of a yellowish-green color. The underground or blanched portion of the stem is yellowish-white, and is more tender than that of any other variety. On this account, and also for its large size, it deserves cultivation. The great length of the leaves makes it important that more space should be allowed between the plants than is usually allotted to other varieties.

1. The Gardener's Assistant. By Robert Thompson.

Garden, Farm and Flower Seeds Catalogue 1903, J.M. Philips' Sons

26

The Onion (Allium cepa)

The Onion is a half-hardy biennial plant: the roots and leaves, however, are annual; as they usually perish during the first year. The bulbs, for which the plant is generally cultivated, are biennial, and differ to a considerable extent in their size, form, and color. The flower-stalk, which is developed the second year, is from three to four feet in height, leafless, hollow, swollen just below the middle, and tapers to the top. The flowers are either white or rose-colored, and are produced at the extremity of the stalk in a regular, globular group, or umbel. The seeds ripen in August. They are deep blue-black, somewhat triangular, and similar in size and form in all the varieties. An ounce contains about seventy-five hundred seeds, which retain their vitality two years.

Soil and Cultivation—The Onion requires a light, loamy, mellow soil; and, unlike most kinds of garden or field vegetables, succeeds well when cultivated on the same land for successive years. With the exception of the Top and the Potato Onion, all the varieties are raised from seed. Previous to sowing, the ground should be thoroughly spaded over or deeply ploughed, and the surface made smooth and even. The seed should be sown as early in spring as the soil may be in good working condition. Sow in drills fourteen inches apart, and half an inch in depth. When the plants are three or four inches high, thin them to two inches asunder; and, in the process of culture, be careful not to stir the soil too deeply, or to collect it about the growing bulbs. The onions will ripen in August, or early in September; and their full maturity will be indicated by the perfect decay of the leaves, or tops. The bulbs may be drawn from the drills by the hand, or by the use of a common garden-rake. After being exposed for a few days to the sun for drying, they will be ready for storing or the market.

Preservation—The essentials for the preservation of the bulbs are a low temperature, freedom from frost, dryness, and thorough ventilation.

Seed—For the production of seed, select the ripest, firmest, and best-formed bulbs; and, in April, transplant them to lines two feet and a half or three feet distant, and from nine to twelve inches apart in the lines, sinking the crowns just below the surface of the ground. As the plants advance in height, tie them to stakes for support. The seeds ripen in August: and the heads, or umbels, should be cut off when they assume a brown color; for then the capsules begin to open, and shed their seeds. After being threshed out, the seed should be exposed to the action of the sun until it is thoroughly dried; for, when stored in a damp state, it is extremely liable to generate heat, and consequently to lose its vitality.

Varieties—Few of the numerous varieties are cultivated to any extent in this country. Many of the kinds succeed only in warm latitudes, and others are comparatively unimportant. The Danvers, Large Red, Silver-skin, and the Yellow seem peculiarly adapted to our soil and climate. The annual product of these varieties greatly exceeds that of all the other sorts combined.

Blood-Red

(French Blood-Red; Dutch Blood-Red; St. Thomas)

Bulb middle-sized, or rather large, flattened; skin dull red,—the coating next within glossy, and very dark red. The internal layers are palest at the base; and, except at the top, are only coloured on their outsides. Each layer is paler than the one which surrounds it; till the centre is reached, which is white.

It is a good keeper, but one of the strongest flavored of all varieties. It imparts to soups, or other dishes of which it may be an ingredient, a brownish or blackish color.

Brown Portugal

(Brown Spanish; Cambrai; Oporto)

A medium-sized, roundish, or flattened onion; neck small; skin yellowish-brown,—next interior layer not tinged with red. It is a popular variety in some parts of France; and is remarkable for its productiveness, excellent quality, and keeping properties.

Danvers

(Danvers Yellow)

This comparatively recent variety was obtained by selection from the Common Yellow. It is somewhat above medium size, and inclined to globular in its form. Average bulbs measure three inches in diameter, and two inches and three-fourths in depth. The skin is yellowish-brown, but becomes darker by age, and greenish-brown if long exposed to the sun; the flesh is similar to that of the Yellow,—white, sugary, comparatively mild, and well flavored.

Danvers Onion

The superiority of the Danvers Onion over the last named consists principally, if not solely, in its greater productiveness. When grown under like conditions, it yields, on the average, nearly one-fourth more; and, on this account, the variety is generally employed for field culture.

It is, however, not so good a keeper; and, for shipping purposes, is decidedly inferior to the

Yellow,—its globular form rendering it more liable to decay, from the heat and dampness incident to sea voyages.

When cultivated for the market, the land is thoroughly ploughed, and well enriched with fine decomposed manure. The surface is then harrowed, and next raked free of stones, and lumps of earth. The seed is sown in April, usually by machines, in rows fourteen inches apart, and three-fourths of an inch in depth; three pounds of seed being allowed to an acre. The crop is treated in the usual form during the summer; and ripens the last of August, or early in September. When the tops have entirely withered, the bulbs are raked from the drills, and spread a few days in the sun for drying; after which they are sorted, and barrelled for storing or the market. The yield varies from five to eight hundred bushels per acre.

Deptford

(Brown Deptford)

Very similar to, if not identical with, the English Strasburg. It sometimes exactly agrees with the description of that variety: but it occasionally has a pale-brown skin, without any tinge of red; and, when this is the case, its flavour is milder than that of the last named.

With the exception of its more globular form, the bulb much resembles the Yellow Onion of this country.

Early Silver Nocera

(Early Small Silver Nocera; White Nocera; Blanc Hatif de Nocera)

This is a very small variety of the Early Silver-skin, with a small, occasionally roundish, but generally oblate bulb. The skin is white; but the layers beneath are striped with bright-green lines. The leaves are very small. Sometimes the bulb has only a single leaf, frequently but two; and, if there are more than four, the plant has not its true character.

It is an excellent sort for pickling; and is the smallest and earliest variety known,—being fifteen or twenty days earlier than the Early Silver-skin: but it is very liable to increase in size, and to degenerate. Very little known or cultivated in this country.

Early Red Wethersfield

A sub-variety of the Large Red Wethersfield, and the earliest of the red onions. Form and color nearly the same as the Large Red; bulb small, measuring about two inches and a half in diameter, and about an inch and a half in depth. It is close-grained; mild; a good keeper; forms its bulbs, with few exceptions, and ripens, the last of July; being three or four weeks earlier than the Large Red. Cultivated to a limited extent in various places on the coast of New England, for early consumption at home, and for shipment to the South and West.

This variety and the Intermediate are very liable to degenerate: they tend to grow larger and later, approaching the original variety; and can be preserved in a pure state only by a careful selection of the bulbs set for seed.

Early Silver-Skin

(Blanc Hatif)

This is a small early variety of the Silver-skin, measuring two inches and three-fourths in diameter, and an inch and three-fourths in depth. The neck is small, and the skin silvery-white. It is much esteemed for its earliness and mild flavor, and is one of the best of all varieties for pickling. When cultivated for the latter purpose, it should be sown and treated as directed for the Silver-skin.

Fusiform, or Cow-Horn

(Corné de Bœuf)

This is a large onion, growing from eight inches to a foot in length. It tapers rather regularly from the base to the top, and is frequently bent or curved in the form of a horn; whence the name. Skin copper-red. It is late, lacks compactness, is very liable to degenerate, decays soon after being harvested, and must be considered more curious than useful.

Intermediate Red Wethersfield

An early variety of the common Large Red. Bulb of medium size, flattened; neck small; color deep purple.

It is rather pungent, yet milder than the Large Red; keeps well; and is grown to a considerable extent, in certain localities in New England, for shipping.

James's Keeping

(James's Long Keeping; De James)

This is an English hybrid, said to have been originated by a Mr. James, an extensive market-gardener in Surrey, Eng. The bulb is pyriform, or pear-shaped; and measures four inches and upwards in depth, and two inches or more at its broadest diameter. Skin copper-yellow,— the coating next under it reddish-brown; flavor strong. It is not early, but is much prized for its long keeping; the bulbs not sprouting so early in spring as those of most varieties.

Large Red

(Wethersfield Large Red)

Wethersfield Large Red Onion

Bulb sometimes roundish, but, when pure, comparatively flat. It is of very large size; and, when grown in favorable soil, often measures five inches or more in diameter, and three inches in depth. Skin deep purplish-red; neck of medium size; flesh purplish-white, moderately fine-grained, and stronger flavored than that of the Yellow and earlier Red varieties. It is very productive; one of the best to keep; and is grown to a large extent, in many places on the seacoast of New England, for shipping to the South and West. It is almost everywhere seen in vegetable markets; and, with perhaps the exception of the Yellow or Danvers, is the most prominent of the sorts employed for commercial purposes. It derives its name from Wethersfield, Conneticut; where it is extensively cultivated, and where it has the reputation of having originated.

A sub-variety of the foregoing is cultivated in some localities, with nearly the same variation in form that exists between the Danvers and Common Yellow. It will probably prove somewhat more productive; but it is neither better flavored, nor to be preferred for its superior keeping properties.

Madeira

(Large Globe Tripoli; Romain; De Madère Rond; De Belle Garde)

This is a roundish, obovate onion, of remarkable size, often measuring six inches and a half in depth, and six inches in diameter; neck thick and large; skin reddish-brown,—the layer next within, pale red.

The variety is much prized for its extraordinary size, and for its mild, sugary flavor. The plants, however, often fail to form good bulbs; and, even when well matured, the latter are liable to decay soon after being harvested. It requires a long, warm season for its greatest perfection. The seed should be sown early, in drills sixteen inches apart; and the plants should be thinned to eight inches apart in the rows.

Not suited to New England or the cooler sections of the United States.

New Deep Blood-Red.

(Brunswick Deep Blood-Red; Rouge Très Foncé de Brunswick)

Bulb very small, flattened,—two inches and a quarter in diameter, and an inch and a half in depth; neck small; skin deep violet-red, approaching black. A half early variety, remarkable for its intense purplish-red color.

Pale Red

(Rouge Pale, de Niort)

Bulb roundish, flattened on the upper side, but not so much so as the Blood-red, of which this may be considered a variety; size medium, two inches and a half in diameter, one inch and three-quarters in depth; neck small; skin copper-red, much paler than that of the Blood-red. Compared with the last named, it is earlier and of milder flavor. This and the Blood-red are much esteemed by some for their extreme pungency and for their diuretic properties.

Paris Straw-Colored

(Jaune des Vertus)

A large, somewhat flattened variety, much cultivated about Paris; skin fine russet-yellow; neck small. It is not early, but very productive, and of excellent quality.

Pear-Shaped

Bulb pyriform, measuring four inches and a half in depth, and two inches in diameter at the broadest part; neck small; skin copper-red. It is quite late, but is of good quality, and keeps well.

Potato Onion

(Underground Onion)

Bulb flattened, from two and a half to three inches in diameter, and about two inches in depth; skin copper-yellow; flavour sugary, mild, and excellent. It does not keep so well as many other varieties; but remains sound longer, if the leaves are cut two or three inches above the top of the bulb at the time of harvesting.

The Potato Onion produces no seeds, neither small bulbs upon its stalks, in the manner of many of the species of the Onion family; but, if a full-grown bulb be set in spring, a number of bulbs of various sizes will be formed, beneath the surface of the ground, about the parent bulb. By means of these it is propagated, and an abundant supply often secured in localities where the varieties raised from seed frequently wholly fail, either from the maggot, effects of climate, or other causes.

Like the other kinds of onions, it requires a rich, deep soil, well manured, and dry at the bottom. This should be deeply and thoroughly stirred, and then raised in ridges of moderate height, fifteen inches apart. In April, select the large bulbs, and set them on the ridges, ten inches apart, with the crown of the bulbs just below the surface of the ground. The subsequent culture consists in keeping them clean from weeds, and gathering a little earth about them from time to time in the process of cultivation. As soon as the tops are entirely dead, they will be ready for harvesting.

It is very prolific, yielding from four to six fold. Such of the crop as may be too small for the table should be preserved during the winter, to be set in the following spring; planting them out in April, in drills one foot apart and three inches from each other in the drills, and sinking the crowns just below the surface of the ground. They attain their full size by September.

Silver-Skin

(White Portugal, of New England)

Bulb of medium size, flattened,—average specimens measuring about three inches in diameter, and an inch and a half or two inches in thickness; neck very small; skin silvery-white. After the removal of the outer envelope, the upper part of the bulb is often veined and clouded with green, while the portion produced below ground is generally clear white. Flesh white, fine-grained, sugary, and remarkably mild flavored.

It forms its bulb early and regularly, ripens off well, and is quite productive; an average yield being about four hundred bushels per acre. It is a very poor keeper; and this is its most serious objection. It is always preserved through the winter with much difficulty, and almost invariably decays if kept from light and exposed to dampness. The best method for its preservation is to spread the roots in a dry, light, and airy situation.

The Silver-skin Onion is much esteemed in the middle and southern sections of the United States, and is cultivated to a considerable extent in New England. It is well adapted for sowing in

August, or the beginning of September, for early use, and for marketing during the ensuing spring. Where the winter are mild, the crop, with slight protection, will sustain no injury in the open ground. In Europe it is much esteemed, and extensively grown for pickling, as its "white color, in contrast with the fine green veins, or lines, gives it a very agreeable appearance. For pickling, the seed should be sown very thickly, then slightly covered with fine soil, and afterwards rolled. If the seed is covered more deeply, the bulb, from not being quite on the surface, has a larger and thicker neck; so that it loses its finely rounded form, and is, moreover, less compact."

This variety, erroneously known in New England as the "White Portugal," is unquestionably the true Silver-skin, as described both by English and French authors. The application of the term "Silver-skin" to the common Yellow Onion, as very extensively practised by seedsmen and market-men in the Eastern States, is neither pertinent nor authorised.

Strasburg

(Yellow Strasburg; Flanders; Dutch; Essex)

This is the variety most generally cultivated in Great Britain. Its form varies from flat to globular, or oval; bulb large, three inches wide, and full two inches in depth; outside coating brown, of firm texture. Divested of this, the color is reddish-brown, tinged with green. Flavor comparatively mild. It is a very hardy sort, succeeds in cold localities, and keeps well.

The Strasburg and Deptford Onions much resemble the common Yellow Onion of New England; and the difference between the sorts is not great, when English-grown bulbs of the first-named varieties are compared with the bulbs of the Yellow Onion, American-grown: but seeds of the Strasburg or Deptford, raised in England and sown in this country, almost invariably fail to produce plants that form bulbs so generally or so perfectly as American-grown seeds of the Yellow Onion.

Top or Tree Onion

(Egyptian)

Bulb large, a little flattened; producing, instead of seeds, a number of small bulbs, or onions, about the size of a filbert, which serve as a substitute for seeds in propagation. The flesh is coarse; and the bulbs are very liable to decay during winter, unless kept in a cool and dry situation. The variety has been considered rather curious than useful.

Planting and Culture—Either the bulbs formed in the ground, or the small ones upon the stems, may be planted out in April or May. The former are set one foot apart in each direction, and the stem-bulbs four inches apart in rows eight inches asunder. Stems that bear heavily require to be supported. When ripe, the stem-bulbs should be dried, and kept free from damp in a cool place.

Tripoli

(Flat Madeira; De Madère Plat)

This is one of the largest varieties. The bulb tapers abruptly from the middle to the neck, and almost equally so to the base. It is five inches and upwards in diameter; color light reddish-brown, —beneath the skin, pale brownish-red, tinged with green.

It requires the whole season, and in some localities is considered excellent for a late crop. The flesh is soft, and the bulbs soon perish after being taken from the ground. In its season, it is much esteemed for its mild and delicate flavor. Like the Madeira Onion, the plants fail to form bulbs so generally as other varieties. Not adapted to the climate of the Northern States.

Two-Bladed

(Double Tige)

This variety derives its name from the fact that the small bulbs have generally but two leaves. The larger ones have more; rarely, however, exceeding four: but, unless by far the greater portion have only two leaves, either the seed or the cultivation is at fault.

The bulbs are small, flat, light-brown, very firm, and attain maturity early; the neck is small, and the top of the bulb is depressed or hollowed around the stem. It keeps well, and is an excellent variety.

White Globe

Form nearly ovoid, very regular and symmetrical; skin greenish-yellow, marked with rose-colored lines,—the pellicle changing to white on drying. The bulb measures about four inches in depth, and two inches and three-fourths in its largest diameter. It keeps well, and is an excellent variety.

White Globe Onion

Yellow Globe

Nearly allied to the preceding variety; the size and form being the same. Skin reddish-yellow. It is hardy, productive, of good flavor, keeps well, and deserves general cultivation.

White Lisbon

(Lisbon; Early Lisbon; White Florence)

A very large, globular onion, measuring four inches in diameter, and about four inches in depth; neck comparatively thick; skin smooth, thin, clear, and white.

It is a late variety; and, although comparatively hardy, requires a long, warm season for its full development. Under the most favorable conditions, both with regard to soil and exposure, many of the plants fail to form a good bulb. On account of its hardiness, it is a good sort for sowing in the autumn for a supply of young onions for spring salads; or, if these young bulbs be set in the open ground in April, fine, large onions will be formed towards the end of summer. The variety is better suited to the climate of the Middle States than to that of the Northern and Eastern.

White Portugal, or Spanish

(White Spanish; White Reading; Cambridge; Soufre D'Espagne)

A very large, flat onion, measuring three inches and upwards in width by about two inches in depth; skin loose, of a pale-brown or yellowish-brown, falling off spontaneously, and exhibiting the next coating, which is greenish-white. It has a small neck, and is particularly mild flavored. One of the best for early winter use, but early decays.

Very distinct from the White Portugal of the New-England markets.

Yellow Onion

(Silver-skin of New England)

One of the oldest varieties, and, as a market onion, probably better known and more generally cultivated in this coun

try than any other sort. The true Yellow Onion has a flattened form and a very small neck. Its size is rather above medium,—measuring, when well grown, from three inches to three inches and a half in diameter, and from two inches to two inches and a half in depth. Skin yellowish-brown, or copper-yellow,—becoming somewhat deeper by age, or if exposed long to the sun; flesh white, fine-grained, comparatively mild, sugary, and well flavored. It keeps well, and is very prolific: few of the plants, in good soils and seasons, fail to produce good-sized and well-ripened bulbs. For the vegetable garden, as well as for field culture, it may be considered a standard sort.

Yellow Onion

The Danvers Onion, which is but a sub-variety of the common Yellow, may prove somewhat more profitable for extensive cultivation, on account of its globular form; but neither in its flavor nor in its keeping properties can it be said to possess any superiority over the last named.

The term "Silver-skin," by which this onion is very generally though erroneously known throughout New England, has created great confusion between seedsmen and dealers. Much perplexity might be avoided if its application to the Yellow Onion were entirely abandoned. The genuine Silver-skin, as its name implies, has a skin of pure, silvery whiteness; and is, in other respects, very dissimilar to the present variety.

When extensively cultivated for the market, it should be sown and subsequently treated as directed for the Danvers Onion. The yield per acre varies from four to six hundred bushels.

Men, Women and Children Wanted to Plant Good Seeds, Ratekin's Seed House

27

Rocambole (Allium scorodoprasum)

This plant is a half-hardy perennial from Denmark, partaking of the character of both the leek and garlic. Bulbs or cloves similar to those of the common garlic, with much the same flavor, though somewhat milder; leaves large; flower-stalk about two feet high, contorted or coiled towards the top, and producing at its extremity a group of bulbs, or rocamboles, intermixed with flowers.

***Propagation and Culture*—**It is propagated by planting either the underground bulbs, or the small cloves, or bulbs, that are produced upon the stem of the plant. These should be set in April, in drills ten inches apart, and four or five inches asunder in the drills. In the following August they will have attained their full size, and may be used immediately; or they may be taken up, spread to dry, tied in bunches, and housed, for future consumption. All the culture required is the removal of weeds, and the occasional stirring of the soil.

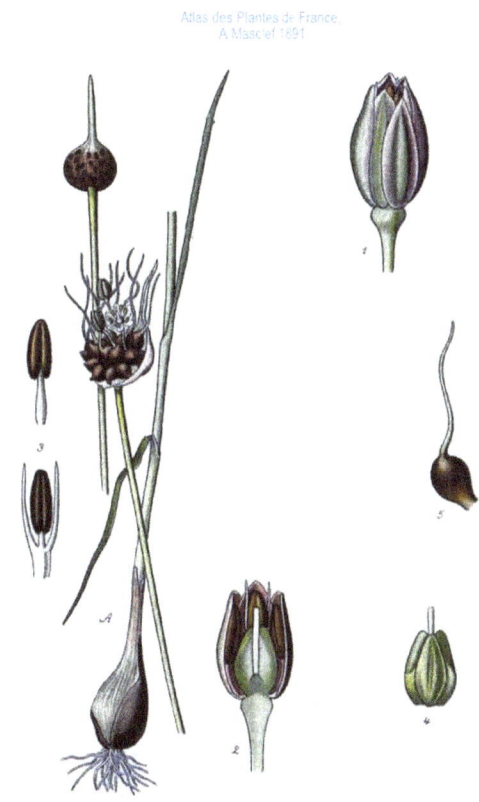

Alliums scorodoprasum L.
Amédée Masclef

Use—The cloves, or small bulbs, as well those from the stem as those beneath the surface of the ground, are used in the manner of shallots and garlics, and nearly for the same purposes.

There is but one variety.

28

Shallot (Allium ascalonicum)

The Shallot (sometimes written Eschalot) is a native of Palestine,—the specific term "Ascalonicum" being derived from Ascalon, a town in Syria: hence also the popular English name, "Scallion."

The root of the plant is composed of numerous small bulbs, united at their base; the whole being enclosed in a thin skin, or pellicle, varying in color in the different varieties. Leaves fistulous, or hollow, produced in tufts, or groups; flowers reddish, in terminal, compact, spherical bunches. The plants, however, very seldom blossom.

Soil—The soil best adapted for growing the Shallot is a light, rich, sandy loam; but, as such soils are scarce, any light, dry soil that has been cultivated and manured a year or two will answer. In wet soils, it is liable to be attacked by the maggot; and such location should, therefore, be avoided.

Propagation and Culture—The roots of the Shallot, which are bulbous, are very readily increased by offsets. The bulbs are oblong, but somewhat irregular in their form, and seldom attain a large size. As they increase into clusters, they do not swell like roots that grow singly.

They are propagated by dividing these clustered roots into separate offsets, and planting the divisions in April, in very shallow drills one foot apart; placing them about six inches apart in the drills, and covering them lightly with earth. Soot mixed with the surface-soil has been found of much service to prevent the maggot from committing extensive depredations upon this plant. The only after-culture required is that of keeping them clean from weeds, and occasionally stirring the ground.

Harvesting—

"As soon as the leaves decay, the bulbs will have attained their growth, and should then be taken up, and spread out in some dry loft; when, after being thoroughly dried and picked, they may be put in bags, boxes, or tied in bundles by the stalks. If kept from frost, they will remain fit for use for several months."

Use—The largest of the bulbs are selected, and employed in the same manner as garlic or onions.

"On account of the mildness of its flavour, when compared with that of other cultivated plants of the Onion family, it is preferred in cookery as a seasoner in soups and stews. It is also much used in the raw state: the cloves, or sections of the root, cut up into small pieces, form an ingredient in French salads; and are also sprinkled over steaks, chops, &c. The true epicure, however, cuts a clove or bulb in two, and, by rubbing the inside of the plate, secures the amount of relish to suit his palate.

"Shallot vinegar is made by putting six cloves, or bulbs, into a quart bottle of that liquid; and, when sealed down, it will keep for years. The Shallot also makes an excellent pickle."[1]

Varieties—

Common or Small Shallot
(Échalote ordinaire)
Bulbs about three-fourths of an inch in diameter at the base, elongated, and enclosed in a reddish-yellow skin, or pellicle; leaves small, ten or twelve inches high.
This variety is early, keeps well, and is one of the best for cultivation.

Jersey
Bulbs of large size, measuring two inches in length, and rather more than an inch in diameter at the base; grouped like the other varieties, and enclosed in a light-brown pellicle, as fine in texture as the skin of an onion, which this Shallot much resembles in form and odor. Compared with the Common Shallot, it is more round, the neck is smaller, and it is also more close or compact. Leaves remarkably glaucous, not tall, but of good substance,—quite distinct in these respects from the Small or the Large sort. It also sometimes produces seeds; which is, perhaps, a recommendation, as these, when sown, frequently produce new varieties. It is one of the earliest of all the sorts; but is comparatively tender, and decays early.

Large Alençon

(Échalote grosse d'Alençon)

Bulb very large, exceeding in size that of the Jersey Shallot; which variety it much resembles in form and color, and in being tender, decaying early, and sometimes running to seed. It is, however, not quite so early; and the leaves are longer and more glaucous. Flavor mild and pleasant.

At the time of harvesting, the bulbs should be long exposed to the sun, in order that they may be thoroughly dried before packing away. "The bulbs are slow in forming, and the worst keepers, as, when stored, they soon begin to sprout."

This variety, and also the Jersey Shallot, closely resemble the Onion. It is possible they may constitute a distinct species.

Large Shallot

(Échalote grosse)

Bulbs about two inches in diameter at the base, elongated, and enclosed in a brownish-yellow skin, or pellicle; leaves fifteen to eighteen inches high.

This variety, in size, much exceeds that of the Common or Small Shallot; and, though later in ripening, is nevertheless the first to be found in the market, as it forms its bulbs early in the season. Its keeping properties are inferior to the last named.

Long Keeping

This resembles the Common Shallot; but is considered superior to that variety in its keeping properties, and in being less subject to the attack of the maggot. It is said that the variety may be kept two years.

---

1. The Book of the Garden. By Charles M'Intosh. 2 vols. Edinburgh and London, 1855.

29

Welsh Onion (Allium fistulosum)

The Welsh Onion is a hardy perennial from Siberia. It is quite distinct from the Common Onion, as it forms no bulbs, but produces numerous elongated, angular, tunicated stems, not unlike scallions, or some of the smaller descriptions of leeks. The flower-stem is about eighteen inches high, swollen near the middle, and terminates in a globular umbel of greenish-white flowers. The seeds are small, black, somewhat irregular in form, and retain their vitality two years. About thirty-six thousand are contained in an ounce.

Sowing and Cultivation—The seeds are sown in drills about half an inch in depth, and the crop subsequently treated as the Common Onion.

There are two varieties:

Common or Red Welsh Onion

Skin, or pellicle, reddish-brown, changing to silvery-white about the base of the leaves; the latter being fistulous, and about a foot in height.

Allium fistulosum

Its principal recommendation is its remarkable hardiness. The seeds are sometimes sown in July and August for the young stems and leaves, which are used during winter and early in spring as salad.

White Welsh Onion

(Early White. Ciboule Blanche Hative)

This is a sub-variety of the Common Red. The skin is rose-white, and, like that of the last named, changes to silvery-white about the upper portion of the stem, or bulb; the leaves are longer, deeper coloured, firmer, and less subject to wither or decay at their extremities, than those of the Common Red. The White is generally considered the better variety; as it is more tender, and milder in flavor, though much less productive.

The Welsh onions are of little value, except in cold latitudes; and are rarely found in the vegetable gardens of this country.

Part Three: Asparaginous Plants

The Artichoke; Asparagus; Cardoon; Hop; Oosung; Phytolacca.

Still Life with a Cardoon on a Shelf
Bernardo Polo

30

The Artichoke (Cynarus scolymus)

The Artichoke is a hardy perennial. The stem is from four to five feet in height, with numerous branches; the leaves are of remarkable size, frequently measuring three feet, and sometimes nearly four feet in length, pinnatifid, or deeply cut on the borders, and more or less invested with an ash-coloured down; the mid-ribs are large, fleshy, and deeply grooved, or furrowed; the flowers are large, terminal, and consist of numerous blue florets, enclosed by fleshy-pointed scales; the seeds (eight hundred and fifty of which are contained in an ounce) are of a greyish color, variegated with deep brown, oblong, angular, somewhat flattened, and retain their vitality five years.

Green Globe Artichoke

Soil—Select a light, rich, and rather moist soil, and trench it well; incorporating in the process a liberal portion of old, well-decomposed compost. Sea-weeds, kelp, rock-weed, and the like, where they can be obtained, are the best fertilizers; but, where these are not accessible, a slight application of salt will be beneficial.

Propagation—Artichokes may be propagated either by seeds, or by slips, or suckers, from established plants. If by slips, they should be taken off in May, when they have grown five or six inches in height, and transplanted four or five inches deep, in rows four feet apart, and two feet apart in the rows. Water freely, if dry weather occurs before the young plants are established. Keep the ground loose by frequent hoeings; and in August or September the heads will be fit for use. Before severe weather, the plants should be covered with straw or stable-litter.

As plants of one year's growth produce but few heads, and are also later in their development of these than established plants, it is the practice of many cultivators to set a few young slips, and to destroy an equal part of the old plantation, yearly.

FIELD AND GARDEN VEGETABLES OF THE LATE NINETEENTH CENTURY | 141

Propagation by Seeds—

"Sow the seeds in April, in a nursery-bed; making the drills a foot apart, and covering the seeds an inch deep. When the plants are three inches high, transplant as before directed. Plants from seeds will seldom flower the first year."[1]

To raise Seeds—Allow a few of the largest central heads to remain; and, just as the flowers expand, bend over the stalk so as to allow the rain to run from the buds, as the seeds are often injured by wet weather. In favorable seasons, they will ripen in September. According to English authority, little dependence can be placed on seedling plants: many produce small and worthless heads, whilst others produce those of large size and of good quality.

Taking the Crop—

"All of the heads should be cut as fast as they are fit for use, whether wanted or not; as allowing them to flower greatly weakens the plants, as does also permitting the stems that produced the heads to remain after the heads are cut off. For pickling whole, the heads should be cut when about two inches in diameter; for other purposes, when they have nearly attained their full size, but before the scales of the calyx begin to open. For what is called 'bottoms,' they should be cut when they are at their largest size, and just as the scales begin to show symptoms of opening, which is an indication that the flowers are about to be formed; after which, the heads are comparatively useless."[2]

Use—The portions of the plant used are the lowest parts of the leaves, or scales, of the calyx; and also the fleshy receptacles of the flower, freed from the bristles and seed-down. The latter are commonly called the "choke," on account of their disagreeable character when eaten.

Sometimes, particularly in France, the central leaf-stalk is blanched, and eaten like cardoons. The bottom, which is the top of the receptacles, is fried in paste, and enters largely into fricassees and ragouts. They are sometimes pickled, and often used in a raw state as a salad. The French also cut them into thin slices; leaving one of the scales, or calyx leaves, attached, by which the slice is lifted, and dipped in oil and vinegar before using. The English present the head whole, or cut into quarters, upon a dry plate; the guests picking off the scales one by one, which have a fleshy substance at the base. These are eaten after being dipped in oil and vinegar.

What is called "artichoke chard" is the tender leaf-stalks blanched, and cooked like cardoons. The Italians and French often eat the heads raw with vinegar, oil, salt, and pepper; but they are generally preferred when boiled.

Varieties—

Dark-Red Spined

Bud very small. The variety is remarkable for the very long spines in which the scales terminate. For cultivation, it is inferior to the other sorts.

Early Purple

(Purple; Purple Globe; Artichaut Violet)

Heads rather small, obtusely conical; scales short and broad, pointed, green at the base, tinged with purplish-red on the outside, towards their extremities, moderately succulent, and of good quality. The variety is early, but not hardy. In France, it is considered excellent in its crude state, served with vinegar and oil; but not so good cooked.

Green Globe

(Large Round-Headed; Globe)

A very large sort, much esteemed, and generally cultivated in England. Heads, or buds, very large, nearly round, and with a dusky, purplish tint. The scales turn in at the top, and the receptacle is more fleshy than that of most varieties. It is generally preferred for the main crop, as the scales, or edible parts, are thicker, and higher flavored, than those of any other artichoke. It is not a hardy variety, and requires ample protection during winter.

Green, or Common

(French)

Bud very large, of a conical or oval form; scales deep-green, thick, and fleshy, pointed at the tips, and turned outwards. Though it has not the same thickness of flesh as the Green Globe Artichoke, it is much hardier, more prolific, and one of the best sorts for cultivation.

Green Provence

Bud large; scales comparatively long and narrow, of a lively green color, erect, fleshy at the base, and terminating in a sharp, brownish spine, or thorn; leaves of the plant deep-green. Most esteemed in its crude state; eaten as a salad in vinegar and oil.

Laon

(Gros vert de Laon)

Similar to the Common Green Artichoke, but of larger size. Scales rather loose and open, very deep-green, fleshy, and pointed. Much cultivated in the vicinity of Paris, and they're considered the best.

Large Flat Brittany
(Artichaut Camus de Bretagne)

Bud of medium size, somewhat globular, but flattened at the top; scales closely set together, green, brownish on the borders,—short, thick, and fleshy at the base. Earlier than the Laon, but not so fleshy. Much grown in Anjou and Brittany.

Purplish-Red

Bud conical; scales green towards their tips, and purplish-red at their base. Not very fleshy, and in no respect superior to the other varieties.

1,2. The Book of the Garden. By Charles M'Intosh. 2 vols. Edinburgh and London, 1855.

Artichokes in Valencia, Spain

31

Asparagus (Asparagus officinalis)

Asparagus is a hardy, perennial, maritime plant. It rises to the height of five feet and upwards, with an erect, branching stem; short, slender, nearly cylindrical leaves; and greenish, drooping flowers. The seeds, which are produced in globular, scarlet berries, are black, somewhat triangular, and retain their germinative powers four years. Twelve hundred and fifty weigh an ounce.

It is indigenous to the shores of various countries of Europe and Asia; and, since its introduction, has become naturalized to a considerable extent in this country. It is frequently seen in mowing-fields upon old farms; and, in some instances, has found its way to the beaches and marshes of the seacoast.

Propagation—It is propagated from seed, which may be sown either in autumn, just before the closing-up of the ground; or in spring, as soon as the soil is in good working condition. The nursery, or seed-bed, should be thoroughly spaded over, the surface levelled and raked smooth and fine, and the seed sown, not very thickly, in drills twelve or fourteen inches apart, and about an inch in depth. An ounce of seed is sufficient for fifty or sixty feet of drill.

When the plants are well up, thin them to three inches asunder; as they will be much stronger, if grown at some distance apart, than if allowed to stand closely together. Cultivate in the usual manner during the summer, and give the plants a light covering of stable-litter during the winter.

Good plants of one year's growth are preferred by experienced growers for setting; but some choose those of two years, and they may be used when three years old.

Soil and Planting—

"A rich, sandy, alluvial soil, impregnated with salt, is naturally best adapted to the growth of Asparagus; and, in such soil, its cultivation is an easy matter. Soils of a different character must be made rich by the application of fertilizing material, and light and friable by trenching. Sand, in wet, heavy, clayey soil, is of permanent benefit.

"The market-gardeners near London are aware of this; for, highly as they manure their ground for crops generally, they procure sand, or sandy mud, from certain parts of the Thames, for Asparagus plantations, where the soil is too heavy.

"The ground should be thoroughly trenched to the depth of two and a half or three feet: and, in order to make it rich, a large quantity of manure should be incorporated, as well at the bottom as near the surface,—using either sandy mud; the scourings of ditches made into compost; rock-weed, or kelp, where they can be procured; decayed leaves, or leaf-mould; the remains of hot-beds, good peat, or almost any other manure not in too crude a state.

"Where the soil is not so deep, and the subsoil coarse and rather gravelly, the ground is not trenched so deep; the bottom of the trench being merely dug over. Above this, however, a large quantity of manure is applied; and by this, with good after-management,—chiefly consisting in making the soil fine and light for the shoots to push through,—excellent crops are produced.

"The ground should be divided into beds either three or five feet wide, with an alley or path of two feet in width between. The reason for having some of the beds so much narrower than the others is, that the narrow ones are sooner heated by the sun's rays, and consequently an earlier production is induced.

"The distance between the rows in the beds may be regulated as follows: When the beds are three feet wide, two rows should be transplanted along them: each row should be a foot from the edge of the bed, and they will consequently be a foot apart. In beds that are five feet wide, three rows should be transplanted, also lengthwise,—one along the middle, and one on each side, a foot from the edge of the bed. The distance from plant to plant in the rows should not be less than one foot; at this distance, good-sized heads will be produced: but, if very large heads are desired for exhibition or competition, the plants should be fifteen, or even eighteen, inches asunder.

"The transplanting may be performed either in April or May. The three-feet beds should be traced out to run east and west, or so as to present the side of the bed to the direct action of the sun's rays when they are most powerful. Asparagus, in beds so formed, pushes earlier in the season than it does in beds running north and south. For all except the earliest beds, the direction is immaterial; and they may run east and west, or north and south, as may be most convenient.

"In proceeding to transplant, the beds, and paths, or alleys, should be marked off at the required distance. A stout stake should be driven at each corner of the beds, and from these the distances for the rows should be measured. There are various ways of transplanting. Some stretch a line, and cut out a trench only deep enough to allow the roots to be laid out without doubling; and they are spread out like a fan perpendicularly against the side of the cut, the crown of the plant being kept two inches below the surface of the ground. Some dig out a trench, and form little hillocks of fine soil, over which the roots are spread, extending like the sticks of an umbrella. Others make a ridge, astride which they set the plants, spreading their roots on each side of the ridge; and, again, some take off a portion of the soil on the bed, and, after the surface has been raked smooth, the roots of the plants are spread out nearly at right angles on the level.

"The first method is the most expeditious, and is generally practised in setting extensive plantations: but, whatever plan be preferred, the crowns of the plants should all be on the same level; otherwise those that are too high would be liable to be injured by the knife in cutting."

During the summer, nothing will be necessary but to keep the plants clear of weeds; and, in doing this, the hoe should be dispensed with as much as possible, to avoid injuring the roots. In the autumn, when the tops have completely withered, they should be cut down nearly level with the surface of the ground, and burned. The beds should then be lightly dug over, and three or four inches of rich loam, intermixed with well-digested compost, and salt at the rate of two quarts to the square rod, should be applied; which will leave the crowns of the roots about five inches below the surface.

Second Year—Early in spring, as soon as the frost leaves the ground, dig over the beds, taking care not to disturb the roots; rake the surface smooth; and, during the summer, cultivate as before directed: but none of the shoots should be cut for use. In the autumn, after the stalks have entirely withered, cut down and burn as in the previous year; stir the surface of the bed, and add an inch of soil and manure, which will bring the crowns six or seven inches below ground,—a depth preferred, by a majority of cultivators, for established plantations.

Third Year—Early in spring, stir the ground as directed for the two previous years. Some cultivators make a slight cutting during this season; but the future strength of the plants will be increased by allowing the crop to grow naturally as during the first and second years. In autumn, cut and burn as before; dig over the surface; add a dressing of manure; and, in the ensuing spring, the beds may be cut freely for use.

Asparagus officinalis
Hans Simon Holtzbecker

Instead of transplanting the roots, asparagus-beds are sometimes formed by sowing the seeds where the plants are to remain. When this method is adopted, the beds should be laid out and trenched, as before directed, and about three inches of soil removed from the entire surface. The seed should then be sown in drills an inch deep, at the distances marked out for the rows, and covered with rich, light soil. When the seedlings are two or three inches high, they should be thinned to nine or twelve inches apart; and, in thinning, the weakest plants should be removed. In the autumn, cut down the plants after they have withered, stir and smooth the surface, and add a dressing of manure. In the spring of the second year, stir the surface again; and, during the summer, cultivate as before. In the autumn, the plants will be ready for the dressing; which consists of the soil previously taken from the bed, with sufficient well-digested compost added to cover the crowns of the roots five or six inches in depth. The after-culture is similar to that of beds from transplanted roots

"Asparagus-beds should be enriched every autumn with a liberal application of good compost containing some mixture of salt; the benefit of which will be evident, not only in the quantity, but in the size and quality, of the produce. The dressing should be applied after the removal of the decayed stalks, and forked in, that its enriching properties may be washed to the roots of the plants by winter rains.

"In general, transplanted Asparagus comes up quite slender the first year; is larger the second; and, the third year, a few shoots may be fit for cutting. It is nearly in perfection the fourth year; and, if properly managed, will annually give an abundant supply during the life of the maker of a bed or plantation."

Cutting—

"The shoots should be cut angularly, from two to three inches below the surface of the ground; taking care not to wound the younger buds. It is in the best condition for cutting when the shoots are four or five inches above ground, and while the head, or bud, remains close and firm.

"It is the practice to cut off all the shoots as they appear, up to the period when it is thought best to leave off cutting altogether. The time for this depends on the climate, season, nature of the soil, and strength of the plants. Where the climate is good, or when the season is an early one, cutting must be commenced early; and of course, in such a case, it ought not to be continued late, as the plants would thereby be weakened."

In the Middle States, the cutting should be discontinued from the 10th to the 15th of June; and from the 15th to the 25th of the same month in the Eastern States and the Canadas.

"If the plants are weak, they should be allowed to grow up as early as possible, to make foliage, and consequently fresh roots, and thus to acquire more vigor for the ensuing year. It is also advisable to leave off at an early period the cutting of some of the best of the beds intended for early produce, in order that the buds may be well matured early in autumn, and thus be prepared to push vigorously early in spring."

Asparagus-beds will continue from twenty to thirty years; and there are instances of beds being regularly cut, and remaining in good condition for more than fifty years.

Seed—Select some of the finest and earliest heads as they make their appearance in the spring; tie them to stakes during the summer, taking care not to drive the stake through the crown of the plant. If for the market, or to be sent to a distance, wash out the seeds in autumn, and dry thoroughly; if for home-sowing, allow the seeds to remain in the berries till used.

Use—The young shoots are boiled twenty minutes or half an hour, until they become soft; and are principally served on toasted bread, with melted butter. It is the practice of some to boil the shoots entire; others cut or break the sprout just above the more tough or fibrous part, and cook only the part which is tender and eatable. This is snapped or cut into small sections, which are boiled, buttered, seasoned, and served on toast in the usual form. The smaller sprouts are sometimes cut into pieces three-eighths of an inch long, and cooked and served as green pease. The sprouts are also excellent when made into soup.

It is one of the most productive, economical, and healthful of all garden vegetables.

Varieties—

"The names of numerous varieties occur in the catalogues of seedsmen: but there seems to be little permanency of character in the plants; such slight variations as appear from time to time being caused, to a considerable extent, by the nature of the soil, or by the situation in which the plants are grown. What are called the Red-topped and Green-topped may perhaps be somewhat distinct, and considered as varieties."[1]

Soil and location have unquestionably much influence, both as respects the quality and size of the sprouts. A bed of asparagus in one locality produced shoots seldom reaching a diameter of half an inch, and of a very tough and fibrous character; while a bed in another situation, formed of plants taken from the same nursery-bed, actually produced sprouts so large and fine as to obtain the prize of the Massachusetts Horticultural Society.

If any variety really exists peculiar in size, form, color, or quality, it cannot be propagated by seed. Large sprouts may afford seeds, which, as a general rule, will produce finer asparagus than seeds from smaller plants; but a variety, when it occurs, can be propagated only by a division of the roots.

Mr. Thompson[2] states, that on one part of Mr. Grayson's extensive plantation, on the south side of the Thames, near London, the so-called Grayson's Giant was produced; and in another section, the common sort: but, when both were made to change places, the common acquired the dimensions of the Giant, whilst the latter diminished to the ordinary size.

Seeds of the following named and described sorts may be obtained of seedsmen, and will undoubtedly, in nearly all cases, afford fine asparagus; but they will not produce plants which will uniformly possess the character of the parent variety:

Battersea

Battersea is famed for producing fine asparagus, and the name is applied to the particular variety there grown. The heads are large, full, and close, and the tops tinted with a reddish-green color. It is probably intermediate between the Green and Purple-topped.

Gravesend

Originated and named under like circumstances with the Battersea. The top is greener, and not generally so plump and close; but it is considered finer flavored. Both varieties are, however, held in great estimation.

Grayson's Giant.

This variety, as also the Deptford, Mortlake, and Reading, all originated and were named under the same conditions as the varieties before described. All are fine sorts; but the difference between them, and indeed between all of the kinds, if important, is certainly not permanent, so long as they are offered in the form of seeds for propagation.

Mr. Grayson, the originator of this variety, produced a hundred sprouts, the aggregate weight of which was forty-two pounds,—the largest ever raised in Britain.

German

(Asperge d'Allemagne)

This variety very nearly resembles the Giant Purple-Topped. It is, however, considered a little earlier, and the top is deeper colored.

Giant Purple-Top.

(Dutch; Red-Top)

Sprout white; the top, as it breaks ground, purple; size very large, sometimes measuring an inch and three-fourths in diameter, but greatly affected by soil and cultivation.

A hundred sprouts of this variety have been produced which weighed twenty-five pounds.

Green-Top

This variety, when grown under the same conditions as the Giant Purple-top, is generally smaller or more slender. The top of the sprout, and the scales on the sides, are often slightly tinged with purple. The plant, when full grown, is perceptibly more green than that of the Giant Purple-top. From most nursery-beds, plants of both varieties will probably be obtained, with every intervening grade of size and color.

1. Glenny.

2. The Gardener's Assistant. By Robert Thompson.

Asparagus officinalis

32

Cardoon (Cynara cardunculus)

In its general character and appearance, the Cardoon resembles the Artichoke. Its full size is not attained until the second year, when it is "truly a gigantic herbaceous plant," of five or six feet in height. The flowers, which are smaller than those of the artichoke, are produced in July and August of the second year, and are composed of numerous small blue florets, enclosed by somewhat fleshy, pointed scales. The seeds are oblong, a little flattened, of a grayish or grayish-green color, spotted and streaked with deep brown; and, when perfectly grown, are similar in size and form to those of the apple. About six hundred are contained in an ounce; and they retain their vitality seven years.

Soil, Propagation, and Culture—The best soil for the Cardoon is a light and deep but not over-rich loam. It is raised from seed; which, as the plant is used in the first year of its growth and is liable to be injured by the winter, should be sown annually, although the Cardoon is really a perennial. It succeeds best when sown where the plants are to remain; for, if removed, the plants recover slowly, are more liable to run to seed, and, besides, seldom attain the size of those that have not been transplanted.

The seed should be sown as early in spring as the weather becomes warm and settled, in drills three feet apart, an inch and a half in depth, and the young plants afterwards thinned to twelve inches asunder in the drills. The leaves are blanched before being used.

It is sometimes raised and blanched as follows: Sow the seed at the bottom of trenches made about six inches deep, twelve inches wide, three feet apart, and of a length according to the supply required. At the bottom of the trench, thoroughly mix a small quantity of well-digested compost, and sow the seeds in small groups, or collections (three or four seeds together), at about twelve or fifteen inches apart, and cover them an inch or an inch and a half deep. When the young plants have acquired three or four leaves, they should be thinned out to single plants. During the summer, keep them free from weeds; and, as they require much moisture, it is well to water frequently, if the weather is very dry. In September, the plants will have attained their growth for the season, and be ready for blanching; which should be done in a dry day, and when the plants are entirely free from dampness. It is thus performed: The leaves of each plant are carefully and lightly tied together with strong matting; keeping the whole upright, and the ribs of the leaves closely together.

The plant is then bound with twisted hay-bands, or bands of straw, about an inch and a half in diameter; beginning at the root, and continuing the winding until two-thirds or three-fourths of the height is covered. If there is no heavy frost, the leaves will blanch quickly and finely without further pains: but, if frosty weather occurs, it will be necessary to earth up about the plants, as is practised with celery; but care should be taken not to raise the earth higher than the hay-bands.

One method of blanching is simply to tie the leaves together with matting, and then to earth up the plants from time to time like celery; beginning early in September, and adding gradually every week until they are sufficiently covered. Those, however, blanched by the banding process, are superior, both in respect to color and in the greater length of the parts blanched.

Another practice is to earth up a little about the base of the plant, tie the leaves together with thread or matting, and then envelop the whole quite to the top with a quantity of long, clean wheat or rye straw, placed up and down the plant, and tied together with small cord or strong matting. The leaves will thus blanch without being earthed up, and speedily become white. This process is a good one, is economical, and presents a neat appearance.

"In either of the methods, it is very necessary to be careful that the plants are perfectly dry before they are enveloped in their covering: they will otherwise rot." In about three weeks after being tied up, the cardoons will be fit for use.

Harvesting—When the stems and midribs of the leaves are thoroughly blanched, they are ready for use. Until the occurrence of severe weather, the table may be supplied directly from the garden: but, before the closing-up of the ground, "the plants should be taken up, roots and leaves entire, and removed to the cellar; where they should be packed in sand, laying the plants down in rows, and packing the sand around them, one course over another, till finished. In this way, they not only keep well, but become more perfectly blanched."

To Raise Seed—Allow two or three plants to remain unblanched, and leave them in the ground during the winter, protected by straw or other convenient material. They will grow to the height, and flower and seed, as before described. One plant will afford sufficient seed for any common garden.

Use—

"The stems of the leaves, as well as the mid-ribs, when blanched, are used for soups, stews, and even for salads, in autumn and winter. The longer these parts of the plant are, and the more rapidly they are grown, the more they are esteemed, on account of their greater crispness, tenderness, and color."

The "Gardener's Chronicle"[1] gives the following directions for dressing them:

"When a Cardoon is to be cooked, the solid stalks of the leaves are to be cut in pieces about six inches long, and boiled, like any other vegetable, in pure water (not salt and water), till they are tender. They are then to be carefully deprived of the slime and strings that will be found to cover them; and, having been thus thoroughly cleansed, are to be plunged in cold water, where they must remain until they are wanted for the table. They are then taken out, and heated with white sauce, or marrow. The process just described is for the purpose of rendering them white, and of depriving them of a bitterness which is peculiar to them. If this is neglected, the cardoons will be black, not white, as well as disagreeable."

M'Intosh[2] remarks, that, when skilfully prepared, they form an excellent and wholesome dish, deserving far more general notice.

In France, the flowers are gathered, and dried in the shade; and, when so preserved, are used as a substitute for rennet, to coagulate milk.

Varieties—

Common, or Large Smooth

(Smooth Large Solid; Plein Inerme)

This kind grows from four to five feet high. The leaves are large and strong, though somewhat smaller than those of the Tours or Prickly Cardoon. They are of a shining-green color, with little appearance of hoariness on the upper surface, and generally destitute of spines; though some of the plants occasionally have a few small ones at the base of the leaflets.

The Cardon *Plein Inerme* of the French, which is described in the "Bon Jardinier" as a novelty, corresponds nearly with the Large Smooth or Common Cardoon.

Large Spanish

(D'Espagne)

Stem five or six feet high. The divisions of the leaflets are rather narrower, and somewhat more hoary, than those of the Common Cardoon. The ribs are longer, and the whole plant stronger and generally more spiny; though, on the whole, comparatively smooth. It is not, however, always very readily distinguished from the Common or Large Smooth Cardoon. It runs up to seed quicker than the other varieties.

Puvis

(Artichoke-leaved; Lance-Leaved; Puvis de Bourg)

The Puvis Cardoon is remarkable for its strong growth, the large size it attains, and the thickness of the mid-ribs of the leaves, which are almost solid. The leaves are thick, and not at all prickly, or very slightly so. The terminal lobe is very large, and lance-formed: whence the name. It is a fine variety, and of more tender substance than the Tours Cardoon.

Red

(Blood-ribbed; Red-stemmed; Large Purple)

The leaves of this variety are green, without any hoariness; long, narrow, and more sharply pointed than those of most of the other kinds. The ribs are large, solid, and tinged with red. A recent sort, excellent in quality, but wanting in hardiness.

Tours

(Large Tours Solid; Cardon de Tours)

The leaves of this variety are very hoary on the upper surface; the divisions are broad, sharply pointed, and terminate with rigid, sharp spines. Spines also grow, in clusters of from three to five, at the base of the leaflets; and are very strong, and of a yellowish color. This variety is not so tall as the Spanish or Large Smooth. The ribs are large and solid.

The Tours Cardoon is cultivated by the market-gardeners around Paris; and, notwithstanding the inconvenience arising from its numerous and rigid spines, it is considered by them as the best, because of its thick, tender, and delicate ribs.

1. The Gardener's Chronicle. Weekly. By Prof. Lindley.

2. The Book of the Garden. By Charles M'Intosh. 2 vols. Edinburgh and London, 1855.

Cardoons taken out of the field for blanching and storage in glass-roofed pits.
Jacques Boyer

33

The Hop (Humulus lupulus)

Humulus lupulus Foemina
Flora Batava, Plate Number 90

The Hop is considered a native of this continent, and is found wild in all parts of the United States. The root is perennial, but the stems are annual. The latter are from ten to twenty-five feet in length, angular, rough, and twine from right to left. The leaves are placed opposite each other on the stem, on long, winding footstalks: the smaller ones are heart-shaped; the larger ones three or five lobed, veiny, and rough. The barren and fertile flowers are produced on separate plants: the former being very numerous and paniculated; the latter in the form of an ament, or collection of small scales, which are more or less covered with a fine, yellow powder called "lupulin."

While several distinct sorts of the fertile or hop-bearing plant have been long in cultivation, only one variety of the male or barren plant is known.

Soil and Location—Though it may be cultivated with success in a variety of soils, the Hop prefers a rich, deep loam, which should be thoroughly ploughed, and, if necessary, enriched with well-digested compost. In general, it may be said that "good corn-land is good hop-land." Hops, however, are reputed to be of better quality when raised on comparatively thin soils.

Propagation and Culture—It is propagated by a division of the roots early in spring. When extensively cultivated, the plants are set in hills, five to seven feet apart, and three or four cuttings or slips allowed to a hill; but in garden culture, to procure the young shoots, the plants are set in rows about three feet apart, and one foot from plant to plant in the rows.

Use—The plant is principally cultivated for its flowers, which are largely employed in the manufacture of malt liquors. The young shoots are cut in spring, when they are five or six inches in height, and eaten as salad, or used as asparagus, which they somewhat resemble in taste.

34

Hoosung/Oosung

(Celtuce; Asparagus Lettuce; Aa Choy; *Lactuca sativa* **var.** *augustana, angustata,* **or** *asparagina)¹*

A lettuce-like plant from Shanghai. Stems cylindrical, from two to three feet high, erect, light green, with a green, succulent pith; leaves oblong, tapering to the base, the uppermost clasping; the flowers are small, yellow, in panicles slightly drooping. If sown in April or May, the plants will ripen their seed in August.

Sowing and Cultivation—Sow in a cool frame, in either April or May, or continuously, for a succession, at intervals during May, and transplant into the open ground in the usual manner of treating lettuces; making the rows about eighteen inches apart, and placing the plants about the same distance apart in the rows. The plants will be fit for use early in June.

Celtuce/Aa Choy/Wo Sun, very likely the vegetable referred to.

Use—The succulent stem is the part used. This is divested of its outer rind, and either simply boiled, with a little salt in the water, and dressed as asparagus, or stewed in soy, with salt, pepper, and butter added, or boiled in soup as okra. It is a very agreeable and pleasant addition to the list of vegetable esculents, and worthy of trial.

The plant is very little cultivated; and there are no described varieties.

1. No Latin name was given for this vegetable, however the description indicates that it is this vegetable, which goes by various names, all of which were added by the editors.

35

Pokeweed/Pigeon Berry (Phytolacca decandra)

A hardy, herbaceous, perennial plant, common by roadsides, in waste places, and springing up spontaneously on newly burned pine-lands. It has a branching, purplish stem, five to seven feet in height; and large, oval, pointed, entire leaves. The flowers are produced in July and August, in long clusters; and are of a dull-white color. The fruit consists of a flat, purple, juicy berry; and is sometimes used for dyeing purple.

Soil and Culture—It will thrive in almost any soil or situation; and can be easily propagated from seed, or by dividing the roots. The plant requires little cultivation, and is so abundant in many localities as to afford an ample supply for the mere labor of gathering.

Use—The young shoots are eaten early in the season, as a substitute for asparagus, which they resemble in taste. When treated in the manner of sea-kale, the flavor of the sprouts is scarcely distinguishable from that of asparagus. The root has reputed important medicinal properties; and, when taken internally, acts as a violent emetic.

Annual Phytolacca.
Phytolacca esculenta.
An annual species, with foliage similar to the foregoing. It is much less vigorous and stocky in habit. The seed should be sown in April, in drills fifteen inches apart. The young shoots, or plants, are used in the manner of the species before described.

Phytolacca decandra
Millspaugh, Charles Frederick

Part Four: Curcubitaceous Plants

The Cucumber; Egyptian Cucumber; Globe Cucumber; Gourd, or Calabash; The Melon; Musk-melon; Persian Melons; Water-melon; Papanjay, or Sponge Cucumber; Prickly-fruited Gherkin; Pumpkin; Snake Cucumber; Squash.

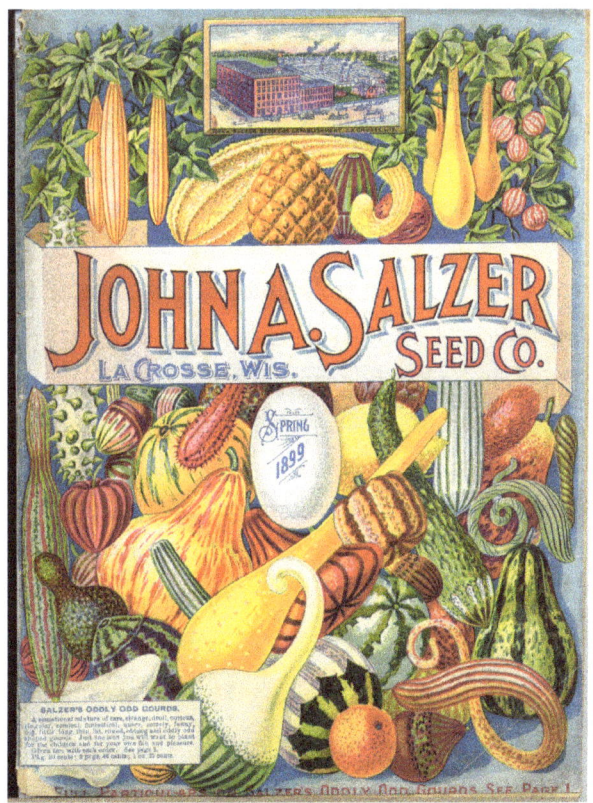

Spring 1899, John A. Salzer Seed Co.

36

The Cucumber (Cucumis sativus)

The Cucumber is a tender, annual plant; and is a native of the East Indies, or of tropical origin. It has an angular, creeping stem; large, somewhat heart-shaped, leaves; and axillary staminate or pistillate flowers. The fruit is cylindrical, generally elongated, often somewhat angular, smooth, or with scattering black or white spines; the flesh is white or greenish-white, and is divided at the centre of the fruit into three parts, in each of which the seeds are produced in great abundance. These seeds are of an elliptical or oval form, much flattened, and of a pale yellowish-white color. About twelve hundred are contained in an ounce; and they retain their vitality ten years.

Soil and Culture—Very dry and very wet soils should be avoided. Cucumbers succeed decidedly best in warm, moist, rich, loamy ground. The essentials to their growth are heat, and a fair proportion of moisture. They should not be planted or set in the open air until there is a prospect of continued warm and pleasant weather; as, when planted early, not only are the seeds liable to decay in the ground, but the young plants are frequently cut off by frost.

The hills should be five or six feet apart in each direction. Make them fifteen or eighteen inches in diameter, and a foot in depth; fill them three-fourths full of thoroughly digested compost, and then draw four or five inches of earth over the whole, raising the hill a little above the level of the ground; plant fifteen or twenty seeds in each, cover half an inch deep, and press the earth smoothly over with the back of the hoe. When all danger from bugs and worms is past, thin out the plants; leaving but three or four of the strongest or healthiest to a hill.

Taking the Crop—As fast as the cucumbers attain a suitable size, they should be plucked, whether required for use or not. The imperfectly formed, as well as the symmetrical, should all be removed. Fruit, however inferior, left to ripen on the vines, soon destroys their productiveness.

Seed—Cucumbers, from their natural proneness to impregnate each other when, grown together, are exceedingly difficult to keep true to their original points of merit; and consequently, to retain any variety in its purity, it must be grown apart from all other sorts. When a few seeds are desired for the vegetable garden, two or three of the finest-formed cucumbers should be selected early in the season, and allowed to ripen on the plants. In September, or when fully ripe, cut them open, take out the seeds, and allow them to stand a day or two, or until the pulp attached to them begins to separate; when they should be washed clean, thoroughly dried, and packed away for future use.

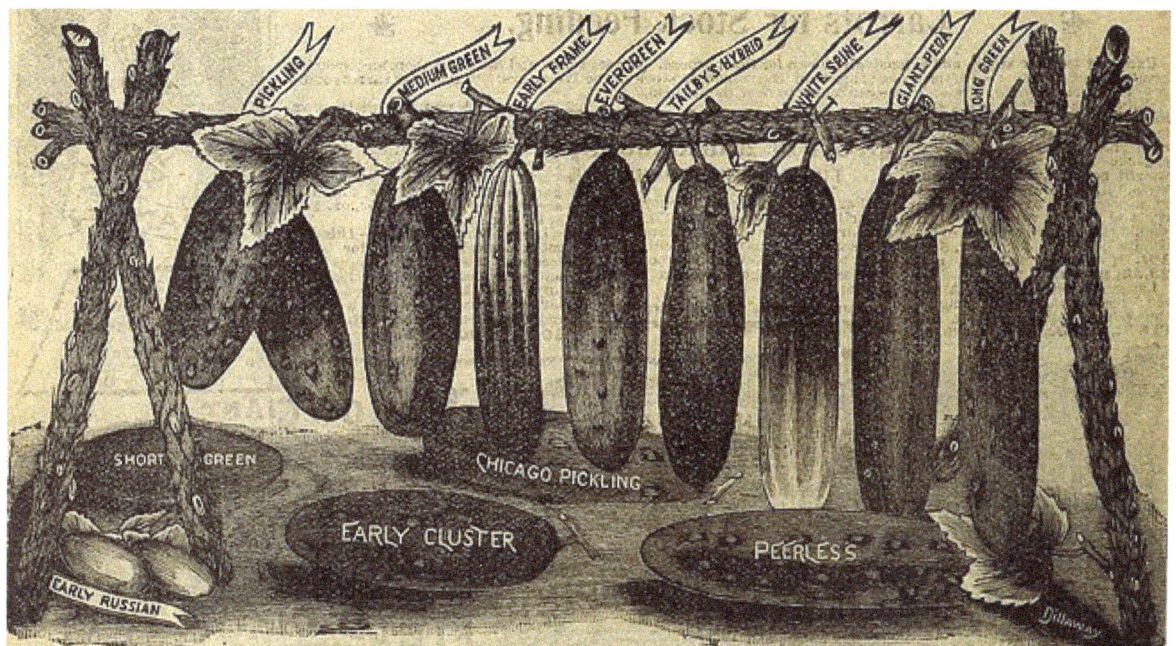

Catalogue of seeds, plants, bulbs & fruits, 1895

For Pickling—The land for raising cucumbers for pickling may be either swarded or stubble; but it must be in good condition, and such as is not easily affected by drought. It should be deeply ploughed, and the surface afterwards made fine and friable by being thoroughly harrowed. The hills should be six feet apart, and are generally formed by furrowing the land at this distance in each direction. Manure the hills with well-digested compost, level off, draw over a little fine earth, and the land is ready for planting.

This may be done at any time from the middle of June to the first week in July. The quantity of seed allowed to an acre varies from three-fourths of a pound, upwards. In most cases, growers seed very liberally, to provide against the depredation of worms and bugs; usually putting six or eight times as many seeds in a hill as will be really required for the crop. When the plants are well established and beyond danger, the field is examined, and the hills thinned to three or four plants; or, where there is a deficiency of plants, replanted.

As fast as the cucumbers attain the proper size, they should be plucked; the usual practice being to go over the plantation daily. In gathering, all the fruit should be removed,—the misshapen and unmarketable, as well as those which are well formed; for, when any portion of the crop is allowed to remain and ripen, the plants become much less productive.

In favorable seasons, and under a high state of cultivation, a hundred and twenty-five thousand are obtained from an acre; while, under opposite conditions, the crop may not exceed fifty thousand. The average price is about a dollar and twenty-five cents per thousand.

Varieties—

Early Cluster

(Early Green Cluster)

A very popular, early cucumber, producing its fruit in clusters near the root of the plant: whence the name. The plant is healthy, hardy, and vigorous; fruit comparatively short and thick. Its usual length is about five inches, and its diameter about two inches; skin prickly, green,—at the blossom-end, often paler, or nearly white,—brownish-yellow when ripe; flesh white, seedy, tender, and well flavored, but less crispy or brittle than that of many other varieties.

It is a good early garden sort, and is very productive; but is not well adapted for pickling, on account of the soft and seedy character of its flesh.

Early Frame

(Short Green)

One of the oldest of the garden sorts, justly styled a standard variety. Plant healthy and vigorous, six to ten feet in length; fruit straight and well formed, five inches and a half long, and two inches and a half in diameter; skin deep-green, paler at the blossom-end, changing to clear yellow as it approaches maturity, and, when fully ripe, of a yellowish, russet-brown color; flesh greenish-white, rather seedy, but tender, and of an agreeable flavor. It is a few days later than the Early Cluster.

The variety is universally popular, and is found in almost every vegetable garden. It is also very productive; succeeds well, whether grown in open culture or under glass; and, if plucked while young and small, makes an excellent pickle.

(Early Russian)

This comparatively new variety resembles, in some respects, the Early Cluster. Fruit from three to four inches in length, an inch and a half or two inches in diameter, and generally produced in pairs; flesh tender, crisp, and well flavored. When ripe, the fruit is deep-yellow or yellowish-brown.

Its merits are its hardiness, extreme earliness, and great productiveness. It comes into use nearly ten days in advance of the Early Cluster, and is the earliest garden variety now cultivated. Its small size is, however, considered an objection; and some of the larger kinds are generally preferred for the main crop.

(London Long Green)

Fruit about a foot in length, tapering towards the extremities; skin very deep-green while the fruit is young, yellow when it is ripe; flesh greenish-white, firm, and crisp; flavour good.

This variety is nearly related to the numerous prize sorts which in England are cultivated under glass, and forced during the winter. There is little permanency in the slight variations of character by which they are distinguished; and old varieties are constantly being dropped from the catalogues, and others, with different names, substituted. Amongst the most prominent of these sub-varieties are the following:—

Carter's Superior—Recently introduced. Represented as one of the largest and finest of the forcing varieties.

Conqueror of the West—Eighteen to twenty inches in length. It is a fine prize sort, and succeeds well in open culture.

Cuthill's Black Spine—Six to nine inches in length, hardy, early, and productive. An excellent sort for starting in a hot-bed. Fruit very firm and attractive.

The Doctor—Sixteen to eighteen inches in length, and contracted towards the stem in the form of a neck. In favorable seasons, it will attain a good size, if grown in the open ground. Crisp, tender, and well flavored.

Eggleston's Conqueror—Very prolific, good for forcing, of fine flavor, hardy, and a really useful sort. Specimens have been grown measuring twenty-eight inches in length, nine inches and a half in circumference, and weighing five pounds.

Flanigan's Prize—An old, established variety; having been grown in England upwards of thirty years. Length fifteen inches.

Hunter's Prolific—Length eighteen inches. Very crisp and excellent, but requires more heat than most other varieties. Spines white; fruit covered with a good bloom, and not liable to turn yellow at the base.

Improved Sion House—This variety has received many prizes in England. Not only is it well adapted for the summer crop, but it succeeds remarkably well when grown under glass.[Pg 175]

Irishman—Length twenty-two to twenty-five inches. Handsome, and excellent for exhibition.

Lord Kenyon's Favourite—Length twelve to eighteen inches. A fine sort for winter forcing.

Manchester Prize—This, like the Nepal, is one of the largest of the English greenhouse prize varieties. It sometimes measures two feet in length, and weighs twelve pounds. In favorable seasons, it will attain a large size in open culture, and sometimes perfect its seed.

Nepal—One of the largest of all varieties; length about twenty-four inches; weight ten to twelve pounds.

Norman's Stitchworth-Park Hero—A recently introduced variety, hardy, long, handsome, very prolific, and fine flavored.

Old Sion House—Length about nine inches. This is a well-tried, winter, forcing variety. Like the Improved Sion House, it also succeeds well in open culture. Quality good, though the extremities are sometimes bitter.

Prize-fighter—Length about sixteen inches. Good for the summer crop or for exhibition.

Rifleman—This variety is described as one of the best prize cucumbers. It has a black spine; always grows very even from stem to point, with scarcely any handle; carries its bloom well; keeps a good fresh color; and is not liable to turn yellow as many other sorts. Length twenty-four to twenty-eight inches. An abundant bearer.

Ringleader—A prominent prize sort, about fifteen inches in length. It succeeds well, whether grown under glass or in the open ground.

Roman Emperor—Length twelve to fifteen inches.

Southgate—This variety has been pronounced the most productive, and the best for forcing, of all the prize sorts. It is not so late as many of the English varieties, and will frequently succeed well if grown in the open ground.

Victory of Bath—Length about seventeen inches. Well adapted for forcing or for the general crop.

Griswold's catalogue of vegetable, flower and field seeds, 1900

Long Green Prickly

(Long Prickly; Early Long Green Prickly)

This is a large-sized variety, and somewhat later than the White-spined. The plant is a strong grower, and the foliage of a deep-green color; the fruit is about seven inches in length, straight, and generally angular; skin dark-green, changing to yellow as the fruit approaches maturity,—when fully ripe, it is reddish-brown, and is often reticulated about the insertion of the stem; prickles black; flesh white, somewhat seedy, but crisp, tender, and well flavored.

The Long Green Prickly is hardy and productive; makes a good pickle, if plucked while young; and is well deserving of cultivation. It differs from the London Long Green and the Long Green

Turkey in its form, which is much thicker in proportion to its length; and also in the character of its flesh, which is more pulpy and seedy.

Long Green Turkey

(Extra Long Green Turkey)

A distinct and well-defined variety; when full grown, sometimes measuring nearly eighteen inches in length. Form long and slender, contracted towards the stem in the form of a neck, and swollen towards the opposite extremity; seeds few, and usually produced nearest the blossom-end. The neck is generally solid. While the fruit is young, the skin is deep-green; afterwards it changes to clear yellow, and finally assumes a rusty-yellow or yellowish-brown. Flesh remarkably firm and crisp; exceeding, in these respects, that of any other variety. Very productive and excellent.

Its remarkably firm and crispy flesh, and the absence of seeds, render it serviceable for the table after it has reached a very considerable size. For the same reasons, it may be pickled at a stage of its growth when other more seedy and pulpy sorts would be comparatively worthless.

Short Prickly

(Short Green Prickly; Early Short Green Prickly)

This variety somewhat resembles the Long Prickly; but it is shorter, and proportionally thicker. Its length, when suitable for use, is about four inches. Skin prickly, green, changing to yellow at maturity; flesh transparent greenish-white, rather seedy, but tender, crisp, and fine flavored.

The variety is very hardy and productive, comes early into fruit, and is one of the best for pickling. It is a few days later than the Early Cluster.

Underwood's Short Prickly

This is an improved variety of the common Short Prickly, and is the best of all the sorts for extensive cultivation for pickling. The plant is hardy and productive. The fruit, when young, is very symmetrical, and of a fine deep-green color. Its flesh is characterized by extraordinary crispness and solidity. When more advanced, the color becomes paler, and the flesh more soft and seedy. The fruit, at maturity, is yellow.

White Spanish

The form of this variety is similar to that of the White-spined. The fruit measures about five inches in length, two inches in diameter, and is generally somewhat ribbed. When suitable for use, the skin is white; a characteristic by which the variety is readily distinguished from all others. The flesh is crisp, tender, and well flavored. At maturity, the fruit is yellow.

White-Spined

(Early White-Spined; New-York Market)

This very distinct variety is extensively grown for marketing, both at the North and South. The plants grow from six to ten feet in length; and, like those of the Early Frame, are of a healthy, luxurious habit. The fruit is of full medium size, straight, and well formed; about six inches in length, and two inches and a half in diameter. Skin deep-green; prickles white; flesh white, tender, crispy, and of remarkably fine flavor. As the fruit ripens, the skin gradually becomes paler; and, when fully ripe, is nearly white: by which peculiarity, in connection with its white spines, the variety is always readily distinguishable.

The White-spined is one of the best sorts for the table; and is greatly prized by market-men on account of its color, which is never changed to yellow, though kept long after being plucked. It is generally thought to retain its freshness longer than any other variety, and consequently to be well fitted for transporting long distances; though, on account of its peculiar color, the freshness may be less real than apparent.

For the very general dissemination of this variety, the public are, in a great degree, indebted to the late I. P. Rand, Esq., of Boston, whose integrity as a merchant, and whose skill as a practical vegetable cultivator and horticulturist, will be long remembered.

1895 Rose Collection (1895), Iowa Seed Co.

37

Egyptian Cucumber (Cucumis chate)

(Hairy Cucumber; Round-leaved Egyptian; Concombre chaté; Cucumis melo var. chate; Carosello Barese Cucumber)[1]

This is a tender, annual plant, with an angular, creeping stem, and alternate, somewhat heart-shaped, leaves. The flowers are axillary, about an inch in diameter, and of a pale-yellow color; the fruit is small, oblong, and very hairy.

It is of little value as an esculent, and is rarely cultivated. The fruit is sometimes eaten in its green state, and also when cooked. According to Duchesne, the Egyptians prepare from the pulp a very agreeable and refreshing beverage.

Plant and cultivate as directed for melons or cucumbers.

1. The identifications *Cucumis melo var. chate* and *Carosello Barese Cucumber* have been added by the editors to help the modern gardener with identification. This variety of 'cucumber' is actually a kind of musk melon that is eaten green, and is suitable for growing in hot climates.

Cucumis melo var. chate

38

Globe Cucumber (Cucumis prophetarum)

(Concombre des prophètes)

A tender annual from Arabia. Stem slender, creeping, and furnished with tendrils, or claspers. The leaves are about three inches in diameter, five-lobed, and indented on the borders; the flowers are axillary, yellow, and nearly three-fourths of an inch in diameter; the fruit is round, and rarely measures an inch in thickness; skin striped with green and yellow, and thickly set with rigid hairs, or bristles; the seeds are small, oval, flattened, and of a yellowish color.

Planting and Culture—The seeds should be planted at the time of planting cucumbers or melons, in hills four or five feet apart, and covered about half an inch deep. Thin to two or three plants to a hill.

Use—The fruit is sometimes eaten boiled; but is generally pickled in its green state, like the common cucumber.

As a table vegetable, it is comparatively unimportant, and not worthy of cultivation.

Cucumis prophetarum
Lukas Hochenleitter und Kompagnie

39

Calabash/Bottle Gourd (Cucurbita lagenaria)

The Calabash, or Common Gourd, is a climbing or creeping annual plant, frequently more than twenty feet in height or length. The leaves are large, round, heart-shaped, very soft and velvety to the touch, and emit a peculiar, musky odor, when bruised or roughly handled. The flowers, which are produced on very long stems, are white, and nearly three inches in diameter. They expand towards evening, and remain in perfection only a few hours; as they are generally found drooping and withering on the ensuing morning. The young fruit is hairy, and quite soft and tender; but, when ripe, the surface becomes hard, smooth, and glossy. The seeds are five-eighths of an inch in length, somewhat quadrangular, of a fawn-yellow color, and retain their vitality five years. About three hundred are contained in an ounce.

Cultivation—The seeds are planted at the same time and in the same manner as those of the Squash. The Gourd succeeds best when provided with a trellis, or other support, to keep the plant from the ground; as the fruit is best developed in a pendent or hanging position.

Use—The fruit, while still young and tender, is sometimes pickled in vinegar, like cucumbers. At maturity, the flesh is worthless: but the shells, which are very hard, light, and comparatively strong, are used as substitutes for baskets; and are also formed into water-dippers, and various other articles both useful and ornamental. The varieties are as follow:

Bottle Gourd

Fruit about a foot in length, contracted at the middle, largest at the blossom-end, but swollen also at the part next the stem.

There is a sub-variety, very much larger; but it is also later.

Hercules Club

(Courge Massue d'Hercule)

Fruit very long. Specimens are frequently produced measuring upwards of five feet in length. It is smallest towards the stem, and increases gradually in size towards the opposite extremity, which is rounded, and near which, in its largest diameter, it measures from four to five inches. Its form is quite peculiar, and is not unlike that of a massive club: whence the name.

It is frequently seen at horticultural and agricultural shows; and, though sometimes exhibited as a "cucumber," has little or no value as an esculent, and must be considered much more curious than useful. It is of a pea-green color while growing, and the skin is then quite soft and tender; but, like the other varieties, the surface becomes smooth, and the skin very hard and shell-like, at maturity.

Powder-Horn

(Courge Poire à Poudre)

Fruit long and slender, broadest at the base, tapering towards the stem, and often more or less curved. In its general form, it resembles a common horn, as implied by the name. Its usual length is twelve or fourteen inches; and its largest diameter, nearly three inches.

Siphon Gourd

(Courge Siphon)

Fruit rounded, and flattened at the blossom-end; then suddenly contracted to a long, slender neck. The latter often bends or turns suddenly at nearly a right angle; and, in this form, the fruit very much resembles a siphon. Pea-green while young, pale-green when mature. Shell thick and hard.

Cucurbita lagenaria
Seikei Zusetsu vol. 26, page 5

40

The Melon (The Musk, Persian & Water Melons)

Of the Melon, there are two species in general cultivation,—the Musk-melon (*Cucumis melo*) and the Water-melon (*Cucurbita citrullus*); each, however, including many varieties. Like the Squash, they are tender, annual plants, of tropical origin, and only thrive well in a warm temperature.

> "The climate of the Middle and Southern States is remarkably favorable for them; indeed, far more so than that of England, France, or any of the temperate portions of Europe. Consequently, melons are raised as field crops by market-gardeners: and, in the month of August, the finest citrons or green-fleshed melons may be seen in the markets of New York and Philadelphia in immense quantities; so abundant, in most seasons, as frequently to be sold at half a dollar per basket, containing nearly a bushel of fruit. The warm, dry soils of Long Island and New Jersey are peculiarly favourable to the growth of melons; and, even at low prices, the product is so large, that this crop is one of the most profitable."[1]

Through the extraordinary facilities now afforded by railroads and ocean steam-navigation, the markets of all the cities and large towns of the northern portions of the United States, and even of the Canadas, are abundantly supplied within two or three days from the time of gathering: and they are retailed at prices so low, as to allow of almost universal consumption; well-ripened and delicious green-fleshed citron-melons being often sold from six to ten cents each.

Soil and Cultivation—Both the Musk and the Water Melon thrive best in a warm, mellow, rich, sandy loam, and in a sheltered exposure. After thoroughly stirring the soil by ploughing or spading, make the hills six or seven feet apart in each direction. Previous to planting, these hills should be prepared as directed for the Squash; making them a foot and a half or two feet in diameter, and twelve or fifteen inches in depth. Thoroughly incorporate at the bottom of the hill a quantity of well-digested compost, equal to three-fourths of the earth removed; and then add sufficient fine loam to raise the hill two or three inches above the surrounding level. On the top of the hill thus formed, plant twelve or fifteen seeds; and, when the plants are well up, thin them out from time to time as they progress in size. Finally, when all danger from bugs and other insect depredators is past, leave but two or three of the most stocky and promising plants to a hill. When the growth is too luxuriant, many practise pinching or cutting off the leading shoots; and, when the young fruit sets in too great numbers, a portion should be removed, both for the purpose of increasing the size and of hastening the maturity of those remaining. Keep the fruit from being injured by lying on the ground; and if slate, blackened shingles, or any dry, dark material, be placed beneath it, by attraction of the sun's rays, the fruit will ripen earlier and better.

The striped bug (*Galereuca vittata*) is the most serious enemy with which the young melon-plants have to contend. Gauze vine-shields, though the most expensive, are unquestionably the most effectual preventive. Boxes either round or square, twelve or fifteen inches in depth, and entirely uncovered at the top, if placed over the hills, will be found useful in protecting the plants. The flight of the bug being generally nearly parallel with the surface of the ground, very few will find their way within the boxes, if of the depth required. Applications of guano, ashes, dilutions of oil-soap, and plaster of Paris, applied while the plants are wet, will be found of greater or less efficacy in their protection. The pungent smell of guano is said to prevent the depredation of the flea-beetle, which, in many localities, seriously injures the plants early in the season, through its attacks on the seed-leaves.

The Musk Melon
(Cucumis melo)

Plant running,—varying in length from five to eight feet; leaves large, angular, heart-shaped, and rough on the upper and under surface; flowers yellow, one-petaled, five-pointed, and about an inch in diameter; seeds oval, flat, generally yellow, but sometimes nearly white, about four-tenths of an inch in length, and three-sixteenths of an inch in breadth,—the size, however, varying to a considerable extent in the different varieties. An ounce contains from nine hundred to eleven hundred seeds; and they retain their germinative properties from eight to ten years.

Varieties—These are exceedingly numerous, in consequence of the great facility with which the various kinds intermix, or hybridize. Varieties are, however, much more easily produced than retained: consequently, old names are almost annually discarded from the catalogues of seedsmen and gardeners; and new names, with superior recommendations, offered in their stead. The following list embraces most of the kinds of much prominence or value now cultivated either in Europe or this country:

Beechwood

Fruit nearly spherical, but rather longer than broad,—usually five or six inches in diameter; skin greenish-yellow, thickly and regularly netted; flesh green, melting, sugary, and excellent. An early and fine variety.

Black-Rock Cantaloupe

A large-fruited, late variety; form variable, but generally round, and flattened at the ends; size large,—ten inches in diameter, eight inches deep, and weighing eight or ten pounds. The skin varies in color from grayish-green to deep-green; becomes yellow at maturity, and is thickly spread with knobby bunches, or small protuberances. Rind very thick; flesh reddish-orange, melting, and sugary. It requires a long season for its full perfection.

Christiana

This variety was originated by the late Capt. Josiah Lovett, of Beverly, Mass. Form roundish; size rather small,—average specimens measuring nearly the same as the Green Citron; skin yellowish-green; flesh yellow, sweet, juicy, and of good quality. Its early maturity is its principal recommendation; the Green Citron, Nutmeg, and many other varieties, surpassing it in firmness of flesh, sweetness, and general excellence.

It would probably ripen at the North, or in short seasons, when other sorts generally fail.[Pg 185]

Citron

(Green-fleshed Citron; Green Citron)

Fruit nearly round, but flattened slightly at the ends,—deeply and very regularly ribbed; size medium, or rather small,—average specimens measuring about six inches in diameter, and five inches and a half in depth; skin green, and thickly netted,—when fully mature, the green becomes more soft and mellow, or of a yellowish shade; flesh green, quite thick, very juicy, and of the richest and most sugary flavor. It is an abundant bearer, quite hardy, and remarkably uniform in its quality. It is deservedly the most popular as a market sort; and for cultivation for family use, every thing considered, has few superiors.

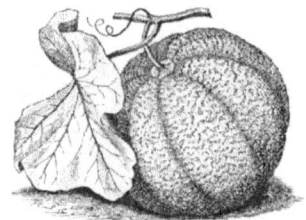
Green Citron Melon

In common with the Carolina Water-melon, the Green Citron is extensively grown at the South for shipping to the northern portions of the United States; appearing in the markets of New York and Boston three or four weeks in advance of the season of those raised in the same vicinity in the open ground.

Early Cantaloupe

This variety possesses little merit aside from its very early maturity. It is a roundish melon, flattened a little at the ends, ribbed, and of comparatively small size; usually measuring about five inches in diameter. Skin yellowish, often spotted with green, and sometimes a little warty; rind quite thick; flesh reddish-orange, sweet, and of good flavor.

It is exceedingly variable in size, form, and colour.

Hardy Ridge

Fruit rather small, round, depressed, strongly ribbed, and irregularly warted all over its surface; skin dull yellow, mottled with dull green; flesh an inch thick, bright orange-red, sweet, and well flavored; rind thick; weight from three to four pounds. Not an early, but a productive variety.

Large-Ribbed Netted

(Common Musk-Melon)

Fruit very large, oval, strongly ribbed; skin yellow, very thickly netted, sometimes so closely as to cover nearly the entire surface; flesh salmon-yellow, remarkably thick and sweet, but not fine-grained or melting, when compared with the more recent and improved varieties.

Large Netted Musk Melon

Hardy and productive. In good soil and favorable seasons, the fruit sometimes attains a length of fifteen inches, and weighs upwards of twenty pounds.

Munroe's Green Flesh

A comparatively new variety. The fruit is nearly spherical, but tapers slightly towards the stem, and is rather regularly as well as distinctly ribbed. Its diameter is about five inches. Cicatrix large; skin greenish-yellow, thickly and finely netted over the entire surface; rind thin; flesh green, remarkably transparent, comparatively thick, very melting, and highly perfumed.

Nutmeg

Fruit oval, regularly but faintly ribbed, eight or nine inches in length, and about six inches in its broadest diameter; skin pale-green, and very thickly netted; rind thin; flesh light-green, rich, sweet, melting, and highly perfumed.

The Nutmeg Melon has been long in cultivation, and is almost everywhere to be found in the vegetable garden, though seldom in a perfectly unmixed state. When the variety is pure, and the fruit perfectly ripened, it is of most delicious excellence, and deservedly ranked as one of "the best."

Orange Cantaloupe

An oval variety, about six inches in length by five inches in diameter, rather prominently ribbed. Skin yellow, marbled with green, thickly netted about the stem, and sparsely so over the remainder of the surface; rind thick; flesh reddish-orange, sweet, highly perfumed, and of good flavor. Very early and productive.

Pine-Apple

Form roundish, inclining to oval, either without ribs or with rib-marking, very faintly defined; size small,—the average diameter being about five inches and a half; skin olive-green, with net-markings more or less abundant; rind thin; flesh green, melting, sweet, and perfumed. Season early.

It is an excellent sort, easily grown, and very productive.

Prescott Cantaloupe

Fruit generally somewhat flattened, but variable in form, deeply ribbed; size large,—well-grown specimens measuring eight or ten inches in diameter, and weighing from seven to nine pounds; skin thickly covered with small tubercles; color varying from grayish-green to clear-green, more or less deep,[Pg 188] changing to yellow at maturity; rind very thick; flesh orange-red, sugary and melting, and of delicious flavor.

There are numerous sub-varieties, as grown by different gardeners, varying somewhat in form, color, and time of maturity; all, however, corresponding nearly with the above description, though known by different names, as the "White," "Gray," "Black," "Prescott," &c. Much esteemed in France, and extensively grown by market-gardeners in the vicinity of Paris.

Skillman's Fine-Netted

This variety much resembles the Pine-apple. Form rounded, flattened slightly at the ends; flesh green, sugary, melting, and excellent. It has been pronounced "the earliest of the green-fleshed sorts."

Victory of Bath

A recently introduced variety of English origin. Fruit egg-shaped, faintly ribbed, rounded at the blossom-end, and slightly contracted towards the stem,—at the insertion of which, it is flattened to a small, plane surface; size medium,—about six inches deep, and five inches in diameter; skin green, clouded with yellow, and sparsely covered with fine net-markings; skin thin; flesh green.

The Persian Melon

These differ remarkably from the varieties commonly cultivated. They are destitute of the thick, hard rind which characterizes the common sorts, and which renders so large a portion of the fruit useless. On the contrary, the Persian melons are protected by a skin so thin and delicate, that they are subject to injury from causes that would produce no perceptible effect on the sorts in general cultivation. As a class, they are not only prolific, but their flesh is extremely tender, rich, and sweet, and flows copiously with a cool juice, which renders them still more grateful. They are, however, not early; and, for their complete perfection, require a long and warm season.

Varieties—

Dampsha

Flesh dark-green near the skin, rather whitish towards the centre, quite melting, and of excellent flavor. The first-produced fruit in the season is somewhat cylindrical, bluntly pointed at both ends; the whole surface being prominently netted, and of a pale-yellow or dark-olive color. The secondary crop has the fruit more pointed and less netted, and the skin becomes much darker. Like the other varieties of winter melons, it may be preserved a long time after being taken from the vines, if suspended in a dry room. Weight four to five pounds.

Daree

This resembles the Geree Melon in color, as well as in many other respects. It is of the same form; but the rind, when netted, exhibits coarser reticulations. The flesh is white, thick, crisp, and melting; when fully ripened, very sweet, but rather insipid if imperfectly matured. It is always, however, cool and pleasant.

Geree

A handsome green fruit. In shape, it is oval, or ovate; and measures eight inches in length by four inches and a half in breadth. The skin is closely mottled with dark sea-green upon a pale ground, and is either netted or not. In the former case, the meshes are very close; by which character, it may be readily distinguished from the Daree. Stalk very short; flesh an inch and a half or two inches thick, bright green, melting, very sweet, and highly flavoured. Though perhaps equally rich, it is not so beautiful or so juicy as the Melon of Keiseng. A good bearer, but requires a warm, long season.

Germek

(Large Germek)

A handsome large-sized, ribbed fruit, shaped like a compressed sphere; usually six inches in length, and from seven to nine inches in diameter. Skin deep-green, closely netted; flesh from an inch and a half to two inches thick, clear green, firm, juicy, and high flavored. This is an excellent variety, an abundant bearer, ripens early, and exceeds in size any of the Persian melons.

Green Hoosainee

A handsome egg-shaped fruit, five inches long by four inches broad: when unripe, of a very deep-green; but, in maturity, acquiring a fine, even, light-green, regularly netted surface, which, on the exposed side, becomes rather yellow. The flesh is pale-greenish white, tender and delicate, full of a highly perfumed, pleasant, sweet juice; the rind is very thin; the seeds are unusually large.

It is a variety of much excellence, a great bearer, and one of the hardiest of the Persian melons.

Green Valencia

A winter sort. Although not rich in flavor, it is firm, saccharine, and juicy; and upon the whole, if fully ripened, a more desirable melon than many of the summer varieties.

Ispahan

(Sweet Ispahan)

This has been pronounced "the most delicious of all melons." The fruit is egg-shaped, varying in length from eight to twelve inches, and weighing from six to eight pounds; skin nearly smooth, of a deep sulphur-yellow; flesh nearly white, extending about half way to its centre, crisp, sugary, and very rich.

It is a variety of much excellence, but is fully perfected only in favorable seasons.

Melon of Keiseng

A beautiful egg-shaped fruit, eight inches long, five inches wide in the middle, six inches wide at the lower extremity; very regularly and handsomely formed. Color pale lemon-yellow; flesh from an inch and a half to two inches and a quarter thick, nearly white, flowing copiously with juice, extremely delicate, sweet, and high flavored, very similar in texture to a well-ripened Beurré pear; rind thin, but so firm that all the fleshy part of the fruit may be eaten.

It differs from the Sweet Ispahan in being closely netted.

Melon of Seen

A fruit of regular figure and handsome appearance, seven inches long by five inches wide. Shape ovate, with a small mamelon at the apex; surface pale dusky yellow, regularly and closely netted, except the mamelon, which is but little marked; rind very thin; flesh from an inch and a half to two inches thick, pale-green, sometimes becoming reddish towards the inside, exceedingly tender and juicy; juice sweet, and delicately perfumed. A good bearer, but requires a long season. Named from Seen, a village near Ispahan; where the variety was procured.

Small Germek

This ripens about a week earlier than the Large Germek, but is not so valuable a fruit. In form, it is a depressed sphere, with about eight rounded ribs. It measures four inches in depth by four inches and a half in width. The skin is even, yellowish, with a little green about the interstices, obscurely netted; the flesh is green, inclining to reddish in the inside, an inch and a half thick, juicy, and high flavored; skin very thin. The pulp in which the seeds are immersed is reddish. It is not a great bearer, and the vines are tender.

Striped Hoosainee

Fruit oval and much netted, dark-green in broad stripes, with narrow intervals of dull white, which become faintly yellow as the fruit ripens; pulp externally green, but more internally pale-red, excessively juicy, and more perfectly melting than that of the famous Ispahan Melon.

It is sweeter and higher flavored than any other Persian variety, but requires a long, warm season for its full perfection.

The Water Melon
Cucurbita citrullus

Plant running,—the length varying from eight to twelve feet; leaves bluish-green, five-lobed, the lobes rounded at the ends; flowers pale-yellow, about an inch in diameter; fruit large, roundish, green, or variegated with different shades of green; seeds oval, flattened, half an inch long, five-sixteenths of an inch broad,—the color varying according to the variety, being either red, white, black, yellowish or grayish brown. An ounce contains from a hundred and seventy-five to two hundred seeds, and they retain their vitality eight years.

The Water-melon is more vigorous in its habit than the Musk-melon, and requires more space in cultivation; the hills being usually made eight feet apart in each direction. It is less liable to injury from insects, and the crop is consequently much more certain. The seed should not be planted till May, or before established warm weather; and but two good plants allowed to a hill. The varieties are as follow:

Apple-Seeded

A rather small, nearly round sort, deriving its name from its small, peculiar seeds; which, in form, size, and color, are somewhat similar to those of the apple. Skin deep, clear-green; rind very thin; flesh bright-red to the centre, sweet, tender, and well-flavored. It is hardy, bears abundantly, seldom fails to ripen perfectly in the shortest seasons, and keeps a long time after being gathered.

Carving One of Our Watermelons
William H. Martin, 1909

Black Spanish

(Spanish)

Form oblong; size large; skin very dark or blackish green; rind half an inch thick; flesh deep-red (contrasting finely with the very deep-green color of the skin), fine-grained, very sugary, and of excellent flavor. The variety is hardy, productive, thrives well, matures its fruit in the Northern and Eastern States, and is decidedly one of the best for general cultivation. Seeds dark-brown, or nearly black.

Bradford

The Bradford is a highly prized, South-Carolina variety; size large; form oblong; skin dark-green, with gray, longitudinal stripes, mottled and reticulated with green; rind not exceeding half an inch in thickness; seed yellowish-white, slightly mottled, and with a yellowish-brown stripe around the edge; flesh fine red to the centre; flavor fine and sugary; quality "best."

Carolina

Fruit of large size, and of an oblong form, usually somewhat swollen towards the blossom-end; skin deep-green, variegated with pale-green or white; flesh deep-red, not fine-grained, but crisp, sweet, and of fair quality; fruit frequently hollow at the centre; seeds black.

This variety is extensively grown in the Southern States for exportation to the North, where it appears in the markets about the beginning of August, and to some extent in July. Many of the specimens are much less marked with stripes and variegations than the true Carolina; and some shipments consist almost entirely of fruit of a uniform deep-green color, but of the form and quality of the Carolina.

Downing mentions a sub-variety with pale-yellow flesh and white seeds.

Citron Water-Melon

Form very nearly spherical; size rather small,—average specimens measuring six or seven inches in diameter; color pale-green, marbled with darker shades of green; flesh white, solid, tough, seedy, and very squashy and unpalatable in its crude state. It ripens late in the season, and will keep until December.

> "It is employed in the making of sweetmeats and preserves, by removing the rind or skin and seeds, cutting the flesh into convenient bits, and boiling in sirup which has been flavored with ginger, lemon, or some agreeable article. Its cultivation is the same as that of other kinds of melons."[2]

Clarendon

(Dark-Speckled)

Size large; form oblong; skin mottled-gray, with dark-green, interrupted, longitudinal stripes, irregular in their outline, and composed of a succession of peninsulas and isthmuses; rind thin, not exceeding half an inch; seed yellow, with a black stripe extending round the edge, and from one to three black spots on each side,—the form and number corresponding on the two sides; flesh scarlet to the centre; flavour sugary and exquisite, and quality "best."

This fine melon originated in Clarendon County, South Carolina; and, when pure, may at all times be readily recognized by the peculiarly characteristic markings of the seeds.

Ice-Cream

A large, very pale-green sort; when unmixed, readily distinguishable from all other varieties. Form nearly round, but sometimes a little depressed at the extremities; rind thicker than in most varieties; flesh white, very sweet and tender, and of remarkably fine flavor; seeds white. It is prolific, and also early; and is remarkably well adapted for cultivation in cold localities, or where the seasons are too short for the successful culture of the more tender and late kinds. Its pale-green skin, white flesh, and white seeds, are its prominent distinctive peculiarities.

Spring 1897, John A. Salzer Seed Co.

Imperial

This variety is said to have been introduced from the Mediterranean. Fruit round, or oblate, and of medium size; skin pale-green, with stripes and variegations of white or paler green; rind thin; flesh pale-red, crisp, sweet, and of excellent flavor; seeds reddish-brown. Very productive, but requiring a warm situation and a long season for its complete perfection.

Mountain Sprout

This variety is similar to the Mountain Sweet. It is of large size, long, and of an oval form. Skin striped and marbled with paler and deeper shades of green; rind thin,—measuring scarcely half an inch in thickness; flesh scarlet, a little hollow at the centre, crisp, sugary, and of excellent flavour.

Like the Mountain Sweet, it is a favorite market sort. It is not only of fine quality, but very productive. Seeds russet-brown.

Mountain Sweet

A large, long, oval variety, often contracted towards the stem in the form of a neck; skin striped and marbled with different shades of green; rind rather thin, measuring scarcely half an inch in thickness; flesh scarlet, and solid quite to the centre; seeds pale russet-brown, but often of greater depth of color in perfectly matured specimens of fruit.

A popular and extensively cultivated variety, quite hardy, productive, and of good quality. "For many years, it was universally conceded to be the best market sort cultivated in the Middle States, but of late has lost some of the properties that recommended it so highly to favor. This deterioration has probably been owing to the influence of pollen from inferior kinds grown in its vicinity."

Odell's Large White

Size very large, sometimes weighing sixty pounds; form round; skin gray, with fine green network spread over its uneven surface; rind nearly three-fourths of an inch in thickness; seeds large, grayish-black, and not numerous; flesh pale-red; flavor fine; quality very good. Productiveness said to exceed that of most other kinds.

This remarkably large melon originated with a negro man on the property of Col. A. G. Sumner, of South Carolina. Its large size, and long-keeping quality after being separated from the vine, will recommend the variety, especially for the market.

Orange

Form oval, of medium size; skin pale-green, marbled with shades of deeper green; rind half an inch in depth, or of medium thickness; flesh red, not fine-grained, but tender, sweet, and of good quality. When in its mature state, the rind separates readily from the flesh, in the manner of the peel from the flesh of an orange.

When first introduced, the variety was considered one of the best quality; but it appears to have in some degree deteriorated, and now compares unfavorably with many other sorts.

Pie-Melon

(California Pie-Melon)

Plant running,—the foliage and general habit resembling the Common Water-melon, but yet distinguishable by its larger size, more hairy stem, and its more stocky and vigorous character; fruit oblong, very large, measuring sixteen inches and upwards in length, and from eight to ten inches in diameter; skin yellowish-green, often marbled with different shades of light-green or pea-green; flesh white, succulent, somewhat tender, but very unpalatable, or with a squash-like flavor, in its crude state. As intimated by the name, it is used only for culinary purposes.

This melon should be cooked as follows: After removing the rind, cut the flesh into pieces of convenient size, and stew until soft and pulpy. Lemon-juice, sugar, and spices should then be added; after which, proceed in the usual manner of making pies from the apple or any other fruit. If kept from freezing, or from dampness and extreme cold, the Pie-melon may be preserved until March.

Ravenscroft

Size large; form oblong; skin dark-green, faintly striped and marked with green of a lighter shade, and divided longitudinally by sutures from an inch and a quarter to two inches apart; rind not more than half an inch in thickness; seed cream-color, tipped with brown at the eye, and having a brown stripe around the edge; flesh fine red, commencing abruptly at the rind, and extending to the centre; flavor delicious and sugary; quality "best."

This valuable water-melon originated with Col. A. G. Sumner, of South Carolina.

Souter

Size large, sometimes weighing twenty or thirty pounds; form oblong, occasionally roundish; skin peculiarly marked with finely reticulated, isolated, gray spots, surrounded by paler green, and having irregular, dark-green, longitudinal stripes extending from the base to the apex; rind thin, about half an inch thick; seed pure cream-white, with a faint russet stripe around the edge; flesh deep-red to the centre; flavor sugary and delicious; quality "best." Productiveness said to be unusually great.

This excellent variety originated in Sumpter District, South Carolina.

1. The Fruit and Fruit-trees of America. By **A. J. Downing**. Revised and corrected by **Charles Downing**, 1858.

2. *New American Cyclopædia.*

1893 Special Advertisement of Burpee's Seeds, W. Atlee Burpee & Co.

41

Luffa/Sponge Cucumber (Luffa acutangula)

***(Papangaye; Luffa aegyptiaca)*[1]**

This is an East-Indian plant, with a creeping stem, and angular, heart-shaped leaves. The flowers (several of which are produced on one stem) are yellow; the fruit is ten or twelve inches in length, about an inch and a half in diameter, deeply furrowed or grooved in the direction of its length, forming ten longitudinal, acute angles; the skin is hard, and of a russet-yellow colour; the seeds are black, rough, and hard, and quite irregular in form,—about five hundred are contained in an ounce.

Use—The fruit is eaten while it is quite young and small; served in the manner of cucumbers, or like vegetable marrow. When fully ripened, it is exceedingly tough, fibrous, and porous, and is sometimes used as a substitute for sponge: whence the name.

Luffa aegyptiaca
Francisco Manuel Blanco

1. *Luffa aegyptiaca is another variety of Luffa very similar to the one described and with the same uses. This variety was not included in the original, but the editors have added it as it may be of use to readers.*

Luffa acutangula
Francisco Manuel Blanco

42

West Indian Gherkin (Cucumis anguria)

(Prickly-Fruited Gherkin; West-Indian Cucumber; Jamaica Cucumber)

This species is said to be a native of Jamaica. The habit of the plant is similar to that of the Globe Cucumber, and its season of maturity is nearly the same. The surface of the fruit is thickly set with spiny nipples, and has an appearance very unlike that of the Common Cucumber. It is comparatively of small size, and of a regular, oval form,—generally measuring about two inches in length by an inch and a third in its largest diameter; color pale-green; flesh greenish-white, very seedy and pulpy. The seeds are quite small, oval, flattened, yellowish-white, and retain their vitality five years.

Concombre sauvage (West Indian Gherkin)
Flore médicale des Antilles, ou, Traité des plantes usuelles

It is somewhat later than the Common Cucumber, and requires nearly the whole season for its full development. Plant in hills about five feet apart; cover the seeds scarcely half an inch deep, and leave three plants to a hill.

The Prickly-fruited Gherkin is seldom served at table sliced in its crude state. It is principally grown for pickling: for which purpose it should be plucked when about half grown, or while the skin is tender, and can be easily broken by the nail. As the season of maturity approaches, the rind gradually hardens, and the fruit becomes worthless. In all stages of its growth, the flesh is comparatively spongy; and, in the process of pickling, absorbs a large quantity of vinegar.

43

The Pumpkin (Curcurbita pepo)

Under this head, on the authority of the late Dr. T. W. Harris, should properly be included "the common New-England field-pumpkin, the bell-shaped and crook-necked winter squashes, the Canada crook-necked, the custard squashes, and various others, all of which (whether rightly or not, cannot now be determined) have been generally referred by botanists to the *Cucurbita pepo* of Linnæus."

The term "pumpkin,"[1] as generally used in this country by writers on gardening and agriculture, and as popularly understood, includes only the few varieties of the Common New-England Pumpkin that have been long grown in fields in an extensive but somewhat neglectful manner; the usual practice being to plant a seed or two at certain intervals in fields of corn or potatoes, and afterwards to leave the growing vines to the care of themselves. Even under these circumstances, a ton is frequently harvested from a single acre, in addition to a heavy crop of corn or potatoes.

The Pumpkin was formerly much used in domestic economy; but, since the introduction of the Crook-necks, Boston Marrow, Hubbard, and other improved varieties of squashes, it has gradually fallen into disuse, and is now cultivated principally for agricultural purposes.

Varieties—The following are the principal varieties, although numerous intermediate sorts occur, more or less distinct, as well as more or less permanent in character:

Canada Pumpkin

(Vermont Pumpkin)

The Canada Pumpkin is of an oblate form, inclining to conic; and is deeply and regularly ribbed. When well grown, it is of comparatively large size, and measures thirteen or fourteen inches in diameter, and about ten inches in depth. Color fine, deep orange-yellow; skin or shell rather thick and hard; flesh yellow, fine-grained, sweet, and well flavored. Hardy, and very productive.

Compared with the common field variety, the Canada is much more flattened in its form, more regularly and deeply ribbed, of a deeper and richer color; and the flesh is generally much sweeter, and less coarse and stringy in its texture. It seems adapted to every description of soil; thrives well in all climates; and is one of the best sorts for agricultural purposes, as well as of good quality for the table.

Cheese Pumpkin

Plant very vigorous; leaves large, deep-green; fruit much flattened, deeply and rather regularly ribbed, broadly dishing about the stem, and basin-like at the opposite extremity. It is of large size; and, when well grown, often measures fifteen or sixteen inches in diameter, and nine or ten inches in depth. Skin fine, deep reddish-orange, and, if the fruit is perfectly matured, quite hard and shell-like; flesh very thick, yellow, fine-grained, sweet, and well flavored. The seeds are not distinguishable from those of the Common Field Pumpkin.

The Cheese Pumpkin is hardy, remarkably productive, and much superior in all respects to most of the field-grown sorts. Whether the variety originated in this country, cannot probably now be determined; but it was extensively disseminated in the Middle States at the time of the American Revolution, and was introduced into certain parts of New England by the soldiers on their return from service. After a lapse of more than seventy-five years,—during which time it must have experienced great diversity of treatment and culture,—it still can be found in its original type; having the same form, color, size, and the same thickness, and quality of flesh, which it possessed at the time of its introduction.

Left: Spaghetti squash (C pepo var β) and gold winter gourd (Cucurbita maxima Duchesne). Right: South gourd/large orange pumpkin variety (Cucurbita pepo var C maxima).
Seikei Zusetsu vol. 27, page 017

Common Yellow Field Pumpkin

Plant of vigorous, stocky habit, extending twelve feet and upwards in length; fruit rounded, usually a little more deep than broad, flattened at the ends, and rather regularly, and more or less prominently, ribbed. Its size is much affected by soil, season, and the purity of the seed. Average specimens will measure about fourteen inches in length, and eleven or twelve inches in diameter.

Color rich, clear orange-yellow; skin, or rind, if the fruit is well matured, rather dense and hard; flesh variable in thickness, but averaging about an inch and a half, of a yellow color, generally coarse-grained, and often stringy, but sometimes of fine texture, dry, and of good quality; seeds of medium size, cream-yellow.

The cultivation of the Common Yellow Field Pumpkin in this country is almost co-eval with its settlement. For a long period, few, if any, of the numerous varieties of squashes, now so generally disseminated, were known; and the Pumpkin was not only extensively employed as a material for pies, but was much used as a vegetable, in the form of squash, at the table. During the struggle for national independence, when the excessively high prices of sugars and molasses prevented their general use, it was the practice to reduce by evaporation the liquid in which the pumpkin had been cooked, and to use the saccharine matter thus obtained as a substitute for the more costly but much more palatable sweetening ingredients. When served at table in the form of a vegetable, a well-ripened, fine-grained pumpkin was selected, divided either lengthwise or crosswise; the seeds extracted; the loose, stringy matter removed from the inner surface of the flesh; and the two sections, thus prepared, were baked, till soft, in a common oven. The flesh was then scooped from the shell, pressed, seasoned, and served in the usual form. By many, it is still highly esteemed, and even preferred for pies to the Squash, or the more improved varieties of pumpkins; but its cultivation at present is rather for agricultural than for culinary purposes.

Connecticut Field Pumpkin

A large, yellow, field variety, not unlike the Common Yellow in form, but with a softer skin, or shell. It is very prolific, of fair quality as an esculent, and one of the best for cultivating for stock or for agricultural purposes.

Long Yellow Field Pumpkin

Plant hardy and vigorous, not distinguishable from that of the Common Yellow variety; fruit oval, much elongated, the length usually about twice the diameter; size large,—well-grown specimens measuring sixteen to twenty inches in length, and nine or ten inches in diameter; surface somewhat ribbed, but with the markings less distinct than those of the Common Yellow; color bright orange-yellow; skin of moderate thickness, generally easily broken by the nail; flesh about an inch and a half in thickness, yellow, of good but not fine quality, usually sweet, but watery, and of no great value for the table.

It is very hardy and productive; well adapted for planting among corn or potatoes; may be profitably raised for feeding out to stock; keeps well when properly stored; and selected specimens will afford a tolerable substitute for the Squash in the kitchen, particularly for pies.

Between this and the Common Yellow, there are various intermediate sorts; and, as they readily hybridize with each other, it is with difficulty that these varieties can be preserved in a pure state. Only one of the sorts should be cultivated, unless there is sufficient territory to enable the cultivator to allow a large distance between the fields where the different varieties are grown.

Nantucket
(Hard-Shell)

Form flattened or depressed, but sometimes oblong or bell-shaped, often faintly ribbed; size medium or rather small; color deep-green, somewhat mellowed by exposure to the sun, or at full maturity; skin or shell thick and hard, and more or less thickly covered with prominent, wart-like excrescences; flesh comparatively thick, yellow, sweet, fine-grained, and of excellent flavor,—comparing favorably in all respects with that of the Sugar Pumpkin. It is a productive sort, and its flesh much dryer and more sugary than the peculiar, green, and warty appearance of the fruit would indicate. When cooked, it should be divided into pieces of convenient size; the seeds, and loose, stringy parts, removed from the inner surface of the flesh, and then boiled or baked in the skin or shell; afterwards scooping out the flesh, as is practised with the Hubbard Squash or other hard-shelled varieties of pumpkins. It is an excellent pie-variety, and selected specimens will be found of good quality when served as squash at the table. It will keep till February or March.

Striped Field Pumpkin

Habit of the plant, and form of the fruit, very similar to the Common Yellow Field Pumpkin. The size, however, will average less; although specimens may sometimes be procured as large as the dimension given for the Common Yellow. Color yellow, striped and variegated with green,—after being gathered, the green becomes gradually softer and paler, and the yellow deeper; flesh yellow, moderately thick, and, though by some considered of superior quality, has not the fine, dry, and well-flavored character essential for table use; seeds similar to the foregoing sorts.

The Striped Field Pumpkin is a hardy sort, and yields well. It is, however, exceedingly liable to hybridize with all the varieties of the family, and is with difficulty preserved in an unmixed condition. It should be grown as far apart as possible from all others, especially when the seed raised is designed for sale or for reproduction at home.

Sugar-Pumpkin
(Small Sugar-Pumpkin)

Sugar Pumpkin

Plant similar in its character and general appearance to the Common Field Pumpkin; fruit small, eight or nine inches at its broadest diameter, and about six inches in depth; form much depressed, usually broadest near the middle, and more or less distinctly ribbed; skin bright orange-yellow when the fruit is well ripened, hard, and shell-like, and not easily broken by the nail; stem quite long, greenish, furrowed, and somewhat reticulated; flesh of good thickness, light-yellow, very fine-grained, sweet, and well flavored; seeds of smaller size than, but in other respects similar to, those of the Field Pumpkin. The variety is the smallest of the sorts usually employed for field cultivation. It is, however, a most abundant bearer, rarely fails in maturing its crops perfectly, is of first-rate quality, and may be justly styled an acquisition. For pies, it is not surpassed by any of the family; and it is superior for table use to many of the garden squashes. The facility with which it hybridizes or mixes with other kinds renders it extremely difficult to keep the variety pure; the tendency being to increase in size, to grow longer or deeper, and to become warty: either of which conditions may be considered an infallible evidence of deterioration.

Varieties sometimes occur more or less marbled and spotted with green; the green, however, often changing to yellow after harvesting.

1. The chapter on squash and marrows contains numerous varieties which in many parts of the English-speaking world are also known as 'pumpkins'.

44

Snake Cucumber (Cucumis flexuosus)

(Armenian cucumber)

Though generally considered as a species of cucumber, this plant should properly be classed with the melons. In its manner of growth, foliage, flowering, and in the odor and taste of the ripened fruit, it strongly resembles the musk-melon. The fruit is slender and flexuous; frequently measures more than three feet in length; and is often gracefully coiled or folded in a serpent-like form. The skin is green; the flesh, while the fruit is forming, is greenish-white,—at maturity, yellow; the seeds are yellowish-white, oval, flattened, often twisted or contorted like those of some varieties of melons, and retain their vitality five years.

Planting and Cultivation—The seeds should be planted in May, in hills six feet apart. Cover half an inch deep, and allow three plants to a hill.

Use—The fruit is sometimes pickled in the manner of the Common Cucumber, but is seldom served at table sliced in its crude state. It is generally cultivated on account of its serpent-like form, rather than for its value as an esculent.

Well-grown specimens are quite attractive; and, as curious vegetable productions, contribute to the interest and variety of horticultural exhibitions.

Armenian cucumber, Cucumis melo var. flexuosus
cones Plantarum Medico-Oeconomico-Technologicarum
(1800-1822)

45

Squash and Marrows (Curcurbita pepo)

All the varieties are tender annuals, and of tropical origin. They only thrive well in a warm temperature: and the seed should not be sown in spring until all danger from frost is past, and the ground is warm and thoroughly settled; as, aside from the tender nature of the plant, the seed is extremely liable to rot in the ground in continued damp and cold weather.

Any good, well-enriched soil is adapted to the growth of the Squash. The hills should be made from eight to ten inches in depth, two feet in diameter, and then filled within three or four inches of the surface with well-digested compost; afterwards adding sufficient fine loam to raise the hill an inch or two above the surrounding level. On this, plant twelve or fifteen seeds; covering about three-fourths of an inch deep. Keep the earth about the plants loose and clean; and from time to time remove the surplus vines, leaving the most stocky and vigorous. Three plants are sufficient for a hill; to which number the hills should ultimately be thinned, making the final thinning when all danger from bugs and other vermin is past. The dwarfs may be planted four feet apart; but the running sorts should not be less than six or eight. The custom of cutting or nipping off the leading shoot of the running varieties is now practised to some extent, with the impression that it both facilitates the formation of fruitful laterals and the early maturing of the fruit. Whether the amount of product is increased by the process, is not yet determined.

In giving the following descriptions, no attempt has been made to present them under scientific divisions; but they have been arranged as they are in this country popularly understood:

Summer Varieties:

Apple Squash

(Early Apple)

Plant running, not of stocky habit, but healthy and vigorous; fruit obtusely conical, three inches broad at the stem, and two inches and a half in depth; skin yellowish-white, thin and tender while the fruit is young, hard and shell-like when ripe; flesh dry and well flavored in its green state, and often of good quality at full maturity.

The fruit is comparatively small; and, on this account, the variety is very little cultivated.

Bush Summer Warted Crookneck

(Early Summer Crookneck; Yellow Summer Warted Crookneck; Cucurbita verrucosa)

Plant dwarfish or bushy in habit, generally about two feet and a half in height or length; fruit largest at the blossom-end, and tapering gradually to a neck, which is solid, and more or less curved; size medium,— average specimens, when suitable for use, measuring about eight inches in length, and three inches in diameter at the broadest part; the neck is usually about two inches in thickness; color clear, bright-yellow; skin very warty, thin, and easily broken by the nail while the fruit is young, and suitable for use,—as the season of maturity approaches, the rind gradually becomes firmer, and, when fully ripe, is very hard and shell-like; flesh greenish-yellow, dry, and well flavored; seeds comparatively small, broad in proportion to the length, and of a pale-yellow color. About four hundred are contained in an ounce.

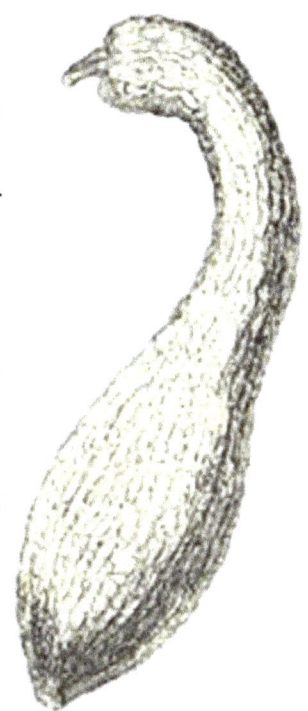

Bush Summer Warted Crookneck Squash

The Bush Summer Crookneck is generally esteemed the finest of the summer varieties. It is used only while young and tender, or when the skin can be easily pierced or broken by the nail. After the fruit hardens, the flesh becomes watery, coarse, strong flavored, and unfit for table use.

On account of the dwarfish character of the plants, the hills may be made four feet apart. Three plants will be sufficient for a hill.

Early White Bush Scolloped

(White Pattypan; Cymbling; White Summer Scolloped; Pattison Blanc)

This is a sub-variety of the Early Yellow Bush Scolloped. The plant has the same dwarf habit, and the fruit is nearly of the same size and form. The principal distinction between the varieties consists in the difference of color.

By some, the white variety is considered a little inferior in fineness of texture and in flavor to the yellow; though the white is much the more abundant in the markets. Both of the varieties are hardy and productive; and there is but little difference in the season of their maturity.

In the month of June, large quantities are shipped from the Southern and Middle States to the North and East, where they anticipate from two to three weeks the products of the home-market gardens; the facilities afforded by steam transportation rendering nearly profitless the efforts of gardeners to obtain an early crop. As the variety keeps well, and suffers little from transportation, the squashes are generally found fresh and in good order on their arrival.

Early Yellow Bush Scolloped

(Cymbling; Pattypan; Yellow Summer Scollop)

Plant dwarf, of rather erect habit, and about two feet and a half in height; leaves large, clear-green; fruit somewhat of a hemispherical form, expanded at the edge, which is deeply and very regularly scolloped. When suitable for use, it measures about five inches in diameter, and three inches in depth; but, when fully matured, the diameter is often ten or twelve inches, and even upwards. Color yellow; skin, while young, thin, and easily pierced,—at maturity, hard and shell-like; flesh pale-yellow, tolerably fine-grained, and well flavored,—not, however, quite so dry and sweet as that of the Summer Crookneck; seeds broader in proportion to their length than the seeds of most varieties, and of comparatively small size. Four hundred and twenty-five weigh an ounce.

Early Yellow Bush Scalloped Squash

This variety has been common to the gardens of this country for upwards of a century; during which period, the form and general character have been very slightly, if at all, changed. When grown in the vicinity of the Bush Summer Crookneck, the surface sometimes exhibits the same wart-like excrescences; but there is little difficulty in procuring seeds that will prove true to the description above given.

Like the Summer Crookneck, the scolloped squashes are used while young or in a green state. After the hardening of the skin, or shell, the flesh generally becomes coarse, watery, strong-flavored, and unfit for the table.

The hills should be made about four feet apart, and three plants allowed to a hill. Season from the beginning of July to the middle or last of August.

Egg-Squash

(Cucurbita ovifera)

An ornamental variety, generally cultivated for its peculiar, egg-like fruit, which usually measures about three inches in length, and two inches or two and a half in diameter. Skin, or shell, white. It is seldom used as an esculent; though, in its young state, the flesh is quite similar in flavor and texture to that of the scolloped varieties. If trained to a trellis, or when allowed to cover a dry, branching tree, it is quite ornamental; and, in its ripened state, is quite interesting, and attractive at public exhibitions. Increase of size indicates mixture or deterioration.

"It has been generally supposed, that the Egg-squash was a native of Astrachan, in Tartary. Dr. Loroche included it in a list of plants not natives of Astrachan, but cultivated only in gardens where it is associated with such exotics as Indian corn, or maize, with which it was probably introduced directly or indirectly from America. We also learn from Loroche that this species varied in form, being sometimes pear-shaped; that it was sometimes variegated in colour with green and white, and the shell served instead of boxes. Here we have plainly indicated the little gourd-like, hard-shelled, and variegated squashes that are often cultivated as ornamental plants.

"From these and similar authorities, it is evident that summer squashes were originally natives of America, where so many of them were found in use by the Indians, when the country began to be settled by Europeans."[1]

Green Bush Scolloped

(Pattison Vert)

Fruit similar in size and form to the Yellow or White Bush Scolloped; skin or shell bottle-green, marbled or clouded with shades of lighter green. It is comparatively of poor quality, and is little cultivated.

Green-Striped Bergen

Plant dwarf, but of strong and vigorous habit; fruit of small size, bell-shaped; colours dark-green and white, striped.

An early but not productive sort, little cultivated at the North or East, but grown to a considerable extent for the New-York market. It is eaten both while green and when fully ripe.

Large Summer Warted Crookneck

A large variety of the Bush or Dwarf Summer Crookneck. Plant twelve feet and upwards in length, running; fruit of the form of the last named, but of much greater proportions,—sometimes attaining a length of nearly two feet; skin clear, bright yellow, and thickly covered with the prominent wart-like excrescences peculiar to the varieties; flesh greenish-yellow, and of coarser texture than that of the Dwarf Summer Crookneck. Hardy and very productive. The hills should be made six feet apart.

Orange

(Cucurbita aurantiaca)

Fruit of the size, form, and color of an orange. Though generally cultivated for ornament, and considered more curious than useful, "some of them are the very best of the summer squashes for table use; far superior to either the scolloped or warted varieties." When trained as directed for the Egg-squash, it is equally showy and attractive.

Variegated Bush Scolloped

(Pattison Panaché)

Pale yellow, or nearly white, variegated with green. Very handsome, but of inferior quality.

Autumn and Winter Varieties:

Autumnal Marrow

(Boston Marrow; Courge de l'Ohio)

Plant twelve feet or more in length, moderately vigorous; fruit ovoid, pointed at the extremities, eight or nine inches in length, and seven inches in diameter; stem very large, fleshy, and contracted a little at its junction with the fruit,—the summit, or blossom-end, often tipped with a small nipple or wart-like excrescence; skin remarkably thin, easily bruised or broken, cream-yellow at the time of ripening, but changing to red after harvesting, or by remaining on the plants after full maturity; flesh rich, salmon-yellow, remarkably dry, fine-grained, and, in sweetness and excellence, surpassed by few varieties. The seeds are large, thick, and pure white: the surface, in appearance and to the touch, resembles glove-leather or dressed goat-skin. About one hundred are contained in an ounce.

In favourable seasons, the Autumnal Marrow Squash will be sufficiently grown for use early in August; and, if kept from cold and dampness, may be preserved till March.

Mr. John M. Ives, of Salem, who was awarded a piece of silver plate by the Massachusetts Horticultural Society for the introduction of this valuable variety, has furnished the following statement relative to its origin and dissemination:

Salem, Mass., Feb. 7, 1858.

Dear Sir,—As requested, I forward you a few facts relative to the introduction of the Autumnal Marrow Squash, the cultivation of which has extended not only over our entire country, but throughout Europe. It succeeds better in England than the Crooknecks; and may be seen in great abundance every season at Covent-Garden Market, in London.

Early in the spring of 1831, a friend of mine from Northampton, in this State, brought to my grounds a specimen of this vegetable, of five or six pounds' weight, which he called "Vegetable Marrow." As it bore no resemblance to the true Vegetable Marrow, either in its form or color, I planted the seeds, and was successful in raising eight or ten specimens. Finding it a superior vegetable, with a skin as thin as the inner envelope of an egg, and the flesh of fine texture, and also that it was in eating early in the fall, I ventured to call it "Autumnal Marrow Squash." Soon a drawing was made, and forwarded, with a description, to the "Horticultural Register" of Fessenden, and also to the "New-England Farmer."

In cultivating this vegetable, I found the fruit to average from eight to nine pounds, particularly if grown on newly broken-up sod or grass land. From its facility in hybridizing with the tribe of pumpkins, I consider it to be, properly speaking, a fine-grained pumpkin. The first indication of deterioration or mixture will be manifested in the thickening of the skin, or by a green circle or coloring of green at the blossom-end.

> More recently, I have been informed, by the gentleman to whom I was indebted for the first specimen, that the seeds came originally from Buffalo, N.Y., where they were supposed to have been introduced by a tribe of Indians, who were accustomed to visit that city in the spring of the year. I have not been able to trace it beyond this. It is, unquestionably, an accidental hybrid.
>
> Yours truly,
> **John M. Ives.**
> Mr. **F. Burr**, Jun.

Canada Crookneck

The plants of this variety are similar in habit to those of the Common Winter Crookneck; but the foliage is smaller, and the growth less luxuriant. In point of size, the Canada Crookneck is the smallest of its class. When the variety is unmixed, the weight seldom exceeds five or six pounds. It is sometimes bottle-formed; but the neck is generally small, solid, and curved in the form of the Large Winter Crooknecks. The seeds are contained at the blossom-end, which expands somewhat abruptly, and is often slightly ribbed. Skin of moderate thickness, and easily pierced by the nail; color, when fully ripened, cream-yellow, but, if long kept, becoming duller and darker; flesh salmon-red, very close-grained, dry, sweet, and fine-flavored; seeds comparatively small, of a grayish or dull-white color, with a rough and uneven yellowish-brown border; three hundred are contained in an ounce.

The Canada is unquestionably the best of the Crooknecked sorts. The vines are remarkably hardy and prolific; yielding almost a certain crop both North and South. The variety ripens early; the plants suffer but little from the depredations of bugs or worms; and the fruit, with trifling care, may be preserved throughout the year. It is also quite uniform in quality; being seldom of the coarse and stringy character so common to other varieties of this class.

Cananda Crookneck Squash

Cashew

(Cushaw Pumpkin)

Somewhat of the form and color of the Common Winter Crookneck. Two prominent varieties, however, occur. The first is nearly round; the other curved, or of the shape of a hunter's horn. The latter is the most desirable. It is not cultivated or generally known in New England or in the northern portions of the United States; for though well suited to Louisiana and other portions of the South, where it is much esteemed, it is evidently too tender for cultivation where the seasons are comparatively short and cool.

In an experimental trial by the late Dr. Harris, specimens raised from seed received from New Jersey "did not ripen well, and many decayed before half ripe."

The Crooknecks of New England "may be distinguished from the Cashew by the want of a persistent style, and by their furrowed and club-shaped fruit-stems."

Cocoa-Nut Squash

Cocoa Squash

Fruit oval, elongated, sixteen to twenty inches in length, eight or ten inches in diameter, and weighing from fifteen to twenty pounds and upwards; skin thin, easily pierced or broken, of an ash-gray color, spotted, and marked with light drab and nankeen-brown,—the furrows dividing the ribs light drab; stem small; flesh deep orange-yellow, of medium thickness; seeds pure white, broader in proportion to their length than those of the Hubbard or Boston Marrow.

The quality of the Cocoa-nut Squash is extremely variable. Sometimes the flesh is fine-grained, dry, sweet, and of a rich, nut-like flavor; but well-developed and apparently well-matured specimens are often coarse, fibrous, watery, and unfit for table use. The variety ripens in September, and will keep till March or April.

Custard Squash

Plant healthy and of vigorous habit, often twenty feet and upwards in length; fruit oblong, gathered in deep folds or wrinkles at the stem, near which it is the smallest, abruptly shortened at the opposite extremity, prominently marked by large, rounded, lengthwise elevations, and corresponding deep furrows, or depressions; skin, or shell, cream-white; flesh pale-yellow, not remarkable for solidity, or fineness of texture, but well flavored; the seeds are yellowish-white, and readily distinguished from those of other varieties by their long and narrow form. Under favorable conditions of soil and season, the Custard Squash attains a large size; often measuring twenty inches and upwards in length, eight or ten inches in diameter, and weighing from eighteen to twenty-five pounds.

It is one of the hardiest and most productive of all varieties. Crops are recorded of fourteen tons from an acre. It is esteemed by some for pies; but, as a table squash, is inferior to most other sorts. Its great yield makes it worthy the attention of agriculturists, as it would doubtless prove a profitable variety to be cultivated for stock.

From the habit of the plant, the form and character of the fruit, and its great hardiness and productiveness, it appears to be allied to the Vegetable Marrow.

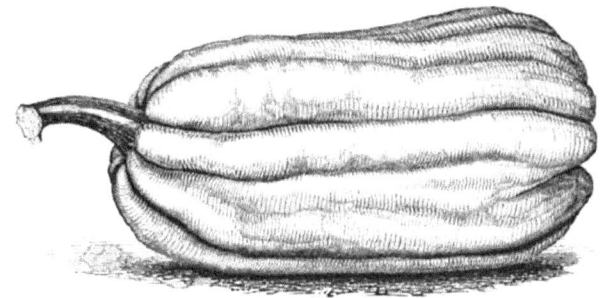
Custard Squash

Egg-Shaped, or Reeves

Fruit large, weighing from fifteen to twenty pounds; but in rich, highly manured soil, and with only a few on each plant, it may be grown to upwards of fifty pounds' weight. It is short, ovate, sometimes tapering rather abruptly. Skin, or shell, hard, of a reddish color; flesh firm, red, excellent in a ripe state cooked as a vegetable, or in any other way in which squashes are prepared. The stems run to a very great length, and bear all along most abundantly. Altogether, it is a sort highly deserving of cultivation.

It was brought into notice by John Reeves, Esq.; who has contributed to horticulture many valuable plants from China, where he resided for many years.

Plant in hills eight feet apart, and thin to two plants to a hill.

Honolulu

Plant twelve feet or more in length, remarkably strong and vigorous; leaves very large,—the leaf-stems often three feet and upwards in length; fruit large, oblate, depressed about the stem, broadly, and sometimes deeply, but in general faintly, ribbed; skin moderately thick, but not shell-like, of an ash-green color, striped and variegated with drab or lighter shades of green; flesh reddish-orange, very thick, of good flavor, but less dry and sweet than that of the Hubbard or Boston Marrow; seeds large, white.

This recently introduced variety is hardy, productive, a good keeper, excellent for pies, and by some esteemed for table use.

Specimens frequently occur of a reddish cream-color, striped and marked with drab or pale-yellow.

Hubbard

Plant similar in character and appearance to that of the Autumnal Marrow; fruit irregularly oval, sometimes ribbed, but often without rib-markings, from eight to ten inches in length, seven or eight inches in diameter, and weighing from seven to nine pounds,—some specimens terminate quite obtusely, others taper sharply towards the extremities, which are frequently bent or curved; skin, or shell, dense and hard, nearly one-eighth of an inch thick, and overspread with numerous small

Hubbard Squash

protuberances; stem fleshy, but not large; color variable, always rather dull, and usually clay-blue or deep olive-green,—the upper surface, if long exposed to the sun, assuming a brownish cast, and the under surface, if deprived of light, becoming orange-yellow; flesh rich salmon-yellow, thicker than that of the Autumnal Marrow, very fine-grained, sweet, dry, and of most excellent flavor,—in this last respect, resembling that of roasted or boiled chestnuts; seeds white,—similar to those of the Autumnal Marrow. Season from September to June; but the flesh is dryest and sweetest during autumn and the early part of winter.

The Hubbard Squash should be grown in hills seven feet apart, and three plants allowed to a hill. It is essential that the planting be made as far as possible from similar varieties, as it mixes, or hybridizes, readily with all of its kind. In point of productiveness, it is about equal to the Autumnal Marrow. "The average yield from six acres was nearly five tons of marketable squashes to the acre."

Mr. J. J. H. Gregory, of Marblehead, Mass., who introduced this variety to notice, and through whose exertions it has become widely disseminated, remarks in the "New-England Farmer" as follows:

"Of its history I know next to nothing, farther than that the seed was given to me by an aged female, about twelve years since, in remembrance of whom I named it; and that the party from whom she received it cannot tell from whence the seed came. I infer that it is of foreign origin, partly from the fact that the gentleman to whom I traced it is a resident of a seaport town, and is largely connected with those who follow the seas."

Italian Vegetable Marrow

(Courge Coucourzelle)

This forms a dwarf bush, with short, reclining stems, and upright leaves, which are deeply five-lobed. The fruits are used when the flowers are about to drop from their ends. They are then from four to five inches long, and an inch and a half to two inches in diameter. When ripe, the fruit is from fifteen to eighteen inches in length, and about six inches in diameter. It is of a pale yellow, striped with green. It should, however, be used in the young, green state; for, when mature, it is not so good as many of the other sorts. It bears very abundantly; and, as it does not run, may be grown in smaller compass than the true Vegetable Marrow.

Mammoth

(Mammoth Pumpkin; Large Yellow Gourd, of the English; Potiron jaune, of the French; Cucurbita maxima)

This is the largest-fruited variety known. In a very rich compost, and under favorable conditions of climate, it grows to an enormous size. Fruit weighing a hundred and twenty pounds is not uncommon; and instances, though exceptional, are recorded of weights ranging from two hundred to nearly two hundred and fifty pounds.

The leaves are very large, and the stems thick, running along the ground to the distance of twenty or thirty feet if not stopped, and readily striking root at the joints.

The fruit is round, or oblate; sometimes flattened on the under side, owing to its great weight; sometimes obtusely ribbed, yellowish, or pale buff, and frequently covered to a considerable extent with a gray netting. Flesh very deep yellow; seeds white.

It is used only in its full-grown or ripe state, in which it will keep for several months; and even during the winter, if stored in a dry, warm situation. The flesh is sweet, though generally coarse-grained and watery. It is used in soups and stews, and also for pies; but is seldom served like squash at the table.

Neapolitan

(Courge pleine de Naples)

Plant running; leaves small, smooth, striped and marked with white along the nerves; fruit nearly two feet in length, and rather more than five inches in its smallest diameter, bent at the middle, and broadly but faintly ribbed,—it increases in size towards the extremities, but is largest at the blossom-end, where it reaches a diameter of eight or ten inches; skin bright green; stem small; flesh bright, clear yellow; the neck is entirely solid, and the seed-end has an unusually small cavity; seeds dull white.

The late Rev. A. R. Pope, in a communication to the Massachusetts Horticultural Society, describes it as follows:

> "New, very heavy; having a large, solid neck, and a small cavity for the seeds. Flesh sweet, dry, and somewhat coarse, but not stringy. Very superior for pies, and a good keeper."

Patagonian

A large, long Squash, prominently ribbed. It differs little in form or size from the Custard. Skin very deep green; flesh pale yellow; seeds of medium size, yellowish-white.

The plant is a vigorous grower, and the yield abundant; but its quality is inferior, and the variety can hardly be considered worthy of cultivation for table use. It may, however, prove a profitable sort for growing for agricultural purposes.

Puritan

Plant running, ten feet and upwards in length; leaves clear green, of medium size; fruit bottle-formed, fourteen or fifteen inches long, and about ten inches in diameter at the broadest part; neck solid, four or five inches in diameter; average weight eight to ten pounds; skin thin, usually white or cream-white, striped and marked with green, though specimens sometimes occur, from unmixed seed, uniformly green; flesh pale yellow, dry, sweet, mild, and well flavored; seeds of medium size, white. Season from August to January.

This variety, long common to gardens in the vicinity of the Old Colony, retains its distinctive character to a very remarkable degree, even when grown under the most unfavorable circumstances. Seeds, obtained from a gardener who had cultivated the variety indiscriminately among numerous summer and winter kinds for upwards of twenty years, produced specimens uniformly true to the normal form color, and quality. It is hardy and productive, good for table use, excellent for pies, and well deserving of cultivation.

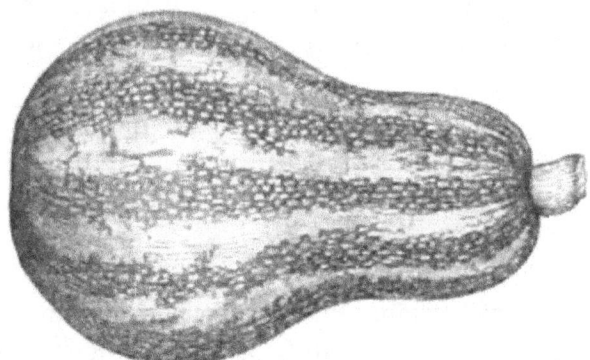

Puritan Squash

Sweet-Potato Squash

Plant very similar in character to that of the Hubbard or Autumnal Marrow; fruit twelve or fourteen inches long, seven or eight inches thick, sometimes ribbed, but usually without rib-markings; oblong, tapering to the ends, which are often bent or curved in the manner of some of the types of the Hubbard; stem of medium size, striated; skin ash-green, with a smooth, polished surface; flesh salmon-yellow, thick, fine-grained, dry, and sweet,—if the variety is pure, and the fruit well matured, its quality approaches that of the Hubbard and Autumnal Marrow; seeds white.

The variety is hardy and productive, keeps well, and is deserving of cultivation. When grown in the vicinity of the last-named sorts, it often becomes mixed, and rapidly degenerates. In its purity, it is uniformly of one color; with perhaps the exception of the under surface, which is sometimes paler or yellowish. It has been suggested that this variety and the Hubbard may have originated under similar circumstances.

Turban

(Acorn; Giraumon Turban; Turk's-cap; Cucurbita piliformis)

Plant running; leaves small, soft, slightly lobed on the borders; fruit rounded, flattened, expanding about the stem to a broad, plain, brick-red surface, of ten or twelve inches in diameter. At the blossom-end, the fruit suddenly contracts to an irregular, cone-like point, or termination, of a greenish color, striped with white; and thus, in form and color, somewhat resembles a turban: whence the name. Flesh orange-yellow, thick, fine-grained, sugary, and well flavored; seeds white, comparatively short, and small.

The Turban Squash is not early, and should have the advantage of the whole season. "Its specific gravity is said to exceed that of any other variety. Its keeping properties are not particularly good; but its flavour, when grown on light, dry soil, will compare well with either the Autumnal Marrow or the Hubbard." It mixes very readily when grown in the vicinity of other varieties, is not an abundant bearer, and cannot be recommended for general cultivation.

Dr. Harris states that "this variety—sometimes called the 'Acorn Squash,' because, when the fruit is small, it resembles somewhat an acorn in its cup—seems to be the *Cucurbita piliformis* of Duchesne." He further adds, that:

> "...it sometimes grows to a large size, measuring fourteen or fifteen inches in transverse diameter, and looks like an immense Turkish turban in shape. Specimens raised in my garden in 1851 were little more than ten inches in diameter, and weighed ten pounds or more; having very thick and firm flesh, and but a small cavity within. They proved excellent for table use,—equal in quality to the best Autumnal Marrows. They keep quite as well as the latter."

Valparaiso
(Porter's Valparaiso; Commodore Porter)

Plant running; leaves large, not lobed, but cut in rounded angles on the borders; fruit oval, about sixteen inches in length, ten or eleven inches in diameter, slightly ribbed, and largest at the blossom-end, which often terminates in a wart-like excrescence; skin cream-white, sometimes smooth and polished, but often more or less reticulated, or netted; flesh comparatively thick, orange-yellow, generally dry, sweet, and well flavored, but sometimes fibrous and watery; seeds rather large, nankeen-yellow, smooth and glossy.

The variety requires the whole season for its perfection. It hybridizes readily with the Autumnal Marrow and kindred sorts, and is kept pure with considerable difficulty. It is in use from September to spring. The variety, if obtained in its purity, will be found of comparative excellence, and well deserving of cultivation. Stripes and clouds of green upon the surface are infallible evidences of mixture and deterioration.

The late Dr. Harris, in a communication to the "Pennsylvania Farm Journal," remarks as follows:

> "The Valparaiso squashes (of which there seem to be several varieties, known to cultivators by many different names, some of them merely local in their application) belong to a peculiar group of the genus *Cucurbita*, the distinguishing characters of which have not been fully described by botanists. The word 'squash,' as applied to these fruits, is a misnomer, as may be shown hereafter. It would be well to drop it entirely, and to call the fruits of this group 'pompions,' 'pumpkins,' or 'potirons.' It is my belief, that they were originally indigenous to the tropical and sub-tropical parts of the western coast of America. They are extensively cultivated from Chili to California, and also in the West Indies; whence enormous specimens are sometimes brought to the Atlantic

States. How much soever these Valparaiso pumpkins may differ in form, size, color, and quality, they all agree in certain peculiarities that are found in no other species or varieties of *Cucurbita*. Their leaves are never deeply lobed like those of other pumpkins and squashes, but are more or less five-angled, or almost rounded and heart-shaped, at base: they are also softer than those of other pumpkins and squashes. The summit, or blossom-end, of the fruit has a nipple-like projection upon it, consisting of the permanent fleshy style. The fruit-stalk is short, nearly cylindrical, never deeply five-furrowed, but merely longitudinally striated or wrinkled, and never clavated, or enlarged with projecting angles, next to the fruit. With few exceptions, they contain four or five double rows of seeds. To this group belong Mr. Ives's Autumnal Marrow Squash (or Pumpkin); Commodore Porter's Valparaiso Squash (Pumpkin); the so-called Mammoth Pumpkin, or *Cucurbita maxima* of the botanists; the Turban or Acorn Squash; *Cucurbita piliformis* of Duchesne; the Cashew Pumpkin; Stetson's Hybrid, called the 'Wilder Squash;' with various others."

Catalogue of Home Grown Seeds (1899), James J.H. Gregory & Son

Vegetable Marrow

(Succade Gourd; Courge à la moëlle, of the French;

Plant twelve feet and upwards in length; leaves deeply five-lobed; fruit about nine inches long, and of an elliptic shape,—but it is sometimes grown to twice that length, and of an oblong form; surface slightly uneven, by irregular, longitudinal, obtuse ribs, which terminate in a projecting apex at the extremity of the fruit. When mature, it is of a uniform pale yellow or straw color. The skin, or shell, is very hard when the fruit is perfectly ripened; flesh white, tender, and succulent, even till the seeds are ripe. It may be used in every stage of its growth. Some prefer it when the flower is still at the extremity of the fruit; others like it older. When well ripened, it will keep well throughout the winter, if stored in a perfectly dry place, out of the reach of frost, and not exposed to great changes of temperature.

To have Vegetable Marrows large and fine for winter, the young fruit should be regularly taken off for use; and, when the plant has acquired strength, a moderate quantity should be allowed to set for maturity. Sufficient for this purpose being reserved, the young fruit that may be subsequently formed should be removed for use in a very young state. The vines, or shoots, may be allowed to run along the surface of the ground; or they may be trained against a wall, or on palings or trellises.

The seed should be planted at the same time and in the same manner as those of the Winter Crookneck or Boston Marrow.

Wilder

(Stetson's Hybrid)

The Wilder Squash was produced about twelve years since, from the Valparaiso and the Autumnal Marrow, by Mr. A. W. Stetson, of Braintree, Mass.; and was named for the Hon. Marshall P. Wilder, a gentleman widely known for his patriotic devotion to the advancement of agricultural and pomological science in the United States.

Winter Crookneck Squash

The plant is a strong grower, and resembles that of the Valparaiso. The fruit is somewhat ovoid, but rather irregular in form, broadly and faintly ribbed (sometimes, however, without rib-markings), and varies in weight from twelve to thirty pounds and upwards; stem very large, striated or reticulated, and often turned at right angles near its connection with the fruit,—the opposite extremity terminates in the wart-like excrescence peculiar to the class; skin reddish-yellow, not unlike that of the Autumnal Marrow; the flesh is remarkably thick, of a salmon-yellow color, sweet and well flavored. In some forms of cookery, and especially for pies, it is esteemed equal, if not superior, to any other variety. When served in the customary manner of serving squash at table, it is inferior to the Hubbard or Autumnal Marrow. The seeds are white.

Winter Crookneck

(Cuckaw)

This is one of the oldest and most familiar of the winter varieties. Plant hardy and vigorous; fruit somewhat irregular in form, the neck solid and nearly cylindrical, and the blossom-end more or less swollen. In some specimens, the neck is nearly straight; in others, sweeping, or circular; and sometimes the extremities nearly or quite approach each other. Size very variable, being affected greatly both by soil and season; the weight ranging from six pounds to forty pounds and upwards. A specimen was raised by Capt. Josiah Lovett, of Beverly, Mass., and exhibited before the Massachusetts Horticultural Society, the weight of which was nearly seventy pounds. Color sometimes green; but, when fully mature, often cream-yellow. The color, like that of the Canada Crookneck, frequently changes after being harvested. If green when plucked, it gradually becomes paler; or, if yellow when taken from the vines, it becomes, during the winter, of a reddish cream-color. Flesh salmon-yellow, not uniform in texture or solidity, sometimes close-grained, sweet, and fine flavored, and sometimes very coarse, stringy, and nearly worthless for the table; seeds of medium size, grayish-white, the border darker, or brownish. About two hundred are contained in an ounce.

It is a very hardy and productive variety; ripens its crop with great certainty; suffers less from the depredations of insects than most of the winter sorts; and, if protected from cold and dampness during the winter months, will keep the entire year.

Winter Striped Crookneck

This is a sub-variety of the common Winter Crookneck. Size large,—the weight varying from six to twenty-five pounds; neck large and solid; seed-end of medium size, and usually smooth; skin thin, very pale-green or light cream-white, diversified with lengthwise stripes and splashes of bright green,—the colors becoming gradually softer and paler after gathering; flesh bright orange, and, like that of the common Winter Crookneck, not uniform in texture or in flavor. Different specimens vary greatly in these respects: some are tough and stringy, others very fine-grained and well flavored. Seeds not distinguishable, in size, form, or color, from those of the Winter Crookneck.

The variety is hardy, grows luxuriantly, is prolific, and keeps well. It is more uniform in shape, and generally more symmetrical, than the Winter Crookneck; though varieties occur of almost every form and colour between this and the last named.

As the plants require considerable space, the hills should not be less than eight feet apart. Two or three plants are sufficient for a hill.

"The 'Crookneck Squash,' as it is commonly but incorrectly called, is a kind of 'pumpkin,'—perhaps a genuine species; for it has preserved its identity, to our certain knowledge, ever since the year 1686, when it was described by Ray. Before the introduction of the Autumnal Marrow, it was raised in large quantities for table use during the winter, in preference to pumpkins, which it almost entirely superseded. Many farmers now use it instead of pumpkins for cattle; the vine being more productive, and the fruit containing much more nutriment in proportion to its size. It varies considerably in form and color. The best kinds are those which are very much curved,—nearly as large at the stem as at the blossom-end,—and of a rich cream-color. It is said to degenerate in the Middle and Southern States; where, probably, the Valparaiso or some kindred variety may be better adapted to the climate."[2]

1. Dr. T. W. Harris, in Pennsylvania Farm Journal.

2. Dr. Harris.

FIELD AND GARDEN VEGETABLES OF THE LATE NINETEENTH CENTURY | 209

Seed & Plant Guide, H.W. Buckbee

Part Five: Brassicaceous Plants

Borecole, or Kale; Broccoli; Brussels Sprouts; Cabbage; Cauliflower; Colewort; Couve Tronchuda, or Portugal Cabbage; Pak-Chöi; Pe-Tsai, or Chinese Cabbage; Savoy; Sea-kale.

Brussel Sprouts
The Encyclopedia of Food by Artemas Ward

46

Borecole/Kale (Brassica oleracea sabellica)

The term "Borecole," or "Kale," is applied to a class of plants, of the Cabbage family, which form neither heads as the common cabbage, nor eatable flowers like the broccoli and cauliflower. Some of the varieties attain a height of six or seven feet; but while a few are compact and symmetrical in their manner of growth, and of good quality for table use, many are "ill-colored, coarse, rambling-growing, and comparatively unpalatable and indigestible." Most of the kinds are either annuals or biennials, and are raised from seeds, which, in size, form, and color, resemble those of the cabbage.

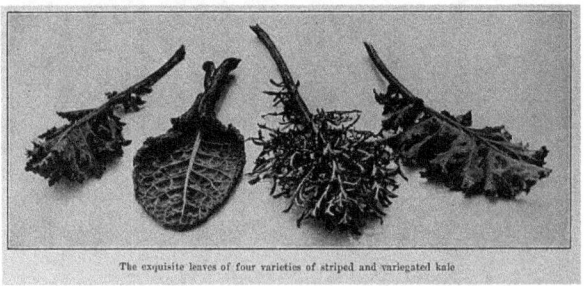

The exquisite leaves of four varieties of striped and variegated kale
The Encyclopedia of Food by Artemas Ward

Sowing—The seeds are sown at the time of sowing the seeds of the cabbage or cauliflower, and in the same manner. Early plants may be started in a hot-bed, or the seeds may be sown in the open ground in April or May. In transplanting, treat the plants like young cabbages; setting them more or less remote, according to the size or habit of the variety.

Though they are extremely hardy, and will endure quite a low temperature, they are generally harvested in autumn, before the closing-up of the ground. If reset in the following spring, they will furnish an abundance of tender sprouts, which, when cooked, are superior in flavour and delicacy to the cabbage, and resemble coleworts or Brussels sprouts.

FIELD AND GARDEN VEGETABLES OF THE LATE NINETEENTH CENTURY | 213

Seeds**—**The plants for seed should be selected from those kept over winter, and in April set rather deeply in a spot well exposed to the sun, and in a sandy rather than stiff soil. The stems should be supported, to prevent breakage by the wind.

E. Teschemacher gives the following directions for culture and use:—

"Sow, the middle or last of May, a small bed on a moderately rich soil, but in a well-exposed situation. Strong plants cannot be obtained from seedlings grown in the shade. When the young plants have six or eight leaves, prepare a piece of well-manured, open soil, plant the young seedlings six or eight inches asunder, water well, and shade for a few days against the hot sun. About a hundred plants are enough for a family. Towards the latter end of July, or middle of August, they should be thick, stocky plants, fit for final transplanting to the spots where they are to remain. They may be planted in the lines from which early crops of pease have been removed. The ground must be well manured, and the plants moved singly and carefully, with as much earth attached to the roots as possible. This last precaution is very necessary in all summer transplanting, as the only means of enabling the plants to bear the hot sun. In a garden, they should be well supplied with water for a few days; but in field-planting, where this is not possible, a moist time should be chosen. They will not show much signs of growth until the cool nights prevail: after that, they will grow rapidly. They will not boil tender or with much flavour until they have been frozen, or have experienced a temperature of about 28° Fahrenheit (-2° Celsius)."

Use—

"The tender, upper part alone is eaten. They are often, but not always, frozen when cut; and, when this is the case, they should be put into a cool cellar or in cold water[Pg 231] until the frost is out of them. It will take one-half to three-quarters of an hour to boil them tender. Put them into the boiling water; to which add a lump of soda. This rather softens them, and causes them to retain their green color. When done, press the water thoroughly out, chop them up with a knife, put them into a vessel to evaporate still more of the water, and serve with melted butter, pepper and salt. In Germany, they frequently boil a few chestnuts, and chop up with the Kale; between which and the stem and stalk of the Kale it is difficult to perceive much difference in taste. The beautiful curled leaves are quite ornamental.

"From one hundred plants, pluckings for the table were made twice a week, from the middle of November to the middle of January; and these fresh from the open garden, although the thermometer in the time had indicated a temperature approaching to zero."[1]

Varieties—The varieties, which are numerous, and in many instances not well marked or defined, are as follow:

Buda Kale, or Borecole

(Russian Kale; Asparagus Kale; Manchester Borecole; Dwarf Feather Kale; Oak-leaved Kale)

The Buda Kale somewhat resembles the Purple; but the stalk is shorter. The leaves are purplish, somewhat glaucous, cut and fringed. The variety is not only hardy and well flavored, but continues to produce sprouts longer than any other sort. It is sometimes blanched like sea-kale.

Cabbaging Kale, or Borecole

(Imperial Hearting)

This is a new variety, and very much resembles the Dwarf Green Curled in the nature, colour, and general appearance of the leaves: the heart-leaves, however, fold over each other, somewhat like those of a cabbage, but, on account of the curls of the margin, not so compactly. The quality is excellent.

Cock's-Comb Kale

(Curled Proliferous Kale; Chou frisé prolifère)

Stalk about twenty inches high. The leaves differ to a considerable extent in size, and are of a glaucous-green color. From the upper surface of the ribs and nerves, and also from other portions of the leaves, are developed numerous small tufts, or fascicles of leaves, which, in turn, give rise to other smaller but similar groups. The foliage thus exhibits a cock's-comb form: whence the name.

The variety is hardy, but more curious than useful.

Cow-Cabbage

(Tree-Cabbage; Cesarean Borecole; Cesarean Cabbage; Chou Cavalier)

This variety generally grows to the height of about six feet; although in some places it is reported as attaining a height of twelve feet, and even upwards. The leaves are large,—measuring from two and a half to nearly three feet in length,—smooth, or but slightly curled.

It is generally grown for stock; but the young sprouts are tender and mild-flavored when cooked. Its value for agricultural purposes appears to have been greatly overrated; for, when tried in this country against other varieties of cabbages, the produce was not extraordinary.

The plants should be set three feet or three feet and a half apart.

Daubenton's Creeping Borecole

(Chou vivace de Daubenton)

Stalk four or five feet in height or length. The leaves are nearly two feet long, deep green; the leaf-stems are long and flexible. It sometimes takes root where the stem rests upon the surface of the ground; and, on this account, has been called perennial.

The variety is hardy, and yields abundantly; though, in this last respect, it is inferior to the Thousand-Headed.

Dwarf Green Curled Borecole

(Dwarf Curled Kale; Green Scotch Kale; Dwarf Curlies; Chou frisé à Pied court; Canada Dwarf Curled)

The Dwarf Green Curled is a very hardy but comparatively low-growing variety; the stems seldom exceeding sixteen or eighteen inches in height. The leaves are finely curled; and the crowns of the plants, as well as the young shoots, are tender and delicate, especially after having been exposed to the action of frost.

The plants may be set eighteen inches apart.

Farmer with Kale, 13 December 1955 (Eerste vorst in ons land, tuinder boerenkoolplu)

Field Cabbage

(Field Kale for mowing; Chou à Faucher)

Leaves sixteen to eighteen inches in length, very dark green, deeply lobed, or lyrate, and hairy, or hispid, on the nerves and borders. The leaf-stems are nearly white.

The variety produces small tufts, or collections of leaves, which are excellent for fodder, and which may be cut several times during the season. It is sometimes cultivated for stock; but, as a table vegetable, is of little value.

Flanders Kale

(Chou Caulet de Flanders)

This is a sub-variety of the Tree-cabbage, from which it is distinguished by the purplish color of its foliage. Its height is nearly the same, and the plant has the same general appearance. It is, however, considered somewhat hardier.

Green Marrow-Stem Borecole

(Chou Moellier)

Stem green, about five feet high, clavate, or club-formed; thickest at the top, where it measures nearly two inches, and a half in diameter. This stem, or stalk, is filled with a succulent pith, or marrow, which is much relished by cattle; and, for this quality, the plant is sometimes cultivated. The leaves are large, and nearly entire on the edges; the leaf-stems are thick, short, white, and fleshy.

It is not so hardy as most of the other varieties. The plants should be grown about three feet apart in one direction, by two feet or two feet and a half in the opposite.

Lannilis Borecole

(Chou de Lannilis; Lannilis Tree-Cabbage)

Stem five feet high, thicker and shorter than that of the Cow or Tree Cabbage; leaves long, entire on the borders, pale-green, and very thick and fleshy. The leaf-stems are also thicker and shorter than those of the last-named varieties.

The stalk is largest towards the top, and has the form of that of the Marrow-stem. It sometimes approaches so near that variety, as to be scarcely distinguishable from it.

Neapolitan Borecole

(Neapolitan Curled Kale; Chou frisé de Naples)

The Neapolitan Borecole is remarkable for its peculiar manner of growth, but is hardly worthy of cultivation as a table vegetable, or even for stock. The stem is short and thick, and terminates in an oval bulb, somewhat in the manner of the Kohl Rabi. From all parts of this bulb are put forth numerous erect, small leaves, finely curled on their edges. The whole plant does not exceed twenty inches in height. The leaves are attached to footstalks six or seven inches long. They are obovate, smooth on the surface, with an extraordinary number of white veins, nearly covering the whole leaf. The fringed edges are irregularly cut and finely curled, and so extended as nearly to conceal the other parts of the leaf. As the plant gets old, it throws out numerous small branches from the axils of the leaves on the sides of the bulb.

The swollen portion of the stem is of a fleshy, succulent character, and is used in the manner of Kohl Rabi; between which and the Cabbage it appears to be intermediate.

Palm Kale

(Palm Borecole; Chou Palmier)

Stalk six feet in height, terminating at the top in a cluster of leaves, which are nearly entire on the borders, blistered on the surface like those of the Savoys, and which sometimes measure three feet in length by four or five inches in width.

As grown in France, the plant is remarkable for its fine appearance, and is considered quite ornamental; though, as an article of food, it is of little value. In England, it is said to have a tall, rambling habit, and to be little esteemed.

The plants should be set three feet and a half apart in each direction.

Purple Borecole
(Red Borecole; Tall Purple Kale; Curled Brown Kale; Chou Frisé Rouge Grand)

With the exception of its color, the Purple Borecole much resembles the Tall Green Curled. As the leaves increase in size, they often change to green; but the veins still retain their purple hue. When cooked, the color nearly or quite disappears.

It is remarkably hardy, and is much cultivated in Germany.

Red Marrow-Stem Borecole
(Red-Stalked Kale)

Stalk purplish-red, four and a half or five feet high, and surmounted by a cluster of large, fleshy leaves, on short, thick stems. The stalk is much larger than that of the Green Marrow-stem, and sometimes measures more than three inches in diameter. It is cultivated in the same manner, and used for the same purposes, as the last-named variety.

Tall Green Curled
(Tall Scotch Kale; Tall Green Borecole; Tall German Greens; Chou Frisé Grand du Nord)

This variety, if unmixed, may be known by its bright-green, deeply lobed, and curled leaves. Its height is two feet and a half and upwards. Very hardy and productive.

The parts used are the crowns of the plants; and also the tender side-shoots, which are produced in great abundance. These boil well, and are sweet and delicate, especially after frost; though the quality is impaired by protracted, dry, freezing weather.

Thousand-Headed Borecole
(Chou Branchu du Poitou; Chou à Mille Têtes)

The Thousand-headed Borecole much resembles the Tree or Cow Cabbage, but is not so tall-growing. It sends out numerous side-shoots from the main stem, and is perhaps preferable to the last-named sort. It is chiefly valuable as an agricultural plant, but may occasionally be grown in gardens on account of its great hardiness; but its flavour is inferior to all other winter greens.

Variegated Borecole
(Variegated Kale; Variegated Canadian Kale; Chou Frisé panaché)

This is a sub-variety of the Purple Borecole, growing about a foot and a half high. The leaves vary much in size, and are lobed and finely curled. They are also beautifully variegated, sometimes with green and yellowish-white or green and purple, and sometimes with bright-red and green.

It is frequently grown as an ornamental plant, is occasionally employed for garnishing, and is sometimes put into bouquets. It is very good cooked after frost, but is not quite so hardy as the Purple Borecole.

Variegated Cock's-Comb Kale

A variety of the Common Cock's-comb Kale, with the leaves more or less variegated with purple and white. It is not of much value as an esculent.

Woburn Perennial Kale

This is a tall variety of the Purple Borecole, with foliage very finely divided or fringed. The plant lasts many years, and may be propagated by cuttings, as it neither flowers readily nor perfects well its seeds. Its produce is stated to have been more than four times greater than that of either the Green or Purple Borecole on the same extent of ground. The weight of produce from ten square yards was a hundred and forty-four pounds ten ounces; but some of the large kinds of cabbages and savoys will exceed this considerably, and prove of better quality. The Woburn Perennial Kale can therefore only be recommended where the climate is too severe for the more tender kinds of the Cabbage tribe.

1. The Magazine of Horticulture, Botany, and Rural Affairs. By **C. M. Hovey**. Boston. Monthly. 1834 to the present time.

47

Broccoli (Brassica oleracea var.)

In its structure and general habit, the Broccoli resembles the Cauliflower. Between these vegetables the marks of distinction are so obscurely defined, that some of the white varieties of Broccoli appear to be identical with the Cauliflower. Botanists divide them as follows:

"The Cauliflower has generally a short stalk, and white-ribbed, oblong leaves. The stem by which the flower is supported unites at the head of the primary branches into thick, short, irregular bundles, in the form of a corymb. It appears to be a degeneration of the *Brassica oleracea costata*, or Portugal Cabbage.

"In the Broccoli, the stalk is more elevated; the leaf-nerves less prominent; the pedicles, or stems, connected with and supporting the flower, or head, less thick and close. They are also longer; so that, on becoming fleshy, they resemble in shape the young shoots of asparagus: hence the name of 'Asparagoides,' given by ancient botanists to Broccoli. It seems to be a degeneration of the 'Chou cavalier,' or tall, open Cabbage.

"Cultivation, by improving the finer kinds of white Broccolis, is narrowing the distinctive marks: but, although so nearly alike, they must ever remain really distinct, inasmuch as they derive their origin from two very distinct types; viz., the Portugal Cabbage and the Tall Curled Kale. The Cauliflower also originated in the south of Europe, and the Broccoli in the north of Europe, either in Germany or Britain."

Seed—Broccoli-seeds are rarely raised in this country; most of the supply being received from France or England. In size, form, and color, they are similar to those of the Cabbage or Cauliflower. An ounce may be calculated to produce about five thousand plants, although it contains nearly twice that number of seeds.

Sowing and Cultivation—In New England, as well as in the Middle and Western States, the seeds of the later sorts should be sown in March or April, in the manner of early cabbages; whilst the earlier varieties may be sown in the open ground, from the middle to the last of May. If the sowing be made in the open ground, prepare a small nursery-bed not too directly exposed to the sun, and sow in shallow drills six or eight inches apart. The last of June, or as soon as the plants have attained sufficient size, transplant them into soil that is well enriched, and has been deeply stirred; setting them at the distance directed for the variety. If possible, the setting should be performed when the weather is somewhat dull, for then the plants become sooner established; but, if planted out in dry weather, they should be immediately and thoroughly watered. If the plants have been started in a hot-bed, they should be set out at the time of transplanting cabbages.

The after-culture consists in hoeing frequently to keep the ground loose and clean, and in earthing up slightly from time to time about the stem.

Some of the early varieties will be fit for use in September; whilst the later sorts, if properly treated, will supply the table till spring.

The difficulties attending the growing of Broccoli in this country arise mainly from the extreme heat and dryness of the summer and the intense cold of the winter. Whatever will tend to counteract these will promote the growth of the plants, and tend to secure the development of large and well-formed heads.

"When the heads of White Broccoli are exposed to light, and especially to the direct influence of the sun, the color is soon changed to a dingy or yellowish hue. It is, therefore, necessary to guard against this as much as possible by frequently examining the plants; and, when any heads are not naturally screened, one or two of the adjoining side-leaves should be bent over the flower-head to shade it from the light, and likewise to protect it from the rain. Some kinds are almost self-protecting; whilst the leaves of others spread, and consequently require more care in shading."[1]

Taking the Crop—

"Broccoli should not be allowed to remain till the compactness of the head is broken, but should always be cut while the 'curd,' as the flowering mass is termed, is entire, or before bristly, leafy points make their appearance through it. In trimming the head, a portion of the stalk is left, and a few of the leaves immediately surrounding the head; the extremities being cut off a little below the top of the latter."[2]

Preservation—

"They are sometimes preserved during winter as follows: Immediately previous to the setting-in of hard frost in autumn, take up the plants on a dry day, with the roots entire, and turn their tops downwards for a few hours, to drain off any water that may be lodged between the leaves. Then make choice of a ridge of dry earth, in a well-sheltered, warm exposure, and plant them down to their heads therein, close to one another; having previously taken off a few of the lower, loose leaves. Immediately erect over them a low, temporary shed, of any kind that will keep them perfectly free from wet, and which can be opened to admit the air in mild, dry weather. In very severe freezing seasons, an extra covering of straw, or other description of dry litter, should be applied over and around the shed; but this should be removed on the recurrence of moderate weather."

They will keep well in a light, dry cellar, if set in earth as far as the lower leaves.

Broccoli
Scott Foresman

Seeds—The seeds of Broccoli are not distinguishable from those of the Cauliflower. They, however, rarely ripen well in this country, and seedsmen are generally supplied from abroad.

Use—The heads, or flowers, are cooked and served in all the forms of the Cauliflower.

Varieties—These are exceedingly numerous; although the distinctions, in many instances, are neither permanent nor well defined.

In 1861-62, a hundred and three nominally distinct sorts were experimentally cultivated at the Chiswick Gardens, near London, Eng., under the direction of Robert Hogg, Esq. In reporting the result, he says:

"It is quite evident that the varieties of Broccoli, as now grown, are in a state of great confusion. The old varieties, such as Grange's and the Old Early White, have entirely disappeared, or lost their original character; whilst the distinctive names of Early White and Late White seem now to be possessed of no value, as, in some cases, the one is used for the other, and *vice versâ*."

The kinds catalogued by seedsmen, and recommended for cultivation, are the following:

Ambler's Early White

Similar to Mitchinson's Penzance, but easily distinguished by its winged leaves; those of the last named being interrupted. It is remarkably hardy, and produces a large, creamy-white head, very uniform in size.

Chappell's Large Cream-Coloured

(Chappell's New Cream-Coloured)

A very large and fine sort, earlier than the Portsmouth; flower cream-yellow. Sow in the open ground in May, and transplant three feet apart in each direction.

Danish, or Late Green

(Late Danish; Siberian)

The leaves of this variety are long, narrow, and much undulated; the leaf-stems are tinged with purple; the heads are of medium size, compact, exposed, and of a greenish colour. It is one of the latest and hardiest of all varieties.

Dwarf Brown Close-Headed

This variety resembles the Sulphur-colored; from which it probably originated. It is, however, earlier, and differs in the form, as well as in the color, of the flower. The leaves are small, not much waved, dark-green, with white veins: they grow erect, and afford no protection to the head. Most of the crowns are green at first; but they soon change to large, handsome, brown heads.

The plants should be set two feet apart in each direction.

Early Purple

(Early Purple Sprouting)

An excellent kind, of a deep-purple color. When the variety is unmixed, it is close-headed at first; afterwards it branches, but is liable to be too much branched, and to become green. The plant is from two to three feet high, and

Brocoli branchu violet.
Réd. au huitième: pousse détachée, demi-grandeur.

Purple Sprouting Broccoli
Vilmorin-Andrieux, 1904

a strong grower; the leaves are comparatively short, spreading, and of a purplish-green color; the head is quite open from the leaves. Small leaves are sometimes intermixed with the head, and the plant produces sprouts of flowers from the alæ of the leaves.

It succeeds best in rich soil, and the plants should be set three feet apart.

Early Sprouting

(Asparagus Broccoli; North's Early Purple; Italian Sprouting; Early Branching)

A strong-growing, hardy sort, from two to three feet high. The leaves are spreading, much indented, and of a purplish-green color. The flower is close-headed, and, in the genuine variety, of a rich purple on its first appearance. It is, however, liable to lose its colour, and to become greenish; and sometimes produces numerous small, green leaves, intermixed with the flower, particularly if grown in soil too rich.

The variety is extensively grown by the market-gardeners in the vicinity of London.

Elletson's Gigantic Late White

(Elletson's Mammoth)

One of the largest and latest of the white broccolis. Leaves spreading; stem short.

Fine Early White

(Early White; Devonshire White; Autumn White)

Plant tall, with erect, dark-green, nearly entire leaves. The heads are very white and close.

This variety, in common with a few others, is sometimes cut in considerable quantities by market-gardeners previous to heavy frost, and preserved in cellars for the supply of the market.

White Sprouting Broccoli

Frogmore Protecting

Head pure white, scarcely distinguishable from the finest cauliflower; size large,—when well formed, measuring from seven to nine inches in diameter.

A recently introduced sort, promising to be one of the best. The plants are extremely hardy and vigorous, and rarely fail to develop a large and fine head, having a rich, curdy appearance, and, as before observed, similar to a well-grown cauliflower. It is of dwarf growth; and the outer leaves, closing over the large head of flowers, protect it from the action of severe weather.

Gillespie's Broccoli

A fine, white, early autumn variety, much grown about Edinburgh.

Grange's Early Cauliflower Broccoli

(Grange's Early White; Hopwood's Early White; Marshall's Early White; Bath White; Invisible)

This is an old variety, and, when pure, still stands in high estimation; having a head nearly as large and as white as a cauliflower. The leaf-stems are long and naked; the leaves are somewhat ovate, lobed at the base, very slightly waved, and, incurving a little over the flower, defend it from frost and wet. It is not a large grower; and, being upright in habit, may be grown at two feet distant.

Hardy, and well deserving of cultivation. The London market-gardeners cultivate four varieties, of which this is the principal.

Green Cape

(Autumnal Cape; Maher's Hardy Cape)

Leaves long and narrow; the veins and midribs green; the head is greenish, and generally covered by the leaves.

This variety and the Purple Cape often become intermixed, and are liable to degenerate. They are, however, quite distinct, and, when pure, very beautiful.

Green Close-Headed Winter

(Late Green; Siberian; Dwarf Roman)

This new and excellent Broccoli is apparently a seedling from the Green Cape. The plants are dwarf; the leaves are large and numerous, with white veins. The flower grows exposed, is not of large size, and resembles that of the Green Cape. Its season immediately follows that of the last-named variety.

Hammond's White Cape

An excellent, pure white variety, obtained in England by cultivation and selection.

Kent's Late White

A remarkably hardy, dwarf-growing variety, with very dark-green foliage. Bouquet white, of good size, and well protected.

Kidderminster

Head large and handsome, of pure whiteness, and much exposed. It is evidently a form of "Willcove," and has, undoubtedly, emanated from that variety; but it is somewhat earlier.

Knight's Protecting

(Early Gem; The Gem; Lake's Gem; Waterloo Late White; Dilliston's Late White; Hampton Court; Invisible Late White)

When pure, this variety is of a dwarfish habit of growth, with long, pointed, and winged leaves, which have a spiral twist about the head, and turn in closely over it, so as effectually to protect it from the effect of frost, and preserve it of a fine white color.

It is remarkably hardy; and as the plants are of small size, with comparatively large heads, a great product is realized from a small piece of ground.

Late Dwarf Purple

(Dwarf Swedish; Italian Purple; Dwarf Danish)

This is the latest purple Broccoli. The plants seldom rise above a foot in height. The flower, at first, shows small and green; but soon enlarges, and changes to a close, conical, purple head. The leaves are short and small, dark-green, with white veins, much sinuated, deeply indented, and form a regular radius round the flower. The whole plant presents a singular and beautiful appearance.

Miller's Late White

(Miller's Dwarf)

This is an old variety; but is considered by some to be the best late sort, if it can be obtained true. Hardy. Transplant two feet apart.

Mitchell's Ne Plus Ultra

Hardy, and of dwarf habit; leaves smooth, glaucous, protecting the head, which is cream-colored, large, and compact. Transplant two feet apart.

Mitchinson's Penzance

(Early White Cornish; Mitchinson's Early White)

One of the best of the Spring Whites. The leaves are much waved on the margin, and enclose large and fine heads, which are nearly of a pure white color. Very hardy.

Portsmouth

(Cream-Coloured; Southampton; Maher's New Dwarf)

Leaves large, broad, with white veins, spreading; although the central ones partially cover the flower, or head, which is buff, or cream-colored. It is a hardy sort; and the flower, which is produced near the ground, is said to exceed in size that of any other variety. The plants should be set three feet apart.

Purple Cape

(Early Purple Cape; Purple Silesian; Howden's Superb Purple; Grange's Early Cape; Blue Cape)

This has a close, compact head, of a purple color, and, in favorable seasons, comes as large as a cauliflower. The plants grow from a foot to a foot and a half in height, with short, erect, concave leaves, regularly surrounding the head. The veins and midribs are stained with purple.[Pg 247] The head is exposed to view in growing; and, as it enlarges, the projecting parts of the flower show a greenish-white mixed with the purple color. When boiled, the whole flower becomes green.

Excellent for general culture, as it is not only one of the finest varieties for the table, but the plants form their heads much more generally than many other kinds. It is the earliest of the purple broccolis.

The seed should not be sown before the middle or last of May, and the plants will require a space of two feet and a half in each direction.

Snow's Superb White Winter

(Gill's Yarmouth White)

This variety is of dwarfish habit. The leaves are broad, with short stems; the heads are large, white, very compact, well protected by the incurved leaves, and equal in quality to those of the Cauliflower. By many it is considered superior to Grange's Early Cauliflower Broccoli.

Snow's Spring White or Cauliflower Broccoli

(Naples White; Early White; Adam's Early White; Neapolitan White; Imperial Early White; Grange's Cauliflower; Covent-Garden Market)

Plant about two feet high, robust, and a strong grower. The leaves are large, thick-veined, flat, and narrow; and generally compress the head, so as to render it invisible when ready for cutting, and thus protect it from rain and the effects of frost. Head large, perfectly white.

Sulphur or Brimstone Broccoli

(Late Brimstone; Fine Late Sulphur; Edinburgh Sulphur)

Leaves with long stems; heads large, compact, somewhat conical, sulphur-coloured, sometimes tinged with purple. Hardy.

Walcheren Broccoli

Comparatively new, and so closely resembling a cauliflower as to be scarcely distinguishable from it. The leaves, however, are more curled, and its constitution is of a hardier nature, enduring the cold, and also withstanding heat and drought better. Much esteemed in England, where, by successive sowings, it is brought to the table at every season of the year.

Ward's Superb

This is a form of Knight's Protecting, but is from two to three weeks later. It is of a dwarfish habit of growth, closely protected by the spirally compressed leaves, with a good-sized and perfectly white head. One of the best of the late White Broccolis.

White Cape

Heads of medium size, white, and compact.

Willcove

Late Willcove

The true Willcove is a variety perfectly distinct from every other of its season. The heads are very large, firm, even, and fine, and of a pure whiteness. They are fully exposed, and not protected by the leaves as most other broccolis are. On this account, the variety is more liable to be injured by the weather than any other late sort; and therefore, in severe seasons, it must be regarded as deficient in hardiness.

It derives its name from a small village near Devonport, England; where it originated, and where the Broccoli is said to be grown in great perfection.

1,2. The Gardener's Assistant. By Robert Thompson.

Garden - Field - and Flower (1888), I.V. Faust

48

Brussel Sprouts (Brassica oleracea var.)

(Thousand-Headed Cabbage)

In its general character, this vegetable is not unlike some of the varieties of Kale or Borecole. Its stem is from a foot to four feet in height, and from an inch and a half to upwards of two inches in diameter. It is remarkable for the production of numerous small axillary heads, or sprouts, which are arranged somewhat in a spiral manner, and which are often so closely set together as entirely to cover the sides of the stem.

> "These small heads are firm and compact like little cabbages, or rather like hearted savoys in miniature. A small head, resembling an open savoy, surmounts the stem of the plant, and maintains a circulation of sap to the extremity. Most of the original side-leaves drop off as these small buds, or heads, enlarge."[1]

Culture—The plant is always raised from seeds, which, in size, form, or color, are scarcely distinguishable from the seeds of the Common Cabbage. These should be sown at the time and in the manner of the Cabbage, either in hot-beds in March or April, or in the open ground in April or May. When three or four inches high, transplant two feet apart in each direction, and cultivate as directed for cabbages and cauliflowers. In September, the early plantings will be fit[Pg 250] for gathering; whilst the later plants will afford a succession that will supply the table during the winter. For the latter purpose, they should be harvested before severe freezing weather, and preserved in the cellar as cauliflowers and broccolis. They are quite hardy, easily grown, thrive well in New England or in the Middle States, and deserve more general cultivation.

To Raise Seeds—In the autumn, select two or three of the finest plants; keep them in the cellar, or out of the reach of frost, during winter; and in the spring set them in the open ground, two feet apart, and as far as possible from all flowering plants of the Cabbage family. Cut off the top shoot, and save the branches of pods that proceed from the finest of the small heads on the sides of the main stem.

Use—The small heads are boiled and served in the manner of cabbages. They are also often used in the form of the cauliflower, boiled until soft, then drained, and afterwards stewed with milk, cream, or butter.

Varieties—Two varieties are enumerated by gardeners and seedsmen:

Dwarf Brussels Sprouts

A low-growing sort, usually from eighteen inches to two feet in height. It differs from the following variety principally in size, though it is somewhat earlier. The dwarf stems are said to produce heads which are more tender and succulent when cooked than those obtained from taller plants.

Tall or Giant Brussels Sprouts

Stem nearly four feet in height; plant healthy and vigorous, producing the small heads peculiar to its class in great abundance. It is somewhat hardier than the foregoing variety; and, on account of its greater length of stalk, much more productive.

There is, however, very little permanency to these sorts. Much of the seed found in the market will not only produce plants corresponding with both of the varieties described, but also numerous intermediate kinds.

Brussel Sprouts

---

1. The Gardener's Assistant. By Robert Thompson.

49

Cabbage (Brassica oleracea capitata)

The Cabbage is a biennial plant; and, though comparatively hardy,—growing at all seasons unprotected in England,—will not withstand the winters of the Northern States in the open ground.

When fully developed, it is from four to five feet in height. The flowers are cruciform, generally yellow, but sometimes white or yellowish-white. The seeds, which ripen in July and August of the second year, are round, reddish-brown or blackish-brown, and retain their vitality five years. About ten thousand are contained in an ounce.

Soil and Situation—Though not particularly nice as to soil or situation, cabbages do best when grown in well-manured ground. In such soil, they are generally earlier than when raised in cold and stiff ground. But manure need not be profusely applied, if the ground is naturally of a fertile and open kind; for the flavour is generally better in such soil than where a great quantity of fertiliser is used.

Propagation—All of the varieties are propagated from seed sown annually. For early use, a sowing may be made in a hot-bed in February or March; and, for winter use, the seed may be sown in a nursery-bed in the open ground in May or June. When five or six inches high, transplant to the distance directed in the description of the variety. In the hot-bed or nursery-bed, the plants should not be allowed to stand too thickly together, as this causes them to draw up weak and feeble.

To raise Seed—At the time of harvesting, select a few of the most compact and best-formed heads possessing the characters of the pure variety; and, in the following April, set the plants entire, three feet apart in each direction. As they progress in growth, remove all of the side-shoots, and encourage the main sprout, that will push up through the centre of the head. Seeds from the side-shoots, as well as those produced from decapitated stems, are of little value. No cabbage-seed is really reliable that is not obtained from firm and symmetrical heads; and seed thus cultivated for a few successive seasons will produce plants, ninety per cent of which will yield well-formed and good-sized cabbages.

American-grown seed is generally considered superior to that of foreign growth; and, when it can be obtained from a reliable seedsman or seeds-grower, the purchaser should not be induced by the difference in price to select the nominally cheaper, as there are few vegetables with which the character of the seed is of greater importance.

Varieties—The varieties are numerous, and the distinction, in many instances, well-defined and permanent. Between some of the sorts, however, the variations are slight, and comparatively unimportant.

Atkins's Matchless

This is a variety of the Early York: the head, however, is smaller and more conical, and the leaves are more wrinkled,—somewhat similar to those of the Savoys. It is of tender texture and delicate flavor; and, with the exception of its smaller size, is considered equal, if not superior, to the last-named variety.

It is comparatively a recent sort, and seems to be desirable rather for its precocity and excellent quality than for its size or productiveness.

Transplant to rows fifteen inches apart, and twelve inches asunder in the rows.

Seeds for the Garden, Farm, and Field (1899), Plant Seed Company

Barnes's Early

(Barnes's Early Dwarf)

This variety, in respect to season, size, form, and general habit, seems to be intermediate between, or a hybrid from, the York and Ox-heart. Head ovate, rather compact; texture fine and tender; flavor mild and good.

Set in rows two feet apart, and eighteen inches apart in the rows.

Bergen Drumhead

(Large Bergen; Great American; Quintal; Large German Drumhead)

Head remarkably large, round, flattened at the top, compact; the leaves are of a peculiar, glaucous-green color, of thick texture, firm, and rather erect; the nerves large and prominent; the outer leaves of the head are usually revoluted on the borders; the loose leaves are numerous, and rarely rise above a level with the summit of the head; the stalk is short.

The Bergen Drumhead is one of the largest and latest of all the cabbages; and, when not fully perfected before being harvested, has the reputation, if reset in earth in the cellar, of heading, and increasing in size, during winter. It is a popular market sort; and, notwithstanding its extraordinary proportions, is tender, well flavored, and of more than average quality for family use. The plants should be set three feet apart.

Champion of America

One of the largest of the recently introduced sorts; the whole plant sometimes attaining a weight of forty pounds and upwards. Head very large, flattened, somewhat resembling the Drumheads; outer leaves very few, succulent, and tender; stalk short; quality tender, mild, and well flavored. As a[Pg 254] market variety, it has few, if any, superiors. It heads with great uniformity, and bears transportation well; but its large size is objectionable when required for the use of families numbering but few members.

Early Battersea

(Dwarf Battersea; Early Dwarf Battersea)

The type of the Early Battersea is very old. When fully grown, the four outside or lower leaves are about sixteen inches in diameter; and, when taken off and spread out, their general outline is nearly circular. The stem is dwarfish, and the leaf-stalks come out quite close to each other; so that scarcely any portion of the stem is to be seen between them. The whole cabbage measures about three feet in circumference. The heart is shortly conical, with a broad base; near which it is about two feet in circumference, when divested of the outside leaves. The ribs boil tender.

It is one of the best sorts for the general crop of early cabbages; is not liable to crack; and, when cut close to the stem, often puts forth a number of fresh heads, of fair size and good quality.

Barnes's Early

(Barnes's Early Dwarf)

This variety, in respect to season, size, form, and general habit, seems to be intermediate between, or a hybrid from, the York and Ox-heart. Head ovate, rather compact; texture fine and tender; flavor mild and good.

Set in rows two feet apart, and eighteen inches apart in the rows.

Bergen Drumhead

(Large Bergen; Great American; Quintal; Large German Drumhead)

Head remarkably large, round, flattened at the top, compact; the leaves are of a peculiar, glaucous-green color, of thick texture, firm, and rather erect; the nerves large and prominent; the outer leaves of the head are usually revoluted on the borders; the loose leaves are numerous, and rarely rise above a level with the summit of the head; the stalk is short.

The Bergen Drumhead is one of the largest and latest of all the cabbages; and, when not fully perfected before being harvested, has the reputation, if reset in earth in the cellar, of heading, and increasing in size, during winter. It is a popular market sort; and, notwithstanding its extraordinary proportions, is tender, well flavored, and of more than average quality for family use. The plants should be set three feet apart.

Champion of America

One of the largest of the recently introduced sorts; the whole plant sometimes attaining a weight of forty pounds and upwards. Head very large, flattened, somewhat resembling the Drumheads; outer leaves very few, succulent, and tender; stalk short; quality tender, mild, and well flavored. As a[Pg 254] market variety, it has few, if any, superiors. It heads with great uniformity, and bears transportation well; but its large size is objectionable when required for the use of families numbering but few members.

Early Battersea

(Dwarf Battersea; Early Dwarf Battersea)

The type of the Early Battersea is very old. When fully grown, the four outside or lower leaves are about sixteen inches in diameter; and, when taken off and spread out, their general outline is nearly circular. The stem is dwarfish, and the leaf-stalks come out quite close to each other; so that scarcely any portion of the stem is to be seen between them. The whole cabbage measures about three feet in circumference. The heart is shortly conical, with a broad base; near which it is about two feet in circumference, when divested of the outside leaves. The ribs boil tender.

It is one of the best sorts for the general crop of early cabbages; is not liable to crack; and, when cut close to the stem, often puts forth a number of fresh heads, of fair size and good quality.

FIELD AND GARDEN VEGETABLES OF THE LATE NINETEENTH CENTURY | 231

American-grown seed is generally considered superior to that of foreign growth; and, when it can be obtained from a reliable seedsman or seeds-grower, the purchaser should not be induced by the difference in price to select the nominally cheaper, as there are few vegetables with which the character of the seed is of greater importance.

Varieties—The varieties are numerous, and the distinction, in many instances, well-defined and permanent. Between some of the sorts, however, the variations are slight, and comparatively unimportant.

Atkins's Matchless

This is a variety of the Early York: the head, however, is smaller and more conical, and the leaves are more wrinkled,—somewhat similar to those of the Savoys. It is of tender texture and delicate flavor; and, with the exception of its smaller size, is considered equal, if not superior, to the last-named variety.

It is comparatively a recent sort, and seems to be desirable rather for its precocity and excellent quality than for its size or productiveness.

Transplant to rows fifteen inches apart, and twelve inches asunder in the rows.

Seeds for the Garden, Farm, and Field (1899), Plant Seed Company

Early Cornish

(Penton; Paignton; Pentonville)

This is an intermediate sort, both in respect to size and season; and is said to derive its name from a village in Devonshire, England, where it has been cultivated for ages. The head is of full medium size, somewhat conical in form, and moderately firm and solid. The outside leaves are rather numerous, long, and of a pale or yellowish green color. Its texture is fine and tender, and its flavor mild and agreeable. It is three or four weeks later than the Early York.

If reset in spring, this variety, like the Yanack, will send out from the stalk abundant tender sprouts, which will supply the table with the best of coleworts, or greens, for several weeks of the early part of the season.

The plants are somewhat leafy and spreading, and require full the average space. The rows should be two and a half or three feet apart, and the distance between the plants in the rows full two feet.

Early Drumhead

This is an intermediate variety, about the size of the Early York, and a little later. The head is round, flattened a little at the top, firm and well formed, tender in texture, and well flavored.

It is a good sort for the garden, as it heads well, occupies but little space in cultivation, and comes to the table immediately after the earlier sorts.

The plants should be set in rows two feet apart, and eighteen inches apart in the rows.

Early Dutch Twist

An excellent cabbage of the smallest size. It is very early and delicate, and may be planted almost as close together as a crop of cabbage-lettuce.

The first sowing should be made early; afterwards, sowings should be made at intervals of two or three weeks, which will secure for the table a constant supply of fresh and tender heads from July till winter.

Early Hope

A rather small, solid, oval-headed, early sort, nearly of the season of the Early York. Its color is bright-green, and its leaves rather erect and firm. In quality, it is not unlike the Small Early Ox-heart, and requires the same space in cultivation.

The variety is comparatively new; and, though found on the catalogues of seedsmen, is little disseminated.

Early Low Dutch

(Early Dutch Drumhead)

This well-known and standard variety has a round, medium-sized, solid head, sometimes tinted with brown at its top. The outside and loose leaves are few in number, large, rounded, clasping, blistered, and of a glaucous-green color; the ribs and nerves are small; the stalk is thick and short.

It is rather early, tender, and of good quality; heads well; and is one of the best sorts for growing in a small garden for early table use. The plants should be allowed a space of two feet and a half between the rows, and nearly two feet in the rows.

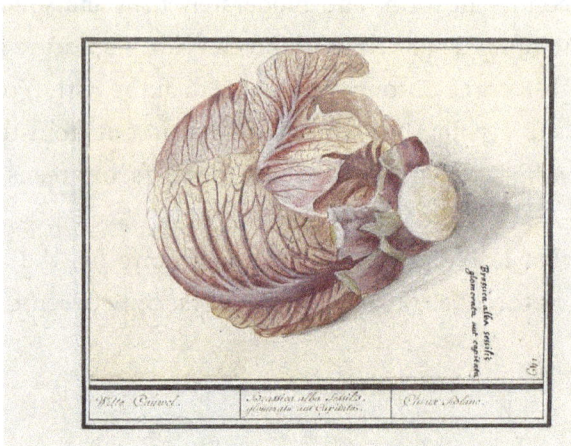

White Cabbage- Brassica alba sessilis
Anselmus Boëtius de Boodt, 1600

Early Nonpareil

Head of medium size, bright-green, rather ovoid or egg-shaped, solid; the leaves are generally erect, roundish, concave, and of thick, firm texture; the stalk is comparatively short, and the spare leaves few in number; flavor mild and pleasant. By some, it is considered the best of the intermediate varieties.

In many respects, it resembles the Small Ox-heart.

Early Sugar-Loaf

The color of this variety, and the form of its head, distinguish it from all others. The plant, when well developed, has an appearance not unlike some of the varieties of Cos lettuces; the head being round and full at the top, and tapering thence to the base, forming a tolerably regular, inverted cone. The leaves are erect, of a peculiar ashy or bluish-green hue, spoon-shaped, and clasp or cove over and around the head in the manner of a hood or cowl.

Though an early cabbage, it is thought to be more affected by heat than most of the early varieties; and is also said to lose some of its qualities, if kept late in the season. Head of medium size, seldom compactly formed; and, when cut and cooked in its greatest perfection, tender and well flavored.

Transplant in rows two feet apart, and from eighteen to twenty-four inches apart in the rows.

Early Wakefield

Head of medium size, generally somewhat conical, but sometimes nearly round, compact; leaves very glaucous; stalk small.

A fine, early variety, heading readily. As the plants occupy but little space, it is recommended as a desirable sort for early marketing.

Early York

According to Rogers,[1] the Early York Cabbage was introduced into England from Flanders, more than a hundred years ago, by a private soldier named Telford, who was there many years in the reign of Queen Anne. On his return to England, he settled as a seedsman in Yorkshire: whence the name and celebrity of the variety.

In this country, it is one of the oldest, most familiar, and, as an early market sort, one of the most popular, of all the kinds now cultivated. The head is of rather less than medium size, roundish-ovoid, close, and well-formed, of a deep or ash-green color, tender, and well flavored. The loose leaves are few in number, often revoluted on the border, and comparatively smooth on the surface; nerves greenish-white. The plants of the true variety have short stalks, occupy but little space, and seldom fail to produce a well-formed, and, for an early sort, a good-sized head. They require a distance of about eighteen inches between the rows, and fifteen or eighteen inches in the row.

Its earliness and its unfailing productiveness make it a favourite with market-gardeners; and it still retains its long-established popularity, notwithstanding the introduction of numerous new sorts, represented as being as early, equally prolific, and surpassing it in general excellence.

East Ham

From East Ham, in Essex, England, It is not a large, but a fine, early sort, not unlike the Ox-heart. The head is of an oval form, compact, and rather regular; the leaves are firm in texture, sometimes reflexed, or curved backward, but generally erect and concave; nerves pale greenish-white; stem very short. It is mild and delicate, and a desirable early variety.

In setting the plants, allow two feet and a half between the rows, and two feet between the plants in the rows.

Green Glazed

(American Green Glazed)

Head large, rather loose and open; the leaves are numerous, large, rounded, waved on the borders, and slightly blistered on the surface; stalk comparatively long. Its texture is coarse and hard, and the variety really possesses little merit; though it is somewhat extensively grown in warm latitudes, where it appears to be less liable to the attacks of the cabbage-worm than any other sort.

A distinguishing characteristic of this cabbage is its deep, shining-green color; the plants being readily known from their peculiar, varnished, or glossy appearance.

Large Late Drumhead

(American Drumhead)

Head very large, round, sometimes flattened a little at the top, close and firm; the loose leaves are numerous, broad, round, and full, clasping, blistered, and of a sea-green color; the ribs and nerves are of medium size, and comparatively succulent and tender; stem short. The variety is hardy, seldom fails to form a head, keeps well, and is of good quality.

In cultivation, it requires more than the average space, as the plants have a spreading habit of growth. The rows should not be less than three feet apart; and two feet and a half should be allowed between the plants in the rows.

There are many varieties of this cabbage, introduced by different cultivators and seedsmen under various names, differing slightly, in some unimportant particulars, from the foregoing description, and also differing somewhat from each other, "but agreeing in being large, rounded, cabbaging uniformly, having a short stem, keeping well, and in being tender and good flavored."

Large York

This is a larger cabbage than the Early York; which variety it somewhat resembles. The head, however, is broader in proportion to its depth, and more firm and solid; the leaves not connected with the head are more erect, of a firmer texture, not quite so smooth and polished, and the surface slightly bullated, or blistered. It also has a shorter stalk, and is two or three weeks later.

The Large York seems to be intermediate between the Early York and the Large Late Drumheads, as well in respect to form and general character as to its season of maturity. It is recommended as being less affected by heat than many other kinds, and, for this reason, well adapted for cultivation in warm climates. It seldom fails in forming its head, and is tender and well flavoured.

Large Ox-Heart

(Large French Ox-heart)

This is a French variety, of the same form and general character as the Small Ox-heart, but of larger size. The stalk is short; the head firm and close, and of a light-green color; the spare leaves are few in number, generally erect, and concave. It is a week or ten days later than the Small Ox-heart, forms its head readily, and is tender and well flavored. One of the best of the intermediate sorts.

The plants should be set two feet apart in each direction.

Marblehead Mammoth Drumhead

One of the largest of the Cabbage family, produced from the Mason, or Stone-mason, by Mr. Alley, and introduced by Mr. J. J. H. Gregory, of Marblehead, Mass.

Heads not uniform in shape,—some being nearly flat, while others are almost hemispherical; size very large, varying from fifteen to twenty inches in diameter,—although specimens have been grown of the extraordinary dimensions of twenty-four inches. In good soil, and with proper culture, the variety is represented as attaining an average weight of thirty pounds. Quality tender and sweet.

Cultivate in rows four feet apart, and allow four feet between the plants in the rows. For early use, start in a hot-bed; for winter, sow in the open ground from the first to the middle of May. Sixty tons of this variety have been raised from a single acre.

Mason

The Mason Cabbage, in shape, is nearly hemispherical; the head standing well out from among the leaves, growing on a small and short stalk. Under good cultivation, the heads[Pg 261]will average about nine inches in diameter and seven inches in depth. It is characterized for its sweetness, and for its reliability for forming a solid head. It is also an excellent variety for cultivation in extreme Northern latitudes, where, from the shortness of the season, or in those sections of the South, where, from excessive heat, plants rarely cabbage well. Under good cultivation, nearly every plant will set a marketable head.

Originated by Mr. John Mason, of Marblehead, Mass.

Pomeranian

This variety is of comparatively recent introduction. The head, which is of medium size, has the form of an elongated cone, and is very regular and symmetrical. It is quite solid, of a pale or yellowish green color, tender and well flavored, and remarkable for the peculiar manner in which the leaves are collected, and twisted to a point, at its top. The loose, exterior leaves are numerous, large, and broad; stalk rather high.

It is not early, but rather an intermediate variety, and excellent either as an autumnal or winter cabbage. As it heads promptly and almost invariably, and, besides, is of remarkable solidity, it makes a profitable market cabbage; keeping well, and bearing transportation with very little injury.

Premium Flat Dutch

(Large Flat Dutch)

Head large, bluish-green, round, solid, broad and flat on the top, and often tinted with red or brown. The exterior leaves are few in number, roundish, broad and large, clasping, blistered on the surface, bluish-green in the early part of the season, and tinged with purple towards the time of harvesting; stalk short.

It is one of the largest of the cabbages, rather late, good for autumn use, and one of the best for winter or late keeping, as it not only remains sound, but retains its freshness and flavor till late in spring. The heads open white and crisp, and, when cooked, are tender and well flavored. It requires a good soil, and should be set in rows not less than three feet apart, and not nearer together than thirty inches in the rows. As a variety for the winter market, the Premium Flat Dutch has no superior. It is also one of the best sorts for extensive culture, as it is remarkably hardy, and seldom fails in forming a good head. An acre of land, well set and cultivated, will yield about four thousand heads.

St. Denis

Head of large size, round, a little flattened, solid; the exterior leaves are numerous, glaucous-green, clasping at their base, and often reflexed at the ends; the ribs and nerves are large and prominent; stem long.

This variety is of good quality, seldom fails to form a head, and yields a large crop in proportion to the quantity of land it occupies. The plants should be set two feet and a half apart in each direction.

Shilling's Queen

A half-early variety, intermediate in form and size between the York and Ox-heart. As a "second early," it is one of the best. It compares favorably with the Early Nonpareil, and is tender, mild, and delicate.

Transplant in rows two feet and a half apart, and eighteen inches apart in the rows.

Small Ox-Heart

(Cœur de Bœuf petit, of the French)

Head below medium size, ovate or egg-shaped, obtuse, broad at the base, compact. The leaves are of the same bright green as those of the York Cabbage, round, of firm texture, sometimes revolute, but generally erect, and concave; the nerves are white, more numerous and less delicate than those of the last-named variety; the stalk is short, and the leaves not composing the head few in number.

The Ox-heart cabbages—with respect to character, and period of maturity—are intermediate between the Yorks and Drumheads; more nearly, however, resembling the former than the latter. The Small Ox-heart is about ten days later than the Early York.

As not only the heads, but the full-grown plants, of this variety are of small size, they may be grown in rows two feet apart, and sixteen inches apart in the rows.

Stone-Mason

An improved variety of the Mason, originated by Mr. John Stone, jun., of Marblehead, Mass. Head larger than that of the original, varying in size from ten to fourteen inches in diameter, according to the strength of the soil and the cultivation given it. The form of the head is flatter than that of the Mason, and but little, if any, inferior to it in solidity. Stem very short and small. Under good culture, the heads, exclusive of the outer foliage, will weigh about nine pounds. Quality exceedingly sweet, tender, and rich. A profitable variety for market purposes; the gross returns per acre, in the vicinity of Boston, Mass., often reaching from two hundred dollars to three hundred and fifty.

The Mason, Stone-mason, and the Marblehead Mammoth, severally originated from a package of seeds received from England, under the name of the "Scotch Drumhead," by Mr. John M. Ives, of Salem, Mass.

Sutton's Dwarf Comb

This is one of the earliest of all the cabbages. It is small and dwarfish in its habit, hearts well early in the season, and will afford a good supply of delicate sprouts throughout a large part of the summer.

The plants require a space of only twelve inches between the rows, and the same distance between the plants in the rows.

The seed of this variety, in common with other dwarfish and early sorts, should be sown more frequently than the larger growing kinds, so as to keep up a succession of young and delicate heads, much after the manner of sowing lettuce.

Vanack

This variety was introduced into England from Holland, more than a century ago, by a wealthy Dutch farmer of the name of Vanack. Though often found upon the catalogues of our seedsmen, it has not been extensively grown in this country, and perhaps is really but little known.

Head somewhat irregular in shape, broad at the base, and terminating in rather a sharp point; color palish-green, the ribs and nerves of the leaves paler. The exterior leaves are large, spreading, deep-green, and strongly veined.

It is tender in texture, sweet and delicate in flavor, cabbages early and uniformly, and, when kept through the winter and reset in spring, pushes abundant and fine sprouts, forming excellent early coleworts, or greens. Lindley pronounces its quality inferior to none of the best cabbages. Transplant to rows two feet and a half apart, and two feet apart in the rows.

Vaugirard Cabbage

(Chou de Vaugirard)

A large, late, but coarse, French variety. The head is generally round; leaves deep-green,—those of the outside having the veins sometimes tinged with red.

The plants should be set three feet apart in each direction.

Waite's New Dwarf

Heads small, but solid and uniform in shape. It has little of the coarseness common to the larger varieties, and the flavor is superior.

One of the finest early cabbages, and one of the best sorts for the market. It occupies but little space compared with some of the older kinds, and a large number of plants may be grown upon a small piece of ground.

Winnigstadt

(Pointed Head)

This is a German variety, somewhat similar to the Ox-heart, but more regularly conical. Head broad at the base, and tapering symmetrically to a point, solid, and of the size of the Ox-heart;

leaves of the head pale or yellowish green, with large nerves and ribs; the exterior leaves are large, short, and rounded, smooth, and of firm texture; the stalk is short.

It is an intermediate sort, immediately following the Early York. A large proportion of the plants will form good heads; and as these are not only of remarkable solidity, but retain their freshness well during winter, it is a good variety for marketing, though rather hard, and somewhat deficient in the qualities that constitute a good table-cabbage.

It requires a space of about eighteen inches by two feet.

Red Varieties—These are comparatively few in number, and generally used as salad or for pickling. When cooked, they are considered less mild and tender than the common varieties, besides retaining a portion of their color; which, by many, is considered an objection.

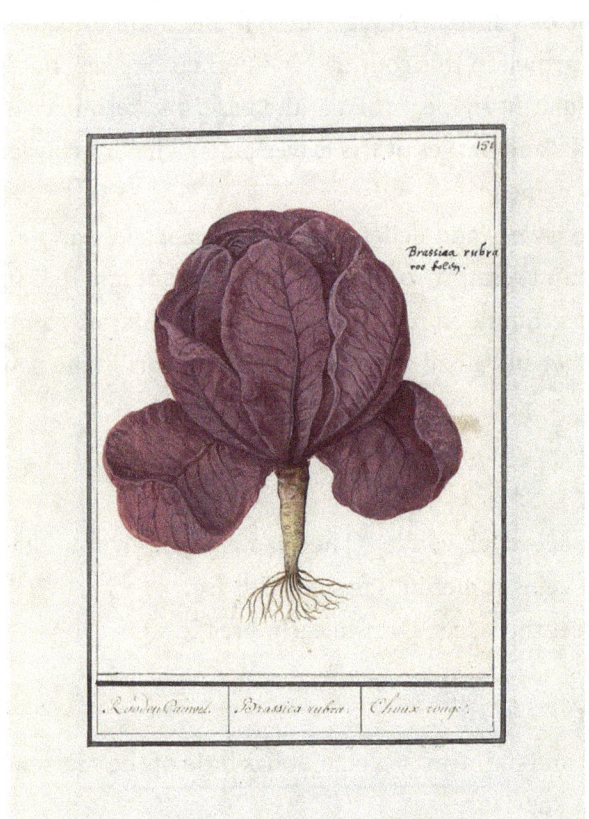

Red Cabbage- Brassica rubra
Anselmus Boëtius de Boodt, 1600

Early Dwarf Red

(Early Blood Red; Small Red)

Head nearly round, generally of a deep-red or dark-purple color. The leaves on the outside of the plant are not numerous, rather rigid or stiff, green, much washed or clouded with red; stalk short.

It is about ten days earlier than the Large Red Dutch, and is quite variable in form and color.

The seed should be sown early; and, when transplanted, the rows should be about two feet apart, and the plants eighteen inches in the rows.

The variety is seldom served at the table, cooked in the manner of other sorts; for, when boiled, it has a dark and unattractive appearance. It is almost invariably shredded, and with the addition of vinegar, olive-oil, mustard, or other seasoning, served as a salad.

Large Red Dutch

The most familiar as well as the most popular of the red varieties. The head is rather large, round, hard, and solid; the leaves composing the head are of an intense purplish-red; the outer leaves are numerous, red, with some intermixture or shades of green, firm in texture, and often petioled at the union with the stalk of the plant, which is of medium height.

On account of its dark color when cooked, it is seldom used in the manner of the common cabbages. It is chiefly used for pickling, or, like the other red sorts, cut in shreds, and served as a salad; though any solid, well-blanched, small-ribbed, white-headed sort will answer for the same purpose, and perhaps prove equally tender and palatable.

The Large Red Dutch is one of the latest of cabbages, and should receive the advantage of nearly the entire season. Make the sowing, if in the open ground, as soon as the soil is in good working condition, and transplant or thin to rows two and a half or three feet apart, and two feet apart in the rows.

The heads may be kept fresh and sound until May.

Superfine Black

Small, like the Utrecht Red, but of a still deeper color. When pickled, however, the dark coloring matter is greatly discharged, so that the substance is left paler than that of others originally not so dark. It is, therefore, not so good for pickling as other sorts which retain their color and brightness.

Utrecht Red

(*Chou noirâtre d'Utrecht*)

A small but very fine dark-red cabbage.

1. The Vegetable Cultivator. By John Rogers. London, 1851.

50

Cauliflower (Brassica oleracea var.)

The Cauliflower, like the Broccoli, is strictly an annual plant; as it blossoms and perfects its seed the year in which it is sown. When fully grown or in flower, it is about four feet in height, and in character and general appearance is similar to the Cabbage or Broccoli at a like stage of growth. The seeds resemble those of the Cabbage in size, form, and colour; although not generally so uniformly plump and fair. From ten to twelve thousand are contained in an ounce, and they retain their germinative properties five years.

Soil— Much of the delicacy and excellence of the Cauliflower depends on the quickness of its growth: therefore, to promote this, the soil cannot be too highly enriched or too deeply cultivated; and, as all the tribe thrive best in new soil, the deeper the ground is dug, and the more new or rested matter that is turned up for the roots, the better.

Sowing and Culture—The seed may be sown in a hot-bed in March, at the same time and in the same manner as early cabbages, and the plants set in the open ground late in May; or the seed may be sown in the open air in April or the beginning of May, in a common nursery-bed, in shallow drills six or eight inches apart; and, when sufficiently grown, the plants may be set where they are to remain. They need not all be transplanted at one time; nor is it important when, except that, as soon as they are large enough, the first opportunity should be improved for beginning the setting.

"Cauliflowers, after transplanting, require no particular skill during summer, and not much labor. The soil, however, must be kept free from weeds, and stirred with the hoe from time to time. As the plants increase in size, a little earth should be drawn about their roots from the middle of the row; and, in continued dry weather, an application of liquid manure will be very beneficial."

The leaves are sometimes gathered, and tied loosely over the tops of the heads, to facilitate the blanching

Taking the Crop—Cauliflowers raised by open culture will generally come to the table in October. Such as have not fully perfected their heads, may, just as the ground is closing, be taken up by their roots, and suspended, with the top downward, in a light cellar, or other place secure from frost; by which process, the heads will increase in size, and be suitable for use the last of December or first of January.

"Cauliflowers are ready for cutting when the heads have attained a good size, and while they are close, firm, and white. They may even be cut before they have attained their full size; but it is always advisable to cut them before the heads begin to open, as the flavor is at this stage much more delicate and agreeable. In taking the crop, the stalks should be cut immediately under the lowest leaves, and the upper parts of these should be cut away near the flower-head.

"It is not size that constitutes a good Cauliflower, but its fine, white, or creamy color, its compactness, and what is technically called its 'curdy' appearance, from its resemblance to the curd of milk in its preparation for cheese. When the flower begins to open, or when it is of a frosty or wart-like appearance, it is less esteemed. In the summer season, it should not be cut long before using."

Use—

"The heads, or flowers, are considered one of the greatest of vegetable delicacies, when served up at the table either plain boiled, to be eaten with meat, like other Brassicæ, or dressed with white sauce, after the French manner. It is much used as a pickle, either by itself, or as forming an ingredient in what is called 'mixed pickles.' It may also be preserved a considerable time when pickled in the manner of 'sour-krout.' It also forms an excellent addition to vegetable soups."[1]

Preservation during Winter—The best way to preserve them during winter is to take them up late in the fall, with as much earth as possible about their roots, and reset them in earth, in a light, dry cellar, or in any other light and dry location secure from frost.

Varieties—These are comparatively few in number; the distinctions, in many instances, being quite unimportant. In the colour, foliage, general habit, and even in the quality, of the entire list, there is great similarity.

Early London Cauliflower

(London Particular; Fitch's Early London)

Stem tall; leaves of medium size. It has a fine, white, compact "curd," as the unexpanded head is termed; and is the sort grown in the vicinity of London for the early crop. It is comparatively hardy, and succeeds well when grown in this country. The plants should be set two feet and a half apart.

Early Paris Cauliflower

Head rather large, white, and compact; leaves large; stalk short. An early sort. In France, it is sown in June, and the heads come to table in autumn.

Erfurt's Early Cauliflower

(Erfurt's Extra Early)

Leaves large, long, waved, and serrated on the borders; stalk of medium height; head large,—measuring from seven to ten inches in diameter,—close, and compact.

From the experience of a single season, this variety promises to be one of the best for cultivation in this country. Specimens exhibited under this name, before the Massachusetts Horticultural Society, measured fully ten inches in diameter; the surface being very close, and the heads possessing the peculiar white, curdy character so rarely attained in the climate of the United States. The plants seldom fail to form a good-sized and symmetrical head, or flower.

Catalogue (1896), Cox Seed and Plant Co.

Large Asiatic Cauliflower

Originally from Holland. It is a fine, large, white, compact variety, taller and later than the Early London Cauliflower; it has also larger leaves. If sown at the same time, it will afford a succession.

Le Normand

Plant about fifteen inches high, with winged leaves, which are broad, and taper abruptly towards the base. They are toothed and waved on the margin, and expose a head which is about nine inches in diameter, and of a creamy color.

It is earlier than the Walcheren, and is readily distinguished from it by the waved and toothed margin of the foliage.

Mitchell's Hardy Early Cauliflower

A new variety. Bouquet not large, but handsome and compact. It is so firm, that it remains an unusual length of time without running to seed or becoming pithy. A desirable sort for private gardens and for forcing.

Stadthold

A new variety, introduced from Holland. Flower fine white, and of large size. Not early.

Waite's Alma Cauliflower

A new variety, represented as being of large size, and firm; surpassing in excellence the Walcheren.

Walcheren Cauliflower

(Early Leyden; Legge's Walcheren Broccoli)

This has been cultivated as a Broccoli for more than ten years; though originally introduced by the London Horticultural Society, under the name of Early Leyden Cauliflower. Stem comparatively short; leaves broad, less pointed and more undulated than those of the Cauliflower usually are. The difference in constitution is, however, important; as it not only resists the cold in winter, but the drought in summer, much better than other cauliflowers. In hot, dry summers, when scarcely a head of these could be obtained, the Walcheren Cauliflower, planted under similar circumstances, formed beautiful heads,—large, white, firm, and of uniform closeness.

Wellington Cauliflower

Messrs. Henderson and Son describe this Cauliflower as the finest kind in cultivation; pure white; size of the head over two feet; in growth, very dwarfish,—the stem not more than two or three inches from the soil. It is one of the hardiest varieties known, and is said to withstand the extreme variations of the climate of the United States. An excellent sort for early planting and for forcing.

1. The Book of the Garden. By Charles M'Intosh. 2 vols. Edinburgh and London, 1855.

Seed and Plant Guide (1899), H.W. Buckbee

51

Colewort/Collards (Brassica oleracea var. viridis.)

The Colewort, strictly speaking, is a plant distinct from the other varieties of Cabbage. It is of small habit, and attains sufficient size for use in a few weeks. It is eatable from the time it has four or six leaves until it has a hard heart. Loudon says the original Colewort seems to be lost, and is now succeeded by what are called "Cabbage Coleworts." These are cabbage-plants in their young state; and, when cooked, are quite as tender and good as the true Colewort. In growing these, all that is necessary is to sow the seed of almost any variety of the common green cabbages in drills a foot apart, and half an inch deep. For a succession, sowings may be made, at intervals of two weeks, from the last of April to the last of August. In the Southern States, the sowings might be continued through the winter.

Collard Leaf Variation

When cultivated for sale, simply allow them to stand till there is enough to be worth bunching and eating. They are boiled and served at table as greens.

Rosette Colewort.

A small but remarkably neat variety; the whole plant, when well grown, measuring twelve inches in diameter, and having the form of a rose not completely expanded,—the head corresponding to the bud still remaining at the heart, or centre; stalk small and short. The plants may be grown twelve inches asunder.

52

Couve Tronchuda/Portugal Cabbage (Brassica oleracea var.)

(Portugal Borecole; Large-Ribbed Borecole; Trauxuda Kale; Couve-Galega)[1]

Though a species of Cabbage, the Couve Tronchuda is quite distinct from the common head varieties. The stalk is short and thick; the outer leaves are large, roundish, of a dark bluish-green, wrinkled on the surface, and slightly undulated on the borders; the mid-rib of the leaf is large, thick, nearly white, and branches into veins of the same color; the plant forms a loose, open head, and, when full grown, is nearly two feet high.

Culture—It should be planted and treated like the Common Cabbage. The seeds may be sown early in frames, and the plants afterwards set in the open ground; or the sowing may be made in the open ground in May. The plants require two feet and a half between the rows, and two feet between the plants in the rows. The seeds, in size, form, and color, resemble those of the Cabbage, and will keep five years. One-fourth of an ounce will produce about a thousand plants.

Portuguese/Galician Cabbage (Couve-Galega)

To raise Seed—In the autumn, before severe weather, remove two or three plants entire to the cellar; and, in April following, reset them about two feet apart. Cut off the lower and smaller

FIELD AND GARDEN VEGETABLES OF THE LATE NINETEENTH CENTURY | 249

side-sprouts as they may appear, and allow only the strong, central shoot to grow. The seeds will ripen in August.

Use—Different parts of the Couve Tronchuda are applicable to culinary purposes. The ribs of the outer and larger leaves, when boiled, somewhat resemble sea-kale in texture and flavor. The heart, or middle of the plant, is, however, the best for use. It is peculiarly delicate, and agreeably flavored, without any of the coarseness which is so often found in plants of the Cabbage tribe.

Woman with a Basket of Portuguese Cabbage on her Head (1948 approx.)
Charles Fenno Jacobs

Dwarf Couve Tronchuda

(Murcianâ; Dwarf Portugal Cabbage; Dwarf Trauxuda Kale)

Much earlier and smaller throughout than the Common Couve Tronchuda. Stem from fifteen to eighteen inches high. The leaves are of medium size, rounded, smooth, and collected at the centre of the plant into a loose heart, or head. When the lower leaves are taken off for use, the plant, unlike the former variety, throws out numerous sprouts, or shoots, from the base of the stem, which make excellent coleworts, or greens. It is, however, wanting in hardiness; and appears to be better adapted for early use than for late keeping.

Soil and Cultivation—Both of the varieties require a well-manured soil. The seeds of the Dwarf Couve Tronchuda may be sown early in frames, and the plants afterwards set in the open ground; or the sowing may be made, in May or June, where the plants are to remain. They should be two feet apart in each direction.

Fringed Tronchuda.

Stem short; leaf-stems thicker and larger than those of the Common Couve Tronchuda, but not so fleshy and succulent. The leaves expand towards their extremities into a spatulate form, the edges being regularly lobed and curled. They are of a glaucous or bluish green color, and form a sort of loose heart, or head, at the centre of the plant. Its only superiority over the common varieties consists in its more hardy character.

The Fringed Tronchuda is, however, very succulent, and of good quality; and is cultivated to some extent in France, particularly in the vicinity of Paris.

White-Ribbed Tronchuda.

(White-Ribbed Avilès Cabbage; White-Ribbed Portugal Cabbage; Chou à Côtes blanches d'Avilès)

This variety nearly resembles the Dwarf Portugal Cabbage, or Dwarf Couve Tronchuda, if it is not identical. It has white ribs, and forms a close heart.

It should be planted, and in all respects treated, as the Dwarf Portugal Cabbage.

---

1. this is actually another variety of Collard (cabbage that does not form a head) that is especially popular in Portugal, Galicia (Spain), and Brazil even in the present day.

53

Pak-Choi/Bok-Choy (Brassica rapa chinensi)

An annual plant, introduced from China. The root-leaves are oval, regular, very smooth, deep-green, with long, naked, fleshy, white stems, somewhat similar to those of the Swiss Chards, or Leaf-beets.

When in blossom, the plant measures about four feet in height, and the stem is smooth and branching. The flowers are yellow; the seeds are small, round, blackish-brown, and, in their general appearance, resemble those of the Turnip or Cabbage. An ounce contains about ten thousand seeds, and they will keep five years.

Sowing and Cultivation—The seed should be sown in April or May, and the plants may be grown in hills or drills. They are usually sown in rows, and thinned to twelve inches apart.

Pak-choi
Vilmorin-Andrieux 1883

Use—The leaves are eaten boiled, like cabbage; but they are much more tender, and of a more agreeable flavor.

Bok-Choy (Brassica rapa chinensis)
Wu Changshuo

54

Pe-Tsai/Chinese Cabbage (Brassica rapa pekinensi)

(Chinese Cabbage; Napa Cabbage)[1]

The Pe-Tsai, like the Pak-Chöi, is an annual plant, originally from China. The leaves are of an oval form, rounded at the ends, somewhat blistered on the surface; and, at the centre, are collected together into a long and rather compact tuft, or head. The plant, when well grown and ready for use, has somewhat the appearance of a head of Cos Lettuce, and will weigh six or seven pounds; though, in its native country, it is said to reach a weight of upwards of twenty pounds.

Chinese Cabbage

Towards the end of the summer, the flower-stalk shoots from the centre of the head to the height of three feet, producing long and pointed leaves, and terminating in loose spikes of yellow flowers. The seeds are small, round, brownish-black, and resemble those of the Common Cabbage. They retain their vitality five years. An ounce contains eight thousand seeds.

Cultivation—Sow in April or May, and thin or transplant to rows eighteen inches apart, and a foot apart in the rows.

Use—It is used like the Common Cabbage, and is sweet, mild-flavored, and easy of digestion. The young plants are also boiled like coleworts or spinach.

1. In the original, the scientific name Brassica chinensis was given. However, Brassica rapa chinensis is actually the name for Bok-choy, the previous entry. The vegetable described here is of the subspecies Pekinensi)

55

Savoy Cabbage (Brassica oleracea var. sabauda)

(Savoy Cabbage; Chou de Milan; Brassica oleracea, var. bullata) [1]

This class of cabbages derives its popular name from Savoy, a small district adjoining Italy, where the variety originated, and from whence it was introduced into England and France more than a hundred and fifty years ago. The Savoys are distinguished from the common head or close-hearted cabbages by their peculiar, wrinkled, or blistered leaves. According to Decandole, this peculiarity is caused by the fact, that the pulp, or thin portion of the leaf, is developed more rapidly than the ribs and nerves.

Savoy Cabbage
Adolphe Millot (1919, approx.)

Besides the distinction in the structure of the leaves, the Savoys, when compared with the common cabbages, are slower in their development, and have more open or less compactly formed heads. In texture and flavor, they are thought to approach some of the broccolis or cauliflowers; having, generally, little of the peculiar musky odor and taste common to some of the coarser and larger varieties of cabbages.

None of the family are hardier or more easily cultivated than the Savoys; and though they will not quite survive the winter in the open ground, so far from being injured by cold and frosty weather, a certain degree of frost is considered necessary for the complete perfection of their texture and flavor.

Soil—They succeed best in strong, mellow loam, liberally enriched with well-digested compost.

Sowing—The first sowing may be made early in a hot-bed, and the plants set in the open ground in May, or as soon as the weather will admit. Subsequent sowings may be made in drills, in the open ground, in May, or early in June. When the seedlings are five or six inches high, thin or transplant to about three feet apart.[Pg 278]

Harvesting—During the autumn, take the heads directly from the garden, whenever they are required for the table; but they should all be taken in before the ground is deeply frozen, or covered with snow. No other treatment will be required during the winter than such as is usually given to the Common Cabbage.

To raise Seed—In April, select a few well-formed, good-sized heads, as near types of the variety as possible; and set them entire, about two feet apart. If small shoots start from the side of the stalk, they should be removed; as only the sprout that comes from the centre of the head produces seed that is really valuable. All varieties rapidly deteriorate, if grown from seeds produced by side-shoots, or suckers.

The seeds, when ripe, in form, size, and color, are not distinguishable from those of the Common Cabbage. An ounce contains ten thousand seeds, which will generally produce about three thousand plants.

Varieties—

Drumhead Savoy

(Cape Savoy)

Head large, round, compact, yellowish at the centre, and a little flattened, in the form of some of the common Drumhead cabbages, which it nearly approaches in size. The exterior leaves of the plant are round and concave, clasping, sea-green or bluish-green, rise above a level with the top of the head, and are more finely and less distinctly fretted or blistered on the surface than the leaves of the Green Globe. Stalk of medium length.

Savoy Cabbage
Patrick Wright (1908)

The Drumhead Savoy seldom fails to heart well, affords a good quantity of produce, is hardy, and, when brought to the table, is of very tender substance, and finely flavored. It is considered one of the best of the large kinds; and, wherever cultivated, has become a standard sort. It keeps well during winter, and retains its freshness late into the spring.

As it requires nearly all of the season for its complete development, the seed should be sown comparatively early.

Transplant to rows at least three feet apart, and allow nearly the same distance between the plants in the row.

Early Dwarf Savoy

(Early Green Savoy)

Head small, flattened, firm, and close; leaves rather numerous, but not large, deep-green, finely but distinctly blistered, broad and rounded at the top, and tapering towards the stalk or stem of the plant, which is short. It is not quite so early as the Ulm Savoy; but it hearts readily, is tender and of good quality, and a desirable sort for early use.

It requires a space of about twenty inches in each direction.

Early Flat Green Curled Savoy

A middle-sized, very dwarf, and flat-headed variety; color deep-green; quality tender and good. The plants should be set fifteen or eighteen inches asunder.

Early Long Yellow Savoy

(Chou de Milan Doré a Tète Longue)

Similar to the Golden Savoy, and, like it, an early sort. It has, however, a longer head, and does not heart so firmly. In flavor and texture, as well as in its peculiar color, there is little difference between the varieties.

Cultivate in rows eighteen inches apart, and fifteen or eighteen inches apart in the rows.

Early Ulm Savoy

(New Ulm Savoy; Earliest Ulm Savoy)

A dwarfish, early sort. Head small, round, solid; leaves rather small, thick, fleshy, and somewhat rigid, of a fine, deep-green, with numerous prominent blister-like elevations. The loose leaves are remarkably few in number; nearly all of the leaves of the plant contributing to the formation of the head.

It very quickly forms a heart, which, though not of large size, is of excellent quality. It is, however, too small a sort for market purposes; but, for private gardens, would, no doubt, be an acquisition. In the London Horticultural Society's garden, it proved the earliest variety in cultivation.

Being one of the smallest of the Savoys, it requires but a small space for its cultivation. If fifteen inches between the rows, and about the same distance in the rows, be allowed, the plants will have ample room for their full development.

Feather-stem Savoy

This curious and useful variety has been in existence for several years, and is said to be a cross between the Savoy and the Brussels Sprouts. It is what may be called a sprouting Savoy; producing numerous shoots, or sprouts, along the stem.

A sowing should be made the last of April, and another from the middle to the 20th of May, and the plants set out as soon as they are of suitable size, in the usual manner of Savoys and other winter greens.

Golden Savoy

(Early Yellow Savoy)

A middle-sized, roundish, rather loose-headed variety; changing during the winter to a clear, bright yellow. The exterior leaves, at the time of harvesting, are erect, clasping, of a pale-green color, and coarsely but not prominently blistered on the surface; stalk short.

The Golden Savoy comes to the table early, hearts readily, is of very tender substance when cooked, and of excellent quality; though its peculiar colour is objectionable to many.

It requires a space of about eighteen inches between the rows, and fifteen to eighteen inches between the plants in the rows.

Green Globe Savoy

(Green Curled Savoy; Large Green Savoy)

One of the best and one of the most familiar of the Savoys; having been long in cultivation, and become a standard sort. The head is of medium size, round, bluish or sea green on the outside, yellow towards the centre, and loosely formed. The interior leaves are fleshy and succulent, with large and prominent midribs,—the exterior leaves are round and large, of a glaucous or sea green color, and, in common with those of the head, thickly and distinctly blistered in the peculiar manner of the Savoys; stalk of medium height.

The variety possesses all the qualities of its class: the texture is fine, and the flavor mild and excellent. On account of its remarkably fleshy and tender character, the inner loose leaves about the head will be found good for the table, and to possess a flavor nearly as fine as the more central parts of the plant.

It is remarkably hardy, and attains its greatest perfection only late in the season, or under the influence of cool or frosty weather. As the plants develop much less rapidly than those of the Common Cabbage, the seed should be sown early. Transplant in rows two and a half or three feet apart, and allow a space of two feet and a half between the plants in the rows.

Long-headed Savoy

(Chou Milan à Tête Longue)

A comparatively small variety, with an oval, long, yellowish-green, but very compact head; leaves erect, inclining to bluish-green, long and narrow, revoluted on the borders, and finely fretted or blistered on the surface; stem rather high.

It is hardy and of excellent quality, but yields less than many other sorts. It is, however, a good kind for gardens of limited size, as it occupies little space, and cabbages well. The plants may be set eighteen inches apart in one direction by about fifteen inches in the opposite.

Marcelin Savoy

A new sort, allied to the Early Ulm, but growing somewhat larger. Though not so early, it is next to it in point of earliness; and, if both sorts are sown at the same time, the Marcelin will form a succession. It is a low grower; the leaves are dark-green, finely wrinkled and curled; the head is round, compact, and of excellent quality. When cut above the lower course of leaves, about four small heads, almost equal in delicacy to Brussels Sprouts, are generally formed. This sort is exceedingly hardy; and, on the whole, must be considered a valuable acquisition.

The plants should be set eighteen inches by twelve inches apart.

Tour's Savoy

(Dwarf Green Curled Savoy; Pancalier de Tourraine)

Head small, loose, and irregular; leaves numerous, bright-green, rigid, concave or spoon-shaped; the nerves and ribs large, and the entire surface thickly and finely covered with the blister-like swellings peculiar to the Savoys.

Chou Milan Pancalier de Touraine
Vilmorin-Andrieux 1904

It has some resemblance to the Early Dwarf Savoy; but is larger, less compact, and slower in its development.

A useful, hardy, smallish sort, adapted to small gardens; requiring only eighteen or twenty inches' space each way. Excellent for use before it becomes fully cabbaged.

Yellow Curled Savoy

(Large Late Yellow Savoy; White Savoy)

Dwarf, middle-sized, round; leaves pale-green at first, but quite yellow in winter; the heart is not so compact as some, but of tender quality, and by many preferred, as it is much sweeter than the other kinds. It is later and hardier than the Yellow Savoys, before described.

---

1. This variety of cabbage is referred to as both Brassica oleracea var. sabauda and Brassica oleracea, var. bullata. The later is the name give in the original, however, for identification purposes, the editors have included both.

56

Sea-Kale (Crambe maritima)

Sea-kale is a native of the southern shores of Great Britain, and is also abundant on the seacoasts of the south of Europe. There is but one species cultivated, and this is perennial and perfectly hardy. The leaves are large, thick, oval or roundish, sometimes lobed on the borders, smooth, and of a peculiar bluish-green color; the stalk, when the plant is in flower, is solid and branching, and measures about four feet in height; the flowers, which are produced in groups, or clusters, are white, and have an odor very similar to that of honey. The seed is enclosed in a yellowish-brown shell, or pod, which, externally and internally, resembles a pit, or cobble, of the common cherry. About six hundred seeds, or pods, are contained in an ounce; and they retain their germinative powers three years. "They are large and light, and, when sold in the market, are often old, or imperfectly formed; but their quality is easily ascertained by cutting them through the middle: if sound, they will be found plump and solid." They are usually sown without being broken.

Sea-Kale

Preparation of the Ground, and Sowing—The ground should be trenched to the depth of from a foot to two feet, according to the depth of the soil, and well enriched through out. The seeds may be sown in April, where the plants are to remain; or they may be sown at the same season in a nursery-bed, and transplanted the following spring. They should be set or planted out in rows three feet apart, and eighteen inches apart in the rows.

Culture—

"After the piece is set, let the plants be kept very clean. The earth should be occasionally stirred, when the rains have run the surface together; and, when the plants come up, let them have their own way the first season. As the plants will blossom the second season if let alone, and the bearing of seed has a tendency to weaken every thing, take off the flower-buds as soon as they appear, and not allow the plants to seed. When the leaves begin to decay in autumn, clear them all off, and dig a complete trench between the rows, and earth up the ridges: that is, all the soil you take out must be laid on the plants, so as to pile or bank up eight inches above the crowns of the roots, thus forming a flat-topped bank a foot across; widening a little downwards, so that the edges shall not break away. In doing this, the piece is formed into alternate furrows and ridges; the plants being under the centre of the ridges.

"As the weather gets warm in the spring, these banks should be watered; and, when the surface is broken by the rising plant, remove the earth, and cut off the white shoots close to their base: for these shoots form the eatable portion; and, being blanched under ground, they are tender and white, and from six to eight inches long. The shoots should be cut as soon as they reach the surface; because, if the shoot comes through, the top gets purple, and the plants become strong-flavored. As all of the shoots will not appear at once, the bed should be looked over frequently, and a shoot cut whenever it has broken the surface of the soil; for, if not taken early, it soon becomes nearly worthless. In the process of cutting the shoots, the earth becomes gradually removed; and the tops of the plants, coming to the surface again, put forth other shoots, which must be allowed to grow the remainder of the summer, only taking off the blossom-shoots as before. When, at the fall of the year, the leaves turn yellow, and decay, earth up again, after clearing the plants of their bad leaves and removing every weed. Before earthing up, fork the surface a little, just to break it up, that the earth may better take hold, and form a regular mass."[1] —*Glenny.*

Pot-forcing and Blanching—

"The ground, once planted, is as good for pot-forcing as for any thing; except that, for pot-forcing, it is usual to plant three plants in a triangle, about nine inches apart. The plants are cleared when the leaves decay, and the ground is kept level instead of being earthed up. Pots and covers (called 'sea-kale pots') are placed over the plants, or patches of plants, and the cover (which goes on and off at pleasure) put on. These pots are of various sizes; usually from ten to fourteen inches in diameter, and from a foot to twenty inches in height. If proper sea-kale pots cannot be procured, large-sized flower-pots will answer as substitutes; the pots being put over the plants as they are wanted, generally a few at a time, so as to keep up a succession. Dung is placed all over them; or, if no dung can be had, leaves are used: and they ferment and give out heat as genial, but not so violent, nor do they command so much influence, as the dung. Some may be placed on in February, and some in March. The dung is removed from the top to admit of seeing if the plant is started; and, by timely examination, it is easily seen when the plant is ready for use. The shoots are as white, when thus treated, as when grown by the other method, because of the total darkness that prevails while they are covered; but there is more air in the empty pots than there possibly could be in the solid earth, and it is considered that the vegetable is not so tender in consequence. However, the greater bulk of Sea-kale is so produced."

Taking the Crop—

"The blanched sprouts should be cut when they are from three to six inches in length, and while stiff, crisp, and compact. They should not be left till they are drawn up so as to bend, or hang down. The soil or other material used for excluding the light should be carefully removed, so as to expose the stem of the sprout; and the latter should be cut just below the base of the petioles or leaf-stem, and just enough to keep these attached."[2]

The Sea-kale season continues about six weeks.

"Cutting too much will finally destroy the plants. With one good cutting the cultivator should be satisfied, and should avoid the practice of covering and cutting a second time. The proper way is to cut the large, fine shoots, and leave the smaller ones that come afterwards to grow stronger during the summer."

Use—

"The young shoots and stalks, when from the length of three to nine inches, are the parts used. These, however, unless blanched, are no better than the coarser kinds of Borecole; but, when blanched, they become exceedingly delicate, and are much prized. The ribs of the leaves, even after they are nearly fully developed, are sometimes used; being peeled and eaten as asparagus. In either state, they are tied up in small bundles, boiled, and served as cauliflowers."[3]

To obtain Seed—

"Select some strong plants, and allow them to take their natural growth, without cutting off their crowns, or blanching. When the seed is ripe, collect the pods, dry them, and put them into open canvas-bags. The seeds keep best in the pods."[4]

1. Glenny.
2. The Gardener's Assistant. By Robert Thompson.
3. The Book of the Garden. By Charles M'Intosh. 2 vols. Edinburgh and London, 1855.
4. The Gardener's Assistant. By Robert Thompson.

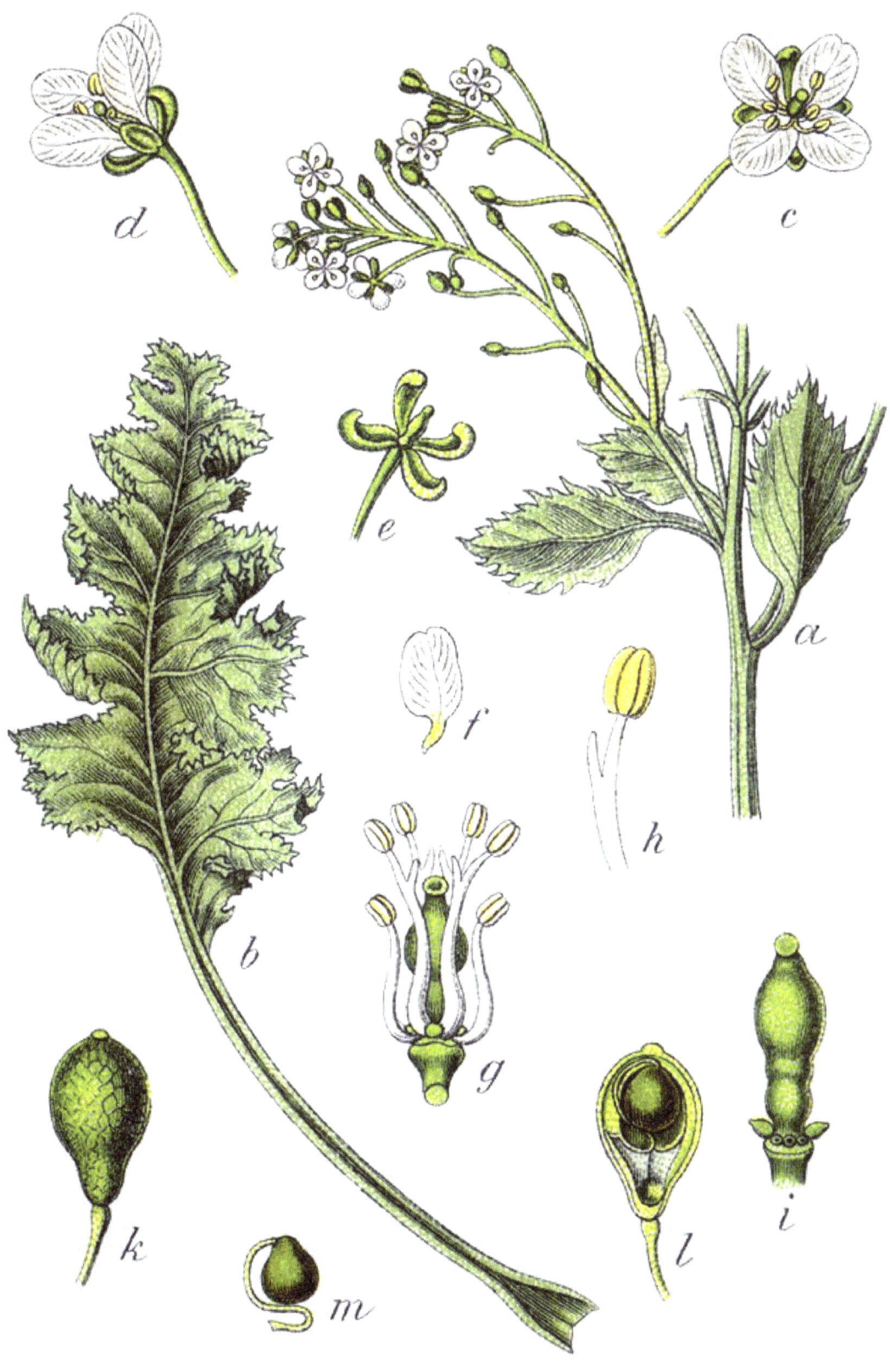

Crambe maritima
E.H.L.Krause

Part Six: Spinaceous Plants

Amaranthus; Black Nightshade; Leaf-beet, or Swiss Chard; Malabar Nightshade; Nettle; New-Zealand Spinach; Orach; Patience Dock; Quinoa; Sea-beet; Shepherd's Purse; Sorrel; Spinach; Wild or Perennial Spinach.

Various Kinds of Chard and Spinach
Seikei Zusetsu vol. 23, page 019

57

Amaranth (Amaranthus)

(Chinese Amaranthus; Chinese Spinach)

A hardy, annual plant, introduced from China; stem three feet in height, much branched, and generally stained with red; leaves variegated with green and red, long, and sharply pointed; the leaf-stems and nerves are red; the flowers, which are produced in axillary spikes, are greenish, and without beauty; the seeds are small, black, smooth, and shining,—twenty-three thousand are contained in an ounce, and they retain their power of germination four or five years.

Soil and Cultivation—Any good garden-soil is adapted to the growth of the Amaranthus. Before sowing, the ground should be thoroughly pulverized, and the surface made smooth and even. The seed may be sown in April, or at any time during the month of May. It should be sown in very shallow drills, fourteen to sixteen inches apart, and covered with fine, moist earth. When the plants are two inches high, thin to five or six inches apart, and cultivate in the usual manner. They will yield abundantly during most of the summer.

Use—The leaves are used in the manner of Spinach, and resemble it in taste.

Red Amaranth
J. Pass, c. 179

Varieties—

Early Amaranthus
(Amarante Mirza)

This plant is a native of the East Indies; and in height, color, and general habit, resembles the Chinese Amaranthus. It is, however, somewhat earlier, and ripens its seed perfectly in climates where the Chinese almost invariably fails. Its uses, and mode of cultivation, are the same.

Hantsi Shanghai Amaranthus
(Amarante Hantsi Shanghai)

Introduced from China by Mr. Fortune, and disseminated by the London Horticultural Society. It differs little from the preceding species; and is cultivated in the same manner, and used for the same purposes. Annual.

Amaranthus tricollor
D. Rabel 1624

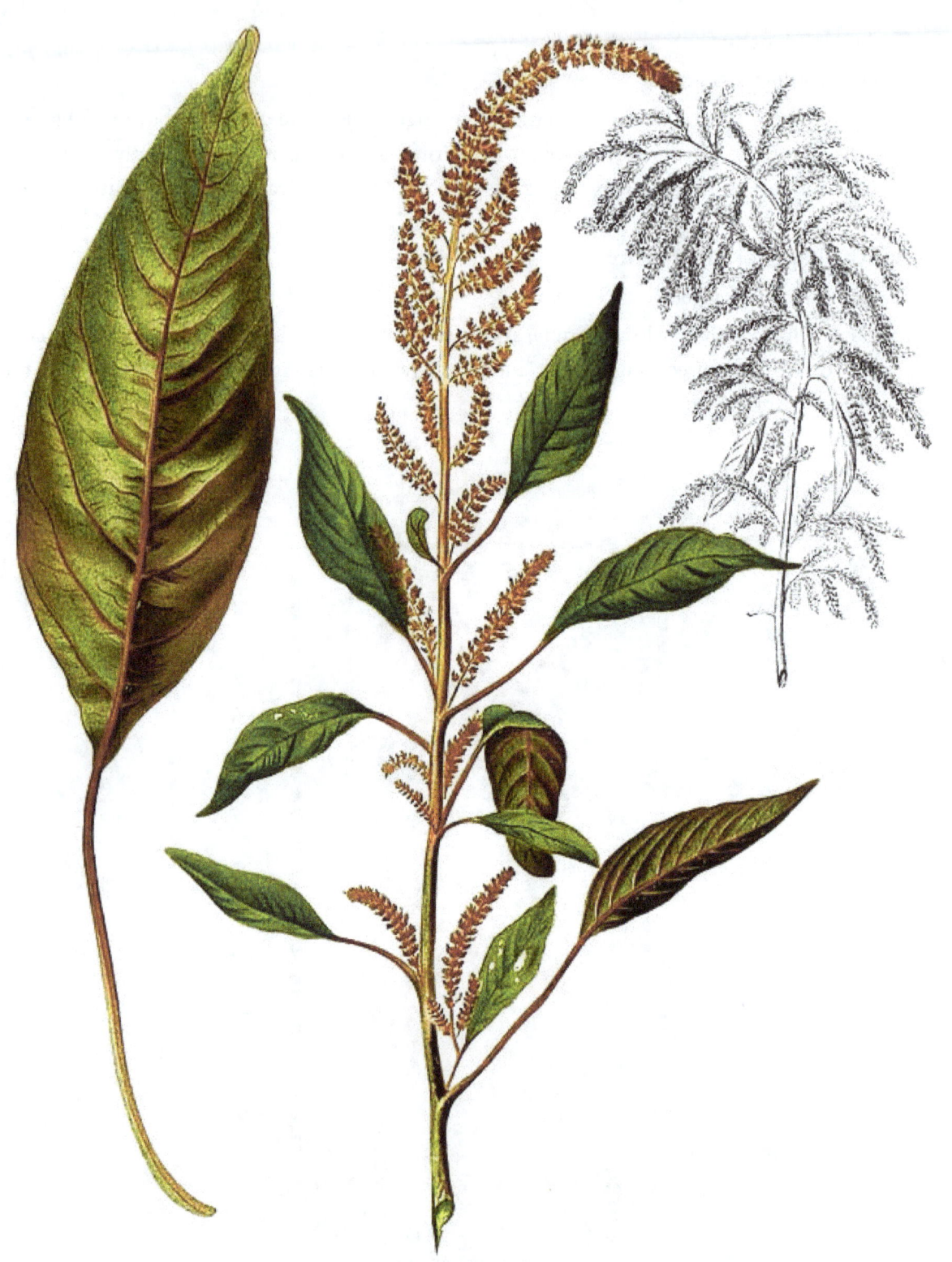

Amaranthus cruentus
Francisco Manuel Blanco

58

Black Nightshade (Solanum nigrum)

(Morelle, of the French)

An unattractive, annual plant, growing spontaneously as a weed among rubbish, in rich, waste places. Its stem is from two to three feet high, hairy and branching; the leaves are oval, angular, sinuate, and bluntly toothed; the flowers are white, in drooping clusters, and are succeeded by black, spherical berries, of the size of a small pea; the seeds are small, lens-shaped, pale yellow, and retain their vitality five years,—twenty-three thousand are contained in an ounce.

Solanum nigrum
Flora Batava, plate number 112

Propagation and Culture—It is raised from seed, which may be sown in April or May, or in autumn. Sow in shallow drills fifteen or eighteen inches apart, and thin to six or eight inches in the drills; afterwards keep the soil loose, and free from weeds, in the usual manner.

Use—The French, according to Vilmorin, eat the leaves in the manner of Spinach; while Dr. Bigelow asserts that **it has the aspect and reputation of a poisonous plant.**[1]

On the authority of American botanists, it was introduced into this country from Europe. By European botanists, it is described as a plant of American origin.

---

1. Although eaten cooked in certain countries, it is generally not recommended to eat any part of this plant due to the high solanine levels. The toxicity levels depend on the strain and proper identification is therefore important. Green berries and raw leaves and berries should never be eaten.

59

Swiss Chard (Beta cicla)

(Leaf-Beet; Sicilian Beet; White Beet; Silverbeet)

The Leaf-beet is a native of the seacoasts of Spain and Portugal. It is a biennial plant, and is cultivated for its leaves and leaf-stalks. The roots are much branched or divided, hard, fibrous, and unfit for use.

Propagation and Cultivation—It is propagated, like other beets, from seed sown annually, and will thrive in any good garden soil. The sowing may be made at any time in April or May, in drills eighteen inches apart, and an inch and a half deep.

> "When the plants are a few inches high, so that those likely to make the best growth can be distinguished, they should be thinned out to nine inches or a foot apart, according to the richness of the soil; more room being allowed in rich ground. Some, however, should be left at half that distance, to make up by transplanting any vacancies that may occur. The ground should be kept clean, and occasionally stirred between the rows; taking care not to injure the roots. In dry weather, plenty of water should be given to promote the succulence of the leaves."[1]

Taking the Crop—

> "The largest and fullest-grown leaves should be gathered first; others will follow. If grown for Spinach, the leaves should be rinsed in clean water, and afterwards placed in a basket to drain dry; if for Chard, or for the leaf-stalks and veins, these should be carefully preserved, and the entire leaves tied up in bundles of six or eight in each."[2]

Seed—During the first season, select a few vigorous plants, and allow them to grow unplucked. Just before the closing-up of the ground in autumn, take up the roots; and, after removing the tops an inch above the crown, pack them in dry sand in the cellar. The following spring, as soon as the ground is in working order, set them out with the crowns level with the surface of the ground, and about two feet and a half apart. As the plants increase in height, tie them to stakes, to prevent injury from wind; and in August, when the seed is ripe, cut off the stems near the ground, and spread them entire, in an airy situation, till they are sufficiently dried for threshing out.

The seed, or fruit, has the appearance peculiar to the family; although those of the different varieties, like the seeds of the Red Beet, vary somewhat in size, and shade of color.

An ounce of seed will sow a hundred feet of drill, or be sufficient for a nursery-bed of fifty square feet.

Use—

"This species of Beet—for, botanically considered, it is a distinct species from *Beta vulgaris*, the Common or Red Beet—is cultivated exclusively for its leaves; whereas the Red Beet is grown for its roots. These leaves are boiled like Spinach, and also put into soups. The midribs and stalks, which are separated from the lamina of the leaf, are stewed and eaten like Asparagus, under the name of "Chard." As a spinaceous plant, the White Beet might be grown to great advantage in the vegetable garden, as it affords leaves fit for use during the whole summer."[3]

The thin part of the leaves is sometimes put into soups, together with sorrel, to correct the acidity of the latter.

The varieties are as follow:

Green or Common Leaf-Beet

Stalks and leaves large, green; the roots are tough and fibrous, and measure little more than an inch in diameter; leaves tender, and of good quality.

If a sowing be made as soon in spring as the frost will permit, another in June, and a third the last of July, they will afford a constant supply of tender greens, nearly or quite equal to Spinach. For this purpose, the rows need be but a foot apart.

Large-Ribbed Curled

(Curled Leaf-Beet)

Stalks white; leaves pale yellowish-green, with broad mid-ribs, large nerves, and a blistered surface like some of the Savoys. It may be grown as a substitute for Spinach, in the manner directed for the Common or Green-leaved variety.

Large-Ribbed Scarlet Brazilian

(Red Stalk Leaf-Beet; Poirée à Carde rouge)

Leaf-stalks bright purplish-red; leaves green, blistered on the surface; nerves purplish-red. A beautiful sort, remarkable for the rich and brilliant color of the stems, and nerves of the leaves.

Large-Ribbed Yellow Brazilian

(Yellow-Stalked Leaf-Beet; Poirée à Carde jaune)

A variety with bright-yellow leaf-stalks and yellowish leaves. The nerves of the leaves are yellow, like the leaf-stalks. The color is peculiarly rich and clear; and the stalks are quite attractive, and even ornamental. Quality tender and good.

Silver-Leaf Beet

(Great White-Leaf Beet; Swiss Chard; Sea-Kale Beet; Large-Ribbed Silver-Leaf Beet)

Stalks very large; leaves of medium size, erect, with strong, white ribs and veins. The leaf-stalks and nerves are cooked and served like Asparagus, and somewhat resemble it in texture and flavor. It is considered the best of the Leaf-beets.

Red Chard
Lisa Redfern

1. The Gardener's Assistant. By Robert Thompson.

2,3. The Book of the Garden. By Charles M'Intosh. 2 vols. Edinburgh and London, 1855.

60

Malabar Spinach (Basella alba)

White/Green Malabar Spinach (Basella alba)

From the East Indies. Though a biennial plant, in cultivation it is generally treated as an annual. Stem five feet and upwards in length, slender, climbing; leaves alternate, oval, entire on the borders, green and fleshy; flowers in clusters, small, greenish; seeds round, with portions of the pulp usually adhering,—eleven to twelve hundred weighing an ounce. They retain their vitality three years.

Basela alba

Large-Leaved Chinese Malabar (Basella cordifolia)

A Chinese species, more vigorous and much stronger in its general habit than the Red or the White. Leaves as large as those of Lettuce,—green, round, very thick, and fleshy; flowers small, greenish; seeds round, nearly of the same form and color as those of the White variety, but rather larger.

The species is slow in developing its flower-stem, and the best for cultivation.

Red Malabar Spinach (Basella rubra)

From China. Properly a biennial plant, but, like the White species, usually cultivated as an annual. It is distinguished from the last named by its color; the whole plant being stained or tinted with purplish red. In the size and color of the seeds, and general habit of the plant, there are no marks of distinction, when compared with the White.

Propagation and Cultivation—All of the species are easily grown from seeds; which may be sown in a hot-bed in March, or in the open ground in May. They take root readily when transplanted; and may be grown in rows like the taller descriptions of pease, or in hills like running beans. Wherever grown, they require a trellis, or some kind of support; otherwise the plants will twist themselves about other plants, or whatever objects may be contiguous. All are comparatively tender, and thrive best, and yield the most produce, in the summer months.

Use—The leaves, which are put forth in great profusion, are used in the form of Spinach. The juice of the fruit affords a beautiful but not permanent purple colour.

Basel rubra
Francisco Manuel Blanco

61

Common Nettle (Urtica dioica)

(Large Stinging Nettle)

The Common Nettle is a hardy, herbaceous perennial, growing naturally and abundantly by waysides and in waste places, "but is seldom seen where the hand of man has not been at work; and may, therefore, be considered a sort of domestic plant." It has an erect, branching, four-sided stem, from three to five feet in height; the leaves are opposite, heart-shaped at the base, toothed on the borders, and thickly set with small, stinging, hair-like bristles; the flowers are produced in July and August, and are small, green, and without beauty; the seeds are very small, and are produced in great abundance,—a single plant sometimes yielding nearly a hundred thousand.

Propagation and Culture—The Nettle will thrive in almost any soil or situation. Though it may be propagated from seeds, it is generally increased by a division of the roots, which may be made in spring or autumn. These should be set in rows two feet apart, and a foot apart in the rows.

Use—

"Early in April, the tops will be found to have pushed three or four inches, furnished with tender leaves. In Scotland, Poland, and Germany, these are gathered, as a pot-herb for soups or for dishes, like Spinach; and their peculiar flavor is by many much esteemed. No plant is better adapted for forcing; and, in winter or spring, it may be made to form an excellent substitute for Cabbage, Coleworts, or Spinach. Collect the creeping roots, and plant them either on a hot-bed or in pots to be placed in the forcing-house, and they will soon send up an abundance of tender tops: these, if desired, may be blanched by covering with other pots. If planted close to a flue in the vinery, they will produce excellent nettle-kale or nettle-spinach in January and February."

Lawson[1] states that:

> "...the common Nettle has long been known as affording a large proportion of fibre, which has not only been made into ropes and cordage, but also into sewing-thread, and beautiful, white, linen-like cloth of very superior quality. It does not, however, appear that its cultivation for this purpose has ever been fairly attempted. The fibre is easily separated from other parts of the stalk, without their undergoing the processes of watering and bleaching; although, by such, the labor necessary for that purpose is considerably lessened. Like those of many other common plants, the superior merits of this generally accounted troublesome weed have hitherto been much overlooked."

1. The Agriculturist's Manual. By Peter Lawson and Son. Edinburgh, 1836.

Urtica dioica

62

New Zealand Spinach (Tetragonia tetragonioides)

(Tetragonia expansa)

This plant, botanically considered, is quite distinct from the common garden Spinach; varying essentially in its foliage, flowers, seeds, and general habit.

It is a hardy annual. The leaves are of a fine green color, large and broad, and remarkably thick and fleshy; the branches are numerous, round, succulent, pale-green, thick and strong,—the stalks recline upon the ground for a large proportion of their length, but are erect at the extremities; the flowers are produced in the axils of the leaves, are small, green, and, except

Tetragonia tetragonioides

that they show their yellow anthers when they expand, are quite inconspicuous; the fruit is of a dingy-brown color, three-eighths of an inch deep, three-eighths of an inch in diameter at the top or broadest part, hard and wood-like in texture, rude in form, but somewhat urn-shaped, with four or five horn-like points at the top. Three hundred and twenty-five of these fruits are contained in an ounce; and they are generally sold and recognized as the seeds. They are, however, really the fruit; six or eight of the true seeds being contained in each. They retain their germinative powers five years.

Propagation and Culture—It is always raised from seed, which may be sown in the open ground from April to July. Select a rich, moist soil, pulverize it well, and rake the surface smooth. Make the drills three feet apart, and an inch and a half or two inches deep; and sow the seed thinly, or so as to secure a plant for each foot of row. In five or six weeks from the planting, the branches will have grown sufficiently to allow the gathering of the leaves for use. If the season should be very dry, the plants will require watering. They grow vigorously, and, in good soil, will extend, before the end of the season, three feet in each direction.

Gathering—The young leaves must be pinched or cut from the branches; taking care not to injure the ends, or leading shoots. These shoots, with the smaller ones that will spring out of the stalks at the points where the leaves have been gathered, will produce a supply until a late period in the season; for the plants are sufficiently hardy to withstand the effects of light frosts without essential injury.

Its superiority over the Common Spinach consists in the fact, that it grows luxuriantly, and produces leaves of the greatest succulency, in the hottest weather.

Anderson, one of its first cultivators, had but nine plants, which furnished a gathering for the table every other day from the middle of June. A bed of a dozen healthy plants will afford a daily supply for the table of a large family.

Seed—To raise seed, leave two or three plants in the poorest soil of the garden, without cutting the leaves. The seeds will ripen successively, and should be gathered as they mature.

Use—It is cooked and served in the same manner as Common Spinach.

There are no described varieties.

63

Orach (Atriplex hortensis)

(Arrach; French Spinach; Mountain Spinach)

Orach is a hardy, annual plant, with an erect, branching stem, varying in height from two to four feet, according to the variety. The leaves are variously shaped, tut somewhat oblong, comparatively thin in texture, and slightly acid to the taste; the flowers are small and obscure, greenish or reddish, corresponding in a degree with the color of the foliage of the plant; the seeds are small, black, and surrounded with a thin, pale-yellow membrane,—they retain their vitality three years.

Soil and Culture—It is raised from seed sown annually. As its excellence depends on the size and succulent character of the leaves, Orach is always best when grown in a rich, deep, and moist soil. The first sowing may be made as soon in spring as the ground is in proper condition; afterwards, for a succession, sowings may be made, at intervals of two weeks, until June.

When the ground has been thoroughly dug over, and the surface made fine and smooth, sow the seed in drills eighteen inches or two feet apart, and cover three-fourths of an inch deep. When the young plants are two or three inches high, thin them to ten or twelve inches apart, and cultivate in the usual manner. Orach is sometimes transplanted, but generally succeeds best when sown where the plants are to remain. In dry, arid soil, it is comparatively worthless.

To raise Seed—Leave a few of the best plants without cutting, and they will afford a plentiful supply of seeds in September.

Use—Orach is rarely found in the vegetable gardens of this country. The leaves have a pleasant, slightly acid taste, and, with the tender stalks, are used boiled in the same manner as Spinach or Sorrel, and are often mixed with the latter to reduce the acidity.

"The stalks are good only while the plants are young; but the larger leaves may be picked off in succession throughout the season, leaving the stalks and smaller leaves untouched, by which the latter will increase in size. The Orach thus procured is very tender, and much esteemed."

A few plants will afford an abundant supply.

Varieties—

Green Orach
(Dark-Green Orach; Deep-green Orach)
The leaves of this variety are of a dark, grass-green colour, broad, much wrinkled, slightly toothed, and bluntly pointed; the stalk of the plant and the leaf-stems are strong and sturdy, and of the same color as the leaves. It is the lowest growing of all the varieties.

Lurid Orach
(Pale-Red Orach)
Leaves pale-purple, tinged with dark-green,—the under surface light-purple, with green veins, slightly wrinkled, terminating rather pointedly, and toothed on the borders only toward the base, which forms two acute angles; the stalk of the plant and the stems of the leaves are bright-red, slightly streaked with white between the furrows,—height three feet and upwards.

Purple Orach
(Dark-Purple Orach)
Plant from three to four feet in height; leaves dull, dark-purple, more wrinkled and more deeply toothed than those of any other variety. They terminate somewhat obtusely, and form two acute angles at the base. The stalk of the plant and the stems of the leaves are deep-red, and slightly furrowed. The leaves change to green when boiled.

Red Orach
(Dark-Red Orach; Bon Jardinier)
Leaves oblong-heart-shaped, somewhat wrinkled, and slightly toothed on the margin: the upper surface is very dark, inclining to a dingy purple; the under surface is of a much brighter color. The stems are deep-red and slightly furrowed; height three feet and upwards.

This is an earlier but a less vigorous sort than the White. The leaves of this variety, as also those of most of the coloured sorts, change to green in boiling.

Red-Stalked Green Orach

Leaves dark-green, tinged with dull-brown, much wrinkled, toothed, somewhat curled, terminating rather obtusely, and forming two acute angles at the base; the stalk and the stems of the leaves are deep-red, and slightly furrowed; the veins are very prominent. It is of tall growth.

Red-Stalked White Orach

(Purple-Bordered Green Orach)

Leaves somewhat heart-shaped, of a yellowish-green, tinged with brown. Their margin is stained with purple, and a little dentated or toothed in some cases, but not in all. The stalk and the stems of the leaves are of a palish-red, and are slightly furrowed, as well as streaked with pale-white between the furrows. The plant is of dwarfish growth.

White Orach

(Pale-Green Orache; White French Spinach; Yellow Orach)

Leaves pale-green or yellowish-green, much wrinkled, with long, tapering points, strongly cut in the form of teeth towards the base, which forms two acute angles; the stalk of the plant and the stems of the leaves are of the same color as the foliage. It is comparatively of low growth.

Triplex hortensis
Anselmus Boëtius de Boodt (1600 Approx.)

64

Patience (Rumex patientia)

(Herb Patience; Patience Dock; Garden Patience)

This plant is a native of the south of Europe. It is a hardy perennial, and, when fully grown, from four to five feet in height. The leaves are large, long, broad, pointed; the leaf-stems are red; the flowers are numerous, small, axillary, and of a whitish-green color,—they are put forth in June and July, and the seeds ripen in August. The latter are triangular, of a pale-brownish color, and will keep three years.

Soil and Cultivation—

Rumex patientia
Jacob Sturm

"The plant will grow well in almost any soil, but best in one that is rich and rather moist. It may easily be raised from seed sown in spring, in drills eighteen inches asunder; afterwards thinning out the young plants to a foot apart in the rows. It may also be sown broadcast in a seed-bed, and planted out; or the roots may be divided, and set at the above distances.

"The plants should not be allowed to run up to flower, but should be cut over several times in the course of the season, to induce them to throw out young leaves in succession, and to prevent seed from being ripened, and scattered about in all directions; for, when this takes place, the plant becomes a troublesome weed."[1]

It is perfectly hardy, and, if cut over regularly, will continue healthy and productive for several years. In the vicinity of gardens where it has been cultivated, it is frequently found growing spontaneously.

Use—

"The leaves were formerly much used as Spinach; and are still eaten in some parts of France, where they are also employed in the early part of the season as a substitute for Sorrel; being produced several days sooner than the leaves of that plant."[2]

Its present neglect may arise from a want of the knowledge of the proper method of using it. The leaves are put forth quite early in spring. They should be cut while they are young and tender, and about a fourth part of Common Sorrel mixed with them. In this way, Patience Dock is much used in Sweden, and may be recommended as forming an excellent spinach dish.

1,2. The Gardener's Assistant. By Robert Thompson.

65

Quinoa (Chenopodium quinoa)

White-seeded Quinoa

An annual plant from Mexico or Peru. Its stem is five or six feet in height, erect and branching; the leaves are triangular, obtusely toothed on the borders, pale-green, mealy while young, and comparatively smooth when old; flowers whitish, very small, produced in compact clusters; seeds small, yellowish-white, round, a little flattened, about a line in diameter, and, on a cursory glance, might be mistaken for those of millet; they retain their vegetative powers three years; about twelve thousand are contained in an ounce.

Sowing and Cultivation—It is propagated from seeds which are sown in April or May, in shallow drills three feet apart. As the seedlings increase in size, they are gradually thinned to a foot apart in the rows. The seeds ripen in September. In good soil, the plants grow vigorously, and produce seeds and foliage in great abundance.

Use—The leaves are used as Spinach or Sorrel, or as greens. In some places, the seeds are employed as a substitute for corn or wheat in the making of bread, and are also raised for feeding poultry.

Black-Seeded Quinoa

The stalks of this variety are more slender, and the leaves smaller, than those of the White-seeded. The plant is also stained with brownish-red in all its parts. Seeds small, grayish-black.

It is sown, and in all respects treated, like the White. The seeds and leaves are used in the same manner.

Red-Seeded Quinoa

(Chenopodium sp.)

This variety, or perhaps, more properly, species, is quite distinct from the White-seeded. It grows to the height of six or eight feet, and even more, with numerous long, spreading branches.

The leaves are more succulent than those of the last named, and are produced in greater abundance. When sown at the same time, it ripens its seeds nearly a month later.

Its foliage and seeds are used for the same purposes as the White. Sow in rows three feet apart, and thin to fifteen inches in the rows.

Quinoa plants near Cachora, Apurímac, Peru
Maurice Chédel

66

Sea-Beet (Beta maritima)

The Sea-beet is a hardy, perennial plant. The roots are not eaten; but the leaves, for which it is cultivated, are an excellent substitute for Spinach, and are even preferred by many to that delicate vegetable. If planted in good soil, it will continue to supply the table with leaves for many years. The readiest method of increasing the plants is by seeds; but they may be multiplied to a small extent by dividing the roots.

The early-produced leaves are the best, and these are fit for use from May until the plants begin to run to flower; but they may be continued in perfection through the whole summer and autumn by cutting off the flower-stems as they arise, and thus preventing the blossoming.

There are two varieties:

English Sea-Beet

The English Sea-beet is a dwarfish, spreading or trailing plant, with numerous angular, leafy branches. The lower leaves are ovate, three or four inches in length, dark-green, waved on the margin, and of thick, fleshy texture; the upper leaves are smaller, and nearly sessile.

Sow in April or May, in rows sixteen or eighteen inches apart, and an inch in depth; thin to twelve inches in the rows. The leaves should not be cut from seedling plants during the first season, or until the roots are well established.

Irish Sea-Beet

This differs from the preceding variety in the greater size of its leaves, which are also of a paler green: the stems are not so numerous, and it appears to be earlier in running to flower. The external differences are, however, trifling; but the flavor of this, when dressed, is far superior to that of the last named.

It requires the same treatment in cultivation as the English Sea-beet.

Beta Maritima
Flora Batava, Plate 233

67

Shephard's Purse (Capsella bursa-pastoris)

(Thlaspi bursa pastoris)

A hardy, annual plant, growing naturally and abundantly about gardens, roadsides, and in waste places. The root-leaves spread out from a common centre, are somewhat recumbent, pinnatifid-toothed, and, in good soil, attain a length of eight or ten inches; the stem-leaves are oval, arrow-shaped at the base, and rest closely upon the stalk. When in blossom, the plant is from twelve to fifteen inches in height; the flowers are small, white, and four-petaled; the seeds are small, of a reddish-brown color, and retain their vitality five years.

Propagation and Cultivation—It is easily raised from seed, which should be sown in May, where the plants are to remain. Sow in shallow drills twelve or fourteen inches apart, and cover with fine mould. Thin the young plants to four inches asunder, and treat the growing crop in the usual manner during the summer. Late in autumn, cover the bed with coarse stable-litter, and remove it the last of February. In March and April, the plants will be ready for the table.

Use—It is used in the manner of Spinach.

Capsella bursa-pastoris
Carl Axel Magnus Lindman

"When boiled, the taste approaches that of the Cabbage, but is softer and milder. The plant varies wonderfully in size, and succulence of leaves, according to the nature and state of the soil where it grows. Those from the gardens and highly cultivated spots near Philadelphia come to a remarkable size, and succulence of leaf. It may be easily bleached by the common method; and, in that state, would be a valuable addition to our list of delicate culinary vegetables."

In April and May it may be gathered, growing spontaneously about cultivated lands; and, though not so excellent as the cultivated plants, will yet be found of good quality.

68

Sorrel (Rumex. sp. et var.)

Sorrel is a hardy perennial. The species, as well as varieties, differ to a considerable extent in height and general habit; yet their uses and culture are nearly alike.

Soil and Cultivation—All of the sorts thrive best in rich, moist soil; but may be grown in almost any soil or situation. The seeds are sown in April or May, in drills fifteen or eighteen inches apart, and covered half an inch in depth. The young plants should be thinned to twelve inches apart; and, in July and August, the leaves will be sufficiently large for gathering.

The varieties are propagated by dividing the roots in April or May; and this method must be adopted in propagating the diœcious kinds, when male plants are required.

> "The best plants, however, are obtained from seed; but the varieties, when sown, are liable to return to their original type. All the care necessary is to hoe the ground between the rows, when needed to fork it over in spring and autumn, and to take up the plants, divide and reset them every three or four years, or less frequently, if they are growing vigorously and produce full-sized leaves."

All of the sorts, whether produced from seeds or by parting the roots, will send up a flower-stalk in summer; and this it is necessary to cut out when first developed, in order to render the leaves larger and more tender.

The plants will require no special protection or care during the winter; though a slight covering of strawy, stable litter may be applied after the forking-over of the bed in the autumn, just before the closing-up of the ground.

Use—It enters into most of the soups and sauces for which French cookery is so famed, and they preserve it in quantities for winter use. It forms as prominent an article in the markets of Paris as does Spinach in those of this country; and it has been asserted, that, amongst all the recent additions to our list of esculent plants, "we have not one so wholesome, so easy of cultivation, or one that would add so much to the sanitary condition of the community, particularly of that class who live much upon salt provisions."

The species and varieties are as follow:

Alpine Sorrel

(Oseille des Neiges; Rumex nivalis)

A new, perennial species, found upon the Alps, near the line of perpetual snow. The root-leaves are somewhat heart-shaped, thick, and fleshy; stem simple, with verticillate branches; flower diœcious.

It is one of the earliest as well as the hardiest of the species, propagates more readily than Alpine plants in general, and is said to compare favorably in quality with the Mountain Sorrel or Patience Dock.

Alpine Sorrel; Rumex nivalis
Anton Hartinger

Common Sorrel

This is a hardy perennial, and, when fully grown, is about two feet in height. The flowers—which are small, very numerous, and of a reddish color—are diœcious, the fertile and barren blossoms being produced on separate plants; the seeds are small, triangular, smooth, of a brownish color, and retain their germinative properties two years. An ounce contains nearly thirty thousand seeds.

Of the Common Sorrel, there are five varieties, as follow:

Belleville Sorrel

(Broad-leaved; Oseille Large de Belleville)

Leaves ten or twelve inches long by six inches in diameter; leaf-stems red at the base. Compared with the Common Garden Sorrel, the leaves are larger and less acid.

The variety is considered much superior to the last-named sort, and is the kind usually grown by market-gardeners in the vicinity of Paris.

It should be planted in rows eighteen inches apart, and the plants thinned to a foot apart in the rows.

Blistered-Leaf Sorrel

Radical leaves nine inches long, four inches wide, oval-hastate or halberd-shaped, growing on long footstalks. The upper leaves are more blistered than those attached to the root; the flower-stems are short. The principal difference between this variety and the Common, or Broad-leaved, consists in its blistered foliage.

It is slow in the development of its flower-stem, and consequently remains longer in season for use. The leaves are only slightly acid in comparison with those of the Common Sorrel. It is a perennial, and must be increased by a division of its roots; for being only a variety, and not permanently established, seedlings from it frequently return to the Belleville, from whence it sprung.

Fervent's New Large Sorrel

(Oseille de Fervent)

An excellent sort, with large, yellowish-green, blistered leaves and red leaf-stems. It is comparatively hardy, puts forth its leaves early, and produces abundantly.

The rows should be eighteen inches apart.

Green or Common Garden Sorrel

Root-leaves large, halberd-shaped, and supported on stems six inches in length. The upper leaves are small, narrow, sessile, and clasping. A hardy sort; but, on account of its greater acidity, not so highly esteemed as the Belleville.

Sow in rows fifteen inches apart, and thin to eight or ten inches in the rows.

Sarcelle Blond Sorrel

(Blond de Sarcelle)

This is a sub-variety of the Belleville, with longer and narrower leaves and paler leaf-stems. It puts forth its leaves earlier in the season than the Common Sorrel, and is of excellent quality. The seed rarely produces the variety in its purity, and it is generally propagated by dividing the roots.

Round-Leaved French Sorrel

(Roman Sorrel; Oseille rond; R. scutatus)

This is a hardy perennial, a native of France and Switzerland. Its stem is trailing, and from twelve to eighteen inches in height or length; the leaves vary in form, but are usually roundish-heart-shaped or halberd-shaped, smooth, glaucous, and entire on the borders; the flowers are hermaphrodite, yellowish; the leaves are more acid than those of the varieties of the preceding species, and for this reason are preferred by many.

The variety is hardy and productive, but not much cultivated.

It requires eighteen inches' space between the rows, and a foot in the rows. There is but one variety.

Rumex scutatus
Anton Hartinger

Mountain Sorrel.

(Oseille verge; R. montanus)

The leaves of this variety are large, oblong, of thin texture, and of a pale-green color; the root-leaves are numerous, about nine inches long and four inches wide, slightly blistered. It is later than the Common Garden Sorrel in running to flower; and is generally propagated by dividing the roots, but may also be raised from seeds. The leaves are remarkable for their acidity.

This is the *Rumex montanus* of modern botanists, though formerly considered as a variety of *R. acetosa*.

Blistered-Leaved Mountain Sorrel

This variety is distinguished from the Green Mountain Sorrel by its larger, more blistered, and thinner leaves. The leaf-stems are also longer, and, as well as the nerves and the under surface of the leaf, finely spotted with red. It starts early in spring, and is slow in running up to flower.

Green Mountain Sorrel.

This is an improved variety of the Mountain Sorrel, and preferable to any other, from the greater size and abundance of its leaves, which possess much acidity. It is also late in running to flower.

The leaves are large, numerous, ovate-sagittate, from ten to eleven inches long, and nearly five inches in width; the radical leaves are slightly blistered, and of a dark, shining green color. It can only be propagated by dividing the roots.

The plants require a space of eighteen inches between the rows, and a foot from plant to plant in the rows.

69

Spinach (Spinacia oleracea)

Spinach is a hardy annual, of Asiatic origin. When in flower, the plant is from two to three feet in height; the stem is erect, furrowed, hollow, and branching; the leaves are smooth, succulent, and oval-oblong or halberd-shaped,—the form varying in the different varieties. The fertile and barren flowers are produced on separate plants,—the former in groups, close to the stalk at every joint; the latter in long, terminal bunches, or clusters. The seeds vary in a remarkable degree in their form and general appearance; those of some of the kinds being round and smooth, while others are angular and prickly: they retain their vitality five years. An ounce contains nearly twenty-four hundred of the prickly seeds, and about twenty-seven hundred of the round or smooth.

Soil and Cultivation —Spinach is best developed, and most tender and succulent, when grown in rich soil. For the winter sorts, the soil can hardly be made too rich.

It is always raised from seeds, which are sown in drills twelve or fourteen inches apart, and three-fourths of an inch in depth. The seeds are sometimes sown broadcast; but the drill method is preferable, not only because the crop can be cultivated with greater facility, but the produce is more conveniently gathered. For a succession, a few seeds of the summer varieties may be sown, at intervals of a fortnight, from April till August.

Taking the Crop—

"When the leaves are two or three inches broad, they will be fit for gathering. This is done either by cutting them up with a knife wholly to the bottom, drawing and clearing them out by the root, or only cropping the large outer leaves; the root and heart remaining to shoot out again. Either method can be adopted, according to the season or other circumstances."[1]

To raise Seed—Spinach seeds abundantly; and a few of the fertile plants, with one or two of the infertile, will yield all that will be required for a garden of ordinary size. Seeds of the winter sorts should be saved from autumn sowings, and from plants that have survived the winter.

Use—The leaves and young stems are the only parts of the plant used. They are often boiled and served alone; and sometimes, with the addition of sorrel-leaves, are used in soups, and eaten with almost every description of meat.

> "The expressed juice is often employed by cooks and confectioners for giving a green color to made dishes. When eaten freely, it is mildly laxative, diuretic, and cooling. Of itself, it affords little nourishment. It should be boiled without the addition of water, beyond what hangs to the leaves in rinsing them; and, when cooked, the moisture which naturally comes from the leaves should be pressed out before being sent to the table. The young leaves were at one period used as a salad."[2]

Varieties—

Flanders Spinach

This is a winter Spinach, and is considered superior to the Prickly or Common Winter Spinach, which is in general cultivation during the winter season in our gardens. It is equally hardy, perhaps hardier.

The leaves are doubly hastate or halberd shaped, and somewhat wrinkled: the lower ones measure from twelve to fourteen inches in length, and from six to eight in breadth. They are not only larger, but thicker and more succulent, than those of the Prickly Spinach. The whole plant grows more bushy, and produces a greater number of leaves from each root; and it is sometimes later in running to seed. The seeds are like those of the Round or Summer Spinach, but larger: they are destitute of the prickles which distinguish the seeds of the Common Winter Spinach.

For winter use, sow at the time directed for sowing the Large Prickly-seeded, but allow more space between the rows than for that variety; subsequent culture, and treatment during the winter, the same as the Prickly-seeded.

Large Prickly-Seeded Spinach

(Large Winter Spinach. Epinard d'Angleterre)

Leaves comparatively large, rounded at the ends, thick and succulent. In foliage and general character, it is similar to some of the round-seeded varieties; but is much hardier, and slower in running to seed. It is commonly known as "Winter Spinach," and principally cultivated for use during this portion of the year. The seeds are planted towards the last of August, in drills a foot apart, and nearly an inch in depth. When well up, the plants should be thinned to four or five inches apart in the drills; and, if the weather is favorable, they will be stocky and vigorous at the approach of severe weather. Before the closing-up of the ground, lay strips of joist or other like material between the rows, cover all over with clean straw, and keep the bed thus protected until the approach of spring or the crop has been gathered for use.

Lettuce-Leaved Spinach

(Epinard à Feuille de Laitue; Epinard Gaudry)

Leaves very large, on short stems, rounded, deep-green, with a bluish tinge, less erect than those of the other varieties, often blistered on the surface, and of thick substance. It is neither so early nor so hardy as some others; but it is slow in the development of its flower-stalk, and there are few kinds more productive or of better quality. The seeds are round and smooth. For a succession, a sowing should be made at intervals of two weeks.

"A variety called 'Gaudry,' if not identical, is very similar to this."

Sorrel-Leaved Spinach

Leaves of medium size, halberd-formed, deep-green, thick, and fleshy. A hardy and productive sort, similar to the Yellow or White Sorrel-leaved, but differing in the deeper color of its stalks and leaves.

Summer or Round-Leaved Spinach

(Round Dutch; Epinard de Hollande)

Leaves large, thick, and fleshy, rounded at the ends, and entire, or nearly entire, on the borders.

Spinacia oleracea
Otto Wilhelm

This variety is generally grown for summer use; but it soon runs to seed, particularly in warm and dry weather. Where a constant supply is required, a sowing should be made every fortnight, commencing as early in spring as the frost leaves the ground. The seeds are round and smooth. Plants from the first sowing will be ready for use the last of May or early in June.

In Belgium and Germany, a sub-variety is cultivated, with smaller and deeper-colored foliage, and which is slower in running to flower. It is not, however, considered preferable to the Common Summer or Round-leaved.

Winter or Common Prickly Spinach

(Epinard ordinaire)

Leaves seven or eight inches long, halberd-shaped, deep-green, thin in texture, and nearly erect on the stalk of the plant; seeds prickly.

From this variety most of the improved kinds of Prickly Spinach have been obtained; and the Common Winter or Prickly-seeded is now considered scarcely worthy of cultivation.

Yellow Sorrel-Leaved Spinach

(White Sorrel-Leaved Spinach; Blond à Feuille d'Oseille)

The leaves of this variety are similar in form and appearance to those of the Garden Sorrel. They are of medium size, entire on the border, yellowish-white at the base, greener at the tips, and blistered on the surface.

New. Represented as being hardy, productive, slow in the development of its flower-stalk, and of good quality.

1. The Vegetable Cultivator. By John Rogers. London, 1851.

2. The Book of the Garden. By Charles M'Intosh. 2 vols. Edinburgh and London, 1855.

70

Wild/Perennial Spinach (Blitum bonus henricus)

(Good King Henry; Tota Bona; Goose-Foot)

A hardy perennial plant, indigenous to Great Britain, and naturalized to a very limited extent in this country. Its stem is two feet and a half in height; the leaves are arrow-shaped, smooth, deep-green, undulated on the borders, and mealy on their under surface; the flowers are numerous, small, greenish, and produced in compact groups, or clusters; the seeds are small, black, and kidney-shaped.

Propagation and Culture—

Chenopodium bonus henricus
Amédée Masclef

"It may be propagated by seed sown in April or May, and transplanted, when the plants are fit to handle, into a nursery-bed. In August or September, they should be again transplanted where they are to remain, setting them in rows a foot apart, and ten inches asunder in the rows, in ground of a loamy nature, trenched to the depth of fifteen or eighteen inches, as their roots penetrate to a considerable depth. The following spring, the leaves are fit to gather for use; and should be picked as they advance, taking the largest first. In this way, a bed will continue productive for several years.

"Being a hardy perennial, it may also be increased by dividing the plant into pieces, each having a portion of the root and a small bit of the crown, which is thickly set with buds, which spring freely on being replanted.

"Most of the species of this genus, both indigenous and exotic, are plants of easy cultivation, and may be safely used as articles of food."[1]

Use—The same as Spinach.

1. The Book of the Garden. By Charles M'Intosh. 2 vols. Edinburgh and London, 1855.

Part Seven: Salad Plants

Alexanders; Brook-lime; Buckshorn Plantain; Burnet; Caterpillar; Celery; Celeriac, or Turnip-rooted Celery; Chervil; Chiccory, or Succory; Corchorus; Corn Salad; Cress, or Peppergrass; Cuckoo Flower; Dandelion; Endive; Horse-radish; Lettuce; Madras Radish; Mallow, Curled-leaf; Mustard; Nasturtium; Garden Picridium; Purslain; Rape; Roquette, or Rocket; Samphire; Scurvy-grass; Snails; Sweet-scented Chervil, or Sweet Cicely; Tarragon; Valeriana; Water-cress; Winter-cress, or Yellow Rocket; Wood-sorrel; Worms.

Water-cress- Nasturtium officinale

71

Alexanders (Smyrnium olusatrum)

(Alisanders)

A hardy, biennial plant, with foliage somewhat resembling that of Celery. Stem three to four feet high, much branched; radical leaves pale-green, compound,—those of the stem similar in form, but of smaller size. The branches of the plant terminate in large umbels, or spherical bunches of yellowish flowers; which are succeeded by roundish fruits, each of which contains two crescent-shaped seeds.

Sowing and Culture—It thrives best in light, deep loam; and is raised from seed sown annually. Make the drills two and a half or three feet apart, and cover the seeds an inch deep. When the plants are two or three inches high, thin to twelve inches apart; or sow a few seeds in a nursery-bed, and transplant.

Blanching—When the plants are well advanced, they should be gradually earthed up about the stems in the process of cultivation, in the manner of blanching Celery or Cardoons; like which, they are also gathered for use, and preserved during winter.

To raise Seed—Leave a few plants unblanched; protect with stable-litter, or other convenient material, during winter; and they will flower, and produce an abundance of seeds, the following summer.

Use—It was formerly much cultivated for its leaf-stalks; which, after being blanched, were used as a pot-herb and for salad. They have a pleasant, aromatic taste and odor; but the plant is now rarely grown, Celery being almost universally preferred.

Perfoliate Alexanders

(Smyrnium perfoliatum)

A hardy, biennial species, from Italy; stem three feet in height, grooved or furrowed, hollow; leaves many times divided, and of a yellowish-green color; flowers, in terminal bunches, yellowish-white; seeds black, of the form of those of the common species, but smaller.

It is considered superior to the last named, as it not only blanches better, but is more crisp and tender, and not so harsh-flavored.

Alexanders (Smyrnium olusatrum): flowering stem, leaf and floral segments
J. Sowerby, 1795

72

Brook-Lime (Veronica beccabunga)

(American Brook-Lime; Marsh Speedwell)

Brook-lime is a native of this country, but is also common to Great Britain. It is a hardy perennial, and grows naturally in ditches, and streams of water, but is rarely cultivated. The stem is from ten to fifteen inches in height, thick, smooth, and succulent, and sends out roots at the joints, by which the plant spreads and is propagated; the leaves are opposite, oval, smooth, and fleshy; the flowers are produced in long bunches, are of a fine blue colour, and stand upon short stems,—they are more or less abundant during most of the summer, and are followed by heart-shaped seed-vessels, containing small, roundish seeds.

Cultivation—It may be propagated by dividing the roots, and setting the plants in wet localities, according to their natural habit. It will thrive well when grown with Water-cress.

Use—The whole plant is used as a salad, in the same manner and for the same purposes as Water-cress. It is considered an excellent antiscorbutic.

Veronica beccabunga
Jacob Sturm

73

Buckshorn Plantain (Plantago coronopus)

(Star of the Earth)

A hardy annual, indigenous to Great Britain, France, and other countries of Europe. The root-leaves are put forth horizontally, and spread regularly about a common centre somewhat in the form of a rosette; the flower-stem is leafless, branching, and from eight to ten inches high; flowers yellow; the seeds are quite small, of a clear, brown color, and retain their power of germination three years,—nearly two hundred and thirty thousand are contained in an ounce.

Soil and Cultivation—It succeeds best in a soil comparatively light; and the seed should be sown in April. Sow thinly, broadcast, or in shallow drills eight inches apart. When the plants are about an inch high, thin them to three or four inches apart.

Use—The plant is cultivated for its leaves, which are used as a salad. They should be plucked while still young and tender, or when about half grown.

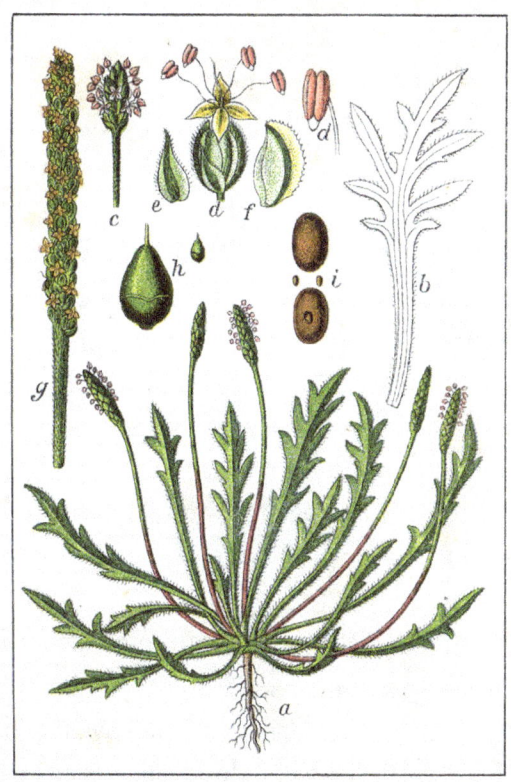

Plantago coronopus
Jacob Sturm

74

Burnet (Poterium sanguisorba)

Burnet is a hardy, perennial plant, indigenous to England, where it is found on dry, upland, chalky soils. When fully developed, it is from a foot and a half to two feet in height. The leaves proceeding directly from the root are produced on long stems, and are composed of from eleven to fifteen smaller leaves, which are of an oval form, regularly toothed, and generally, but not uniformly, smooth. The branches, which are somewhat numerous, terminate in long, slender stems, each of which produces an oval or roundish bunch of purplish-red, fertile and infertile flowers. The fertile flowers produce two seeds each, which ripen in August or September. These are oblong, four-sided, of a yellowish color, and retain their vitality two years. Thirty-five hundred are contained in an ounce.

Sowing and Culture—The plant is easily propagated by seeds, which may be sown either in autumn or spring. Sow in drills ten inches apart, half or three-fourths of an inch deep; and thin, while the plants are young, to six or eight inches in the row. If the seeds are allowed to scatter from the plants in autumn, young seedlings will come up plentifully in the following spring, and may be transplanted to the distances before directed. In dry soil, the plants will continue for many years; requiring no further care than to be occasionally hoed, and kept free from weeds. It may also be propagated by dividing the roots; but, as it is easily grown from seeds, this method is not generally practised.

Use—The leaves have a warm, piquant taste, and, when bruised, resemble cucumbers in odor. They are sometimes used as salad, and occasionally form an ingredient in soups. The roots, after being dried and pulverised, are employed in cases of internal haemorrhage.

It is very little used in this country, and rarely seen in gardens.

Varieties—There are three varieties; the distinctions, however, being neither permanent nor important.

Hairy-Leaved Burnet

Leaves and stems comparatively rough or hairy; in other respects, similar to the Smooth-leaved. Either of the varieties may be propagated by dividing the roots.

Large-Seeded Burnet

This, like the others, is a sub-variety, and probably but a seminal variation.

Smooth-Leaved Burnet

Leaves and stems of the plant comparatively smooth, but differing in no other particular from the Hairy-leaved. Seeds from this variety would probably produce plants answering to both descriptions.

Burnet- Sanguisorba minor
Otto Wilhelm Thomé

75

Caterpillar Plant (Scorpiurus)

(Chenille, of the French)

All of the species here described are hardy, annual plants, with creeping or recumbent stems, usually about two feet in length. The leaves are oblong, entire on the borders, broadest near the ends, and taper towards the stem; the flowers are yellow, and quite small; the seeds are produced in caterpillar-like pods, and retain their vitality five years.

Cultivation—The seeds may be planted in the open ground in April or May; or the plants may be started in a hot-bed, and set out after settled warm weather. The rows should be fifteen inches apart, and the plants twelve or fifteen inches apart in the rows; or the plants may be grown in hills two feet and a half apart, and two or three plants allowed to a hill.

Use—No part of the plant is eatable; but the pods, in their green state, are placed upon dishes of salads, where they so nearly resemble certain species of caterpillars as to completely deceive the uninitiated or inexperienced.

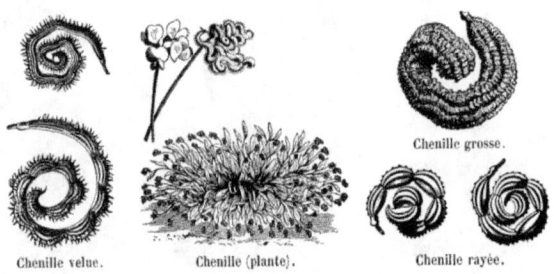

Scorpiurus chenilles: Used in salads as a joke
Vilmorin-Andrieux 1904

Species—The species cultivated are the following;

Common Caterpillar

(Chenille Grosse; Scorpiurus vermiculata)

Pod, or fruit, comparatively large. The interior grooves, or furrows, are indistinct, or quite wanting: the exterior grooves are ten in number, and well defined. Along the summit of these furrows are produced numerous, small, pedicelled tubercles, quite similar to those of some species of worms or caterpillars; and these small tufts, in connection with the brownish-green color and peculiar coiling of the pods, make the resemblance nearly perfect, especially if seen from a short distance. The seeds are large, oblong, flattened at the ends, and of a yellowish color. A well-developed fruit will measure about three-eighths of an inch in diameter; and, when uncoiled, nearly an inch and a half in length.

Furrowed Caterpillar

(Chenille Rayée; Scorpiurus sulcata)

Fruit rather slender, furrowed, grayish-green within the furrows, and brown along the summits. Four of the exterior furrows are surmounted with numerous small, obtuse, or rounded tubercles; and the pods are coiled in the manner peculiar to the class. The seeds resemble those of the Prickly Caterpillar, but are of larger size.

Prickly Caterpillar

(Small Caterpillar; Scorpiurus muricata)

Pod, or fruit, a fourth of an inch in diameter, curved or coiled; longitudinally furrowed, with numerous, small, erect, tufted points, regularly arranged along the surface. It is of a brownish-red color, with shades of green; and, when well grown, bears a remarkable resemblance to some species of hairy worms or caterpillars. The seeds are large, long, wrinkled, and of a yellowish color.

Villous or Hairy Caterpillar

(Chenille Velue; Scorpiurus subvillosa)

This species resembles the Prickly Caterpillar, but is a little larger. The most marked distinction, however, is in the small points, or tubercles, placed along the longitudinal ridges, which in this species are recurved, or bent at the tips. The seeds are larger than those of the foregoing species.

76

Celery (Apium graveolens)

(Smallage)

Celery, or Smallage, is a hardy, umbelliferous, biennial plant, growing naturally "by the sides of ditches and near the sea, where it rises with wedge-shaped leaves and a furrowed stalk, producing greenish flowers in August." Under cultivation, the leaves are pinnatifid, with triangular leaflets; the leaf-stems are large, rounded, grooved, succulent, and solid or hollow according to the variety. The plant flowers during the second year, and then measures from two to three feet in height; the flowers are small, yellowish-white, and are produced in umbels, or flat, spreading groups, at the extremities of the branches; the seeds are small, somewhat triangular, of a yellowish-brown colour, aromatic when bruised, and of a warm, pleasant flavour. They are said to retain their germinative powers ten years; but, by seedsmen, are not considered reliable when more than five years old. An ounce contains nearly seventy thousand seeds.

Soil—Any good garden soil, in a fair state of cultivation, is adapted to the growth of Celery.

Propagation—It is always propagated by seed; one-fourth of an ounce of which is sufficient for a seed-bed five feet wide and ten feet long. The first sowing is usually made in a hot-bed in March: and it may be sown in the open ground in April or May; but, when so treated, vegetates slowly, often remaining in the earth several weeks before it comes up. "A bushel or two of stable manure, put in a hole in the ground against a wall or any fence facing the south, and covered with a rich, fine mould three or four inches deep, will bring the seed up in two weeks." If this method is practised, sprinkle the seed thinly over the surface of the loam, stir the soil to the depth of half an inch, and press the earth flat and smooth with the back of a spade. Sufficient plants for any family may be started in a large flower-pot or two, placed in the sitting-room, giving them plenty of light and moisture.

Cultivation—As soon as the young plants are about three inches high, prepare a small bed in the open air, and make the ground rich and the earth fine. Here set out the plants for a temporary growth, placing them four inches apart. This should be done carefully; and they should be gently watered once, and protected for a day or two against the sun.

"A bed ten feet long and four feet wide will contain three hundred and sixty plants; and, if they be well cultivated, will more than supply the table of a common-sized family from October to May."

"In this bed the plants should remain till the beginning or middle of July, when they should be removed into trenches. Make the trenches a foot or fifteen inches deep and a foot wide, and not less than five feet apart. Lay the earth taken out of the trenches into the middle of the space between the trenches, so that it may not be washed into them by heavy rains; for it will, in such case, materially injure the crop by covering the hearts of the plants. At the bottom of the trench put some good, rich, but well-digested compost manure; for, if too fresh, the Celery will be rank and pipy, or hollow, and will not keep nearly so long or so well. Dig this manure in, and make the earth fine and light; then take up the plants from the temporary bed, and set them out carefully in the bottom of the trenches, six or eight inches apart."[1]

It is the practice of some cultivators, at the time of setting in the trenches, to remove all the suckers, to shorten the long roots, and to cut the leaves off, so that the whole plant shall be about six inches in length. But the best growers in England have abandoned this method, and now set the plants, roots and tops, entire.

Blanching—

"When the plants begin to grow (which they will quickly do), hoe on each side and between them with a small hoe. As they grow up, earth their stems; that is, put the earth up against them, but not too much at a time, and always when the plants are dry; and let the earth put up be finely broken, and not at all cloddy. While this is being done, keep the stalks of the outside leaves close up, to prevent the earth getting between the stems of the outside leaves and inner ones; for, if it gets there, it checks the plant, and makes the Celery bad. When the earthing is commenced, take first the edges of the trenches, working backwards, time after time, till the earth is reached that was taken from the trenches; and, by this time, the earth against the plants will be above the level of the land. Then take the earth out of the middle, till at last the earth against the plants forms a ridge; and the middle of each interval, a sort of gutter. Earth up very often, not putting up much at a time, every week a little; and by the last of September, or beginning of October, it will be blanched sufficient for use."[2]

Another (more recent) method of cultivation and blanching is to take the plants from the temporary bed, remove the suckers, and set them with the roots entire, ten inches apart in the trenches. They are then allowed to grow until they have attained nearly their full size, when the earth for blanching is more rapidly applied than in the previous method.

> "Many plant on the surface,—that is, marking out the size of the bed on ground that has been previously trenched; digging in at least six or eight inches of rich, half-decayed manure, and planting either in single lines four feet apart, or making beds six feet broad, and planting across them, setting the rows fourteen inches apart, and the plants eight inches apart in the lines. They may be earthed up as they advance, or not, until they have attained the height of a foot."[3]

M'Intosh[4] gives the following method, practised by the Edinburgh market-gardeners:

> "Trenches, six feet wide and one foot deep, are dug out; the bottom is loosened and well enriched, and the plants set in rows across the bed, fourteen inches asunder, and the plants nine inches apart in the rows. By this means, space is economized, and the plants attain a fair average size and quality. The same plan is very often followed in private gardens; and, where the new and improved sorts are grown, they arrive at the size most available for family use. This is one of the best methods for amateurs to grow this crop. They should grow their plants in the temporary or nursery beds until they are ten inches or a foot high, before planting in the trenches; giving plenty of water, and afterwards earthing up once a fortnight."

Some allow the plants to make a natural growth, and earth up at once, about three weeks before being required for use. When so treated, the stalks are of remarkable whiteness, crisp, tender, and less liable to russet-brown spots than when the plants are blanched by the more common method.

Taking the Crop—Before the closing-up of the ground, the principal part of the crop should be carefully taken up (retaining the roots and soil naturally adhering), and removed to the cellar; where they should be packed in moderately moist earth or sand, without covering the ends of the leaves.

A portion may be allowed to remain in the open ground; but the hearts of the plants must be protected from wet weather. This may be done by placing boards lengthwise, in the form of a roof, over the ridges. As soon as the frost leaves the ground in spring, or at any time during the winter when the weather will admit, Celery may be taken for use directly from the garden.

Seed—Two or three plants will produce an abundance. They should be grown two feet apart, and may remain in the open ground during the winter. The seeds ripen in August.

Use—The stems of the leaves are the parts of the plant used. These, after being blanched, are exceedingly crisp and tender, with an agreeable and peculiarly aromatic flavor. They are sometimes employed in soups; but are more generally served crude, with the addition of oil, mustard, and vinegar, or with salt only. The seeds have the taste and odor of the stems of the leaves, and are often used in their stead for flavoring soups.

With perhaps the exception of Lettuce, Celery is more generally used in this country than any other salad plant. It succeeds well throughout the Northern and Middle States; and, in the vicinity of some of our large cities, is produced of remarkable size and excellence.

Varieties—

Boston-Market Celery

A medium-sized, white variety; hardy, crisp, succulent, and mild flavored. Compared with the White Solid, the stalks are more numerous, shorter, not so thick, and much finer in texture. It blanches quickly, and is recommended for its hardiness and crispness; the stalks rarely becoming stringy or fibrous, even at an advanced stage of growth. Much grown by market-gardeners in the vicinity of Boston, Mass.

Celery Varieties
Allen's catalogue for 1906

Cole's Superb Red

This is comparatively a new sort, of much excellence, and of remarkable solidity. It is not of large size, but well adapted for cultivation in the kitchen-garden and for family use; not so well suited for marketing or for exhibition purposes. It has the valuable property of not piping or becoming hollow or stringy, and remains long without running to seed. The leaf-stalks are of a fine purple color, tender, crisp, and fine flavored. A well-grown plant will weigh about six pounds.

Cole's Superb White

Much like Cole's Superb Red; differing little, except in color. An excellent sort, hardy, runs late to seed, and is one of the most crisp and tender of the white sorts. Stalks short and thick.

Dwarf Curled White

(Céleri Nain Frisé)

Leaves dark-green, curled, resembling those of Parsley, and, like it, might be employed for garnishing. Leaf-stalks rounded and grooved, comparatively crisp and solid, but not fine flavored. It is quite hardy, and, in moderate winters, will remain in the open ground without injury, and serve for soups in spring. Its fine, curled foliage, however, is its greatest recommendation.

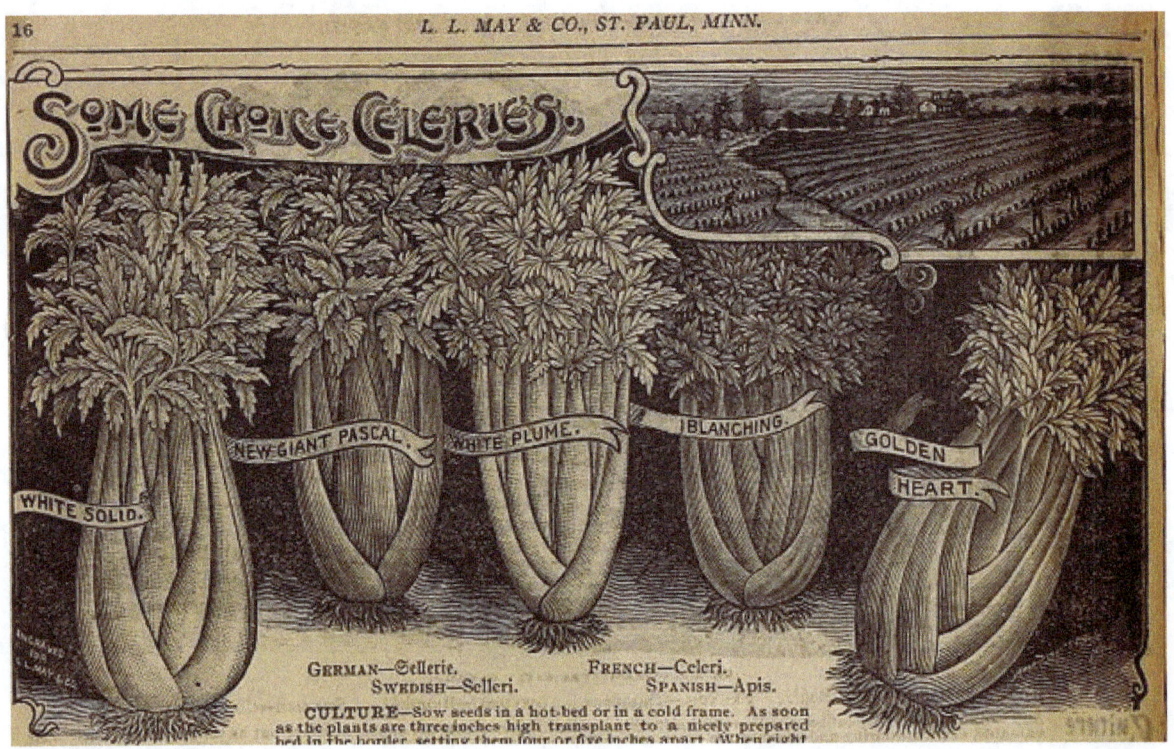

Celery Varieties
Farm and floral guide, 1899

Early Dwarf Solid White

(Céleri plein, blanc, court, hâtif)

Rather dwarf, but thick-stemmed. The heart is remarkably full; the leaf-stalk solid, blanching promptly. There is, in fact, much more finely blanched substance in a plant of this variety than in one of the tall sorts, and the quality is excellent. It comes into use rather early, and is one of the hardiest of the White varieties.

Italian Celery

(Large Upright; Giant Patagonian)

A tall, strong-growing, erect sort; leaf-stems deeply furrowed, sometimes a little hollow; leaves large, deep-green, with coarse, obtuse serratures. It is not so crisp as the Common White Solid; and is suitable only for soups, or where very tall Celery is desirable.

Laing's Improved Mammoth Red Celery

This is considered the largest variety yet produced; specimens having attained, in England, the extraordinary weight of eight or ten pounds, and at the same time perfectly solid. It is nearly perennial in its habit, as it will not run to seed the first year; and is with difficulty started to blossom even during the second, when planted out for the purpose. Colour bright-red; flavour unsurpassed, if equalled.

Manchester Red Celery

(Manchester Red Giant)

This variety scarcely differs from the Red Solid. It has, however, a coarser habit, with a somewhat rounder stalk; and, this being the case, the heart is not so compact. It is grown largely for marketing, and is excellent for soups and stewing.

Nutt's Champion White Celery

Originated with Mr. Nutt, of Sheffield.

It attains, under good management, in good soil, a large size, and, this considered, is of excellent quality; very white, and not apt to run to seed.

Red Solid

(New Large Red; New Large Purple; Tours Purple; Céleri Violet de Tours)

The plant grows to a large size, full-hearted, with a thick stem. Leaf-stalks thick, deeply furrowed, and very solid, of a dark-red or purplish hue where exposed, rose-colored where partially blanched; but the perfectly blanched portion is pure white, more so than the blanched part of the White varieties of Celery. It is also crisp, of excellent flavor, and unquestionably the best variety of Red Celery.

Pink Plume Celery
Peter Henderson & Co. Manual of Everything for the Garden, 1894

Seymour's Superb White

(Seymour's Superb White Solid)

A large-sized, vigorous-growing variety; in good soils, often attaining a height of nearly three feet. The stalks are solid; flat at the base, where they overlap, and form a compact, crisp, and, with ordinary care, a well-blanched heart of excellent quality. It succeeds best, as most other sorts do, in rich, moist soil; and when so grown, and properly blanched, will yield a large proportion of Celery, of a pure white colour, and of the best quality.

It is one of the best sorts for extensive culture for the markets, as it is also one of the best varieties for small gardens for family use. It blanches readily; and, with little care, will supply the table, from the last of September, through most of the winter.

Seymour's White Champion

A variety represented as being superior to Seymour's Superb White. The stalks are broad, flat at the base, and form a compact, well-blanched, crisp heart.

Shepherd's Red

(Shepherd's Giant Red)

Much like the Manchester Red, but has flatter stems: consequently, it is more compact, and blanches sooner and more perfectly, than that variety; to which, for these reasons, it is preferred by growers for competition.

Small Dutch Celery

(Céleri à Couper)

Leaf-stems small, hollow, crisp, and succulent; sprouts, or suckers, abundant. It is seldom blanched; but the leaves are sometimes used for flavoring soups.

The seeds should be sown thickly, and on level beds. The plants often resprout after being cut. Not much cultivated.

Sutton's White Solid

A very large yet solid-growing variety, exceedingly white and crisp.

Turkey or Prussian Celery

(Giant White; Céleri Turc; Turkish Giant Solid)

A remarkably large variety, resembling the Common White Solid. Leaf-stalks long, large, erect, fleshy, and solid; leaves large, with rounded serratures, and of a glossy-green color.

It is one of the largest of the White sorts, and is considered superior to the Common White Solid.

Wall's White Celery

An improved variety of the Italian, esteemed by growers for competition, where quantity, not quality, is the principal consideration.

White Lion's-Paw Celery

(Lion's-Paw)

A short, broad, flat-stalked variety, of excellent quality; crisp and white. Its short, flat, spreading habit gave rise to its name.

White Solid

(Céleri Plein, Blanc; Fine White Solid)

This variety is of strong and rather tall growth; leaf-stalks generally solid, but when grown in rich, highly manured soil, they sometimes become slightly hollow; leaves large, smooth, bright-green; serratures large and obtuse. It blanches readily, is crisp, of excellent quality, and comes into use earlier than the Red sorts. It is generally cultivated in the Northern States, not only on account of its hardiness, but for its keeping qualities. As a market variety, it is one of the best.

1,2. The American Gardener. By William Corbett. Concord, Boston, and New York, 1842.

3,4. The Book of the Garden. By Charles M'Intosh. 2 vols. Edinburgh and London, 1855.

77

Celeriac/Turnip-Rooted Celery (Apium graveolens var. rapaceum)

This variety forms at the base of the leaves, near the surface of the ground, a brownish, irregular, rounded root, or tuber, measuring from three to four inches in diameter. The leaves are small, with slender, hollow stems. In favorable exposures and rich soil, the roots sometimes attain a weight of more than three pounds. It is much hardier than the common varieties of Celery.

Propagation—It is propagated from seeds, which may be sown in the open ground in April or May, in shallow drills six or eight inches apart.

> "When the young plants are three inches high, they should be removed, and set on the surface (not in trenches), in moderately enriched soil. They should be set in rows eighteen inches apart, and a foot from each other in the line. At the time of transplanting, all of the small suckers, or side-shoots, should be rubbed off,—a precaution to be kept in view throughout its growth,—as the energies of the whole plant ought to be directed to the formation of the bulb-like root."[1]

Celeriac Root
Adolphe Millot

Subsequent Cultivation—The growing crop will require no peculiar treatment. When the bulbs are two-thirds grown, they are earthed over for the purpose of blanching, and to render the flesh crisp and tender. Cool and humid seasons are the most favorable to their growth. In warm and dry weather, the bulbs are small, comparatively tough, and strong flavored.

Taking the Crop—Some of the bulbs will be ready for use in September; from which time, till the last of November, the table may be supplied directly from the garden. Before severe weather, the quantity required for winter should be drawn, packed in damp earth or sand, and stored in the cellar.

To save Seed—Give to a few plants, taken up in the autumn, as much light and air as possible during the winter, keeping them cool, but not allowing them to freeze; and, in April, set them in the open ground, eighteen inches apart. The seed will ripen the last of the season. It is often used in the manner of the seed of the Common Celery for seasoning soups.

Use—The root, or bulb, is the part of the plant eaten: the flesh of this is white, and comparatively tender, with the flavor of the stalks of Common Celery, though generally less mild and delicate. It is principally valued for its remarkable hardiness and for its keeping properties. Where the common varieties of Celery are grown or preserved with difficulty, this might be successfully grown, and afford a tolerable substitute. The bulbs are sometimes eaten boiled, and the leaves are occasionally used in soups.

Curled-Leaved Celeriac

(Curled-Leaved; Turnip-Rooted; Céleri-Rave Frisé)

This is a variety of the Common Celeriac, or Turnip-rooted Celery; like which, it forms a sort of bulb, or knob, near the surface of the ground. It is, however, of smaller size; usually measuring about three inches in diameter. The skin is brown, and the flesh white and fine-grained; leaves small, spreading, curled.

It is in no respect superior to the Common Turnip-rooted, and possesses little merit aside from the peculiarity of its foliage. Cultivate, preserve during winter, and use as directed for the common variety.

Early Erfurt Celeriac

(Céleri-Rave d'Erfurt)

A very early variety. Root, or bulb, not large, but regular in form. Its earliness is its principal merit.

1. The Book of the Garden. By Charles M'Intosh. 2 vols. Edinburgh and London, 1855.

78

Chervil (Anthriscus cerefolium)

Common or Plain-Leaved

A hardy, annual plant, from the south of Europe. Stem eighteen inches to two feet in height; the leaves are many times divided, and are similar to those of the Common Plain Parsley; the flowers are small, white, and produced in umbels at the extremities of the branches; the seeds are black, long, pointed, longitudinally grooved, and retain their vitality but two years,—nearly nine thousand are contained in an ounce.

"This is the most common sort; but, except that it is hardier than the Curled varieties, is not worthy of cultivation."

Curled Chervil

A variety of the Common Chervil, with frilled or curled leaves; the distinction between the sorts being nearly the same as that between the Plain-leaved and Curled-leaved varieties of Parsley. The foliage is delicately and beautifully frilled; and, on this account, is much employed for garnishing, as well as for the ordinary purposes for which the plain sort is used.

Being a larger grower, it requires more room for its development; and the plants should stand a foot apart each way. When intended for winter use, it should have the protection of hand-glasses, frames, or branches of trees placed thickly around or amongst it. In very unfavorable situations, it is well to pot a dozen or two plants, and shelter them under glass during the winter.

Frizzled-Leaved or French Chervil

(Double-curled. Cerfeuil frisé)

An improved variety of the Curled Chervil,—even more beautiful; but wanting in hardiness. It succeeds best when grown in the summer months.

Propagation and Cultivation—Chervil is raised from seeds; and, where it is much used, sowings should be made, at intervals of three or four weeks, from April till July. The seeds should be sown thinly, in drills a foot apart, and covered nearly an inch in depth.

Use—It is cultivated for its leaves, which have a pleasant, aromatic taste; and, while young and tender, are employed for flavoring soups and salads.

Chervil- Anthriscus cerefolium
Amédée Masclef

79

Chicory (Cichorium intybus)

(Succory; Wild Endive)

A hardy, perennial plant, introduced into this country from Europe, and often abounding as a troublesome weed in pastures, lawns, and mowing-lands. The stem is erect, stout, and branching, and, in its native state, usually about three feet in height,—under cultivation, however, it sometimes attains a height of five or six feet; the radical leaves are deep-green, lobed, and, when grown in good soil, measure ten or twelve inches in length, and four inches in width; the flowers are large, axillary, nearly stemless, of a fine blue color, and generally produced in pairs; the seeds somewhat resemble those of Endive, though ordinarily smaller, more glossy, and of a deeper-brown color,—they will keep ten years. The plants continue in blossom from July to September; and the seeds ripen from August to October, or until the plants are destroyed by frost.

Soil, Sowing, and Cultivation—As the roots of Chicory are long and tapering, it should be cultivated in rich, mellow soil, thoroughly stirred, either by the plough or spade, to the depth of ten or twelve inches. The seed should be sown in April or May, in drills fifteen inches apart, and three-fourths of an inch deep. When the young plants are two or three inches high, thin them to eight inches apart in the rows; and, during the summer, cultivate frequently, to keep the soil light, and the growing crop free from weeds.

Blanching—Before using as a salad, the plants are blanched, either by covering with boxes a foot in depth, or by strips of boards twelve or fourteen inches wide, nailed together at right angles, and placed lengthwise over the rows. They are sometimes blanched by covering with earth; the leaves being first gathered together, and tied loosely at the top, which should be left exposed to light.

To save Seed—In the autumn, leave a few of the best plants unblanched; let them be about eighteen inches asunder. Protect with stable litter; or, if in a sheltered situation, leave them unprotected during winter, and they will yield abundantly the ensuing summer.

Taking the Crop—When the leaves are properly blanched, they will be of a delicate, creamy white. When they are about a foot high, they will be ready for use; and, as soon as they are cut, the roots should be removed, and others brought forward to succeed them.

"In cutting, take off the leaves with a thin slice of the crown, to keep them together, as in cutting sea-kale. When washed, and tied up in small bundles of a handful each, they are fit for dressing."[1]

Use—It is used as Endive; its flavor and properties being much the same. Though rarely grown in this country, it is common to the gardens of many parts of Europe, and is much esteemed. The blanched leaves are known as *Barbe de Capucin*, or "Friar's Beard."

Varieties—

Improved Chicory, or Succory
(Chicorée Sauvage Améliorée)

Leaves larger than those of the Common Chicory, and produced more compactly; forming a sort of head, or solid heart, like some of the Endives. The plant is sometimes boiled and served in the manner of Spinach.

Variegated or Spotted Chicory

This is a variety of the preceding, distinguished by the color of the leaves, which are veined, and streaked with red. In blanching, the red is not changed, but retains its brilliancy; while the green becomes nearly pure white,—the two colors blending in rich contrast. In this state they form a beautiful, as well as tender and well-flavored, salad.

Improved Variegated Chicory

A sub-variety of the Spotted Chicory, more constant in its character, and more uniform and distinct in its stripes and variegations. When blanched, it makes an exceedingly delicate and beautiful garnish, and a tender and excellent salad.

Either of the improved sorts are as hardy, and blanch as readily, as the Common Chicory.

Large-Rooted or Coffee Chicory
(Turnip-Rooted Chicory)

This variety is distinguished by its long, fleshy roots, which are sometimes fusiform, but generally much branched or divided: when well grown, they are twelve or fourteen inches in length, and about an inch in their largest diameter. The leaves have the form of those of the Common Chicory, but are larger, and more luxuriant.

Though the variety is generally cultivated for its roots, the leaves, when blanched, afford a salad even superior to some of the improved sorts before described.

Vilmorin[2] mentions two sub-varieties of the Large-rooted or Coffee Chicory:

Brunswick Large-Rooted.

Roots shorter than those of the Magdebourg, but of greater diameter; leaves spreading.

Magdebourg Large-Rooted.

Roots long, and comparatively large; leaves erect.

After several years' trial, preference was given to this variety, which proved the more productive.

Sowing and Cultivation—For raising Coffee Chicory, the ground should first be well enriched, and then deeply and thoroughly stirred by spading or ploughing. The seeds should be sown in April or May, in shallow drills a foot apart, and the young plants thinned to three or four inches apart in the rows. Hoe frequently; water, if the weather is dry; and in the autumn, when the roots have attained sufficient size, draw them for use. After being properly cleaned, cut them into small pieces, dry them thoroughly in a kiln or spent oven, and store for use or the market. After being roasted and ground, Chicory is mixed with coffee in various proportions, and thus forms a pleasant beverage; or, if used alone, will be found a tolerable substitute for genuine coffee.

The roots of any of the before-described varieties may be used in the same manner; but as they are much smaller, and consequently less productive, are seldom cultivated for the purpose.

It is an article of considerable commercial importance; large quantities being annually imported from the south of Europe to different seaports of the United States. As the plant is perfectly hardy, of easy culture, and quite productive, there appears to be no reason why the home demand for the article may not be supplied by home production. Of its perfect adaptedness to the soil and climate of almost any section of this country, there can scarcely be a doubt.

Cichorium intybus

1. The Book of the Garden. By Charles M'Intosh. 2 vols. Edinburgh and London, 1855.

2. Description des Plantes Potagères. Par Vilmorin, Andrieux, et Cie. Paris, 1856.

80

Corchorus (Corchorus olitorius)

(Corette Potagère, of the French)

An annual plant from Africa; also indigenous to the West Indies. Stem about two feet high, much branched; leaves deep-green, slightly toothed, varying in a remarkable degree in their size and form,—some being spear-shaped, others oval, and some nearly heart-shaped; leaf-stems long and slender; flowers nearly sessile, small, yellow, five-petaled; seeds angular, pointed, and of a greenish color,—fourteen thousand are contained in an ounce, and they retain their vitality four years.

Soil, Propagation, and Culture—The plant requires a light, warm soil; and should have a sheltered, sunny place in the garden. It is grown from seed sown annually. The sowing may be made in March in a hot-bed, and the plants set in the open ground in May; or the seed may be sown the last of April, or first of May, in the place where the plants are to remain. The drills, or rows, should be fifteen inches apart, and the plants five or six inches apart in the rows. No further attention will be required, except the ordinary labor of keeping the soil loose and the plants clear from weeds.

Corchurus olitorius
William Jackson Hooker

Use—The leaves are eaten as a salad, and are also boiled and served at table in the form of greens or spinach. They may be cut as soon as they have reached a height of five or six inches.

81

Corn-Salad (Valeriana locusta)

(Fetticus; Lamb's Lettuce; Mâche, of the French)

This is a small, hardy, annual plant, said to derive its name from its spontaneous growth, in fields of wheat, in England. It is also indigenous to France and the south of Europe.

When in flower, or fully grown, it is from twelve to fifteen inches in height. The flowers are small, pale-blue; the seeds are rather small, of a yellowish-brown color, unequally divided by two shallow, lengthwise grooves, and will keep six or eight years.

Soil and Culture—It is always grown from seed, and flourishes best in good vegetable loam, but will grow in any tolerably enriched garden soil. Early in April, prepare a bed four feet wide, and of a length according to the quantity of salad required; having regard to the fact, that it is better to sow only a small quantity at a time. Rake the surface of the bed even, make the rows across the bed about eight inches apart, sow the seed rather thinly, and cover about one-fourth of an inch deep with fine, moist soil. If dry weather occurs after sowing, give the bed a

Valeriana locusta
K. Ziarnek

good supply of water. When the young plants are two inches high, thin them to four inches apart, and cut or draw for use as soon as the leaves have attained a suitable size.

As the peculiar value of Corn Salad lies in its remarkable hardiness, a sowing should be made the last of August or beginning of September, for use during the winter or early in spring; but, if the weather is severe, the plants must be protected by straw or some other convenient material. Early in March, or as soon as the weather becomes a little mild, remove the covering, and the plants will keep the table supplied until the leaves from fresh sowings shall be grown sufficiently for cutting.

Seed—To raise seed, allow a few plants from the spring[Pg 340] sowing to remain without cutting. They will grow up to the height and in the manner before described, and blossom, and ripen their seed during the summer. An ounce of seed will sow a row two hundred feet in length, and about five pounds will be required for an acre.

Use—The leaves, while young, are used as a salad; and in winter, or early in spring, are considered excellent. They are also sometimes boiled and served as Spinach.

Varieties—

Common Corn Salad

Root-leaves rounded at the ends, smooth, three or four inches long by about an inch in width. The younger the plants are when used, the more agreeable will be their flavor.

Large Round-Leaved

Leaves larger, of a deeper green, thicker, and more succulent, than those of the foregoing variety. It is the best sort for cultivation. The leaves are most tender, and should be cut for use while young and small.

Large-Seeded Round

This is a sub-variety of the Large Round, and is much cultivated in Germany and Holland. The leaves are longer, narrower, and thinner, and more tender when eaten; but the Large Round is preferred by gardeners for marketing, as it bears transportation better. The seeds are about twice as large.

Italian Corn Salad

(Valerianella eriocarpa)

The Italian Corn Salad is a distinct species, and differs from the Common Corn Salad in its foliage, and, to some extent, in its general habit. It is a hardy annual, about eighteen inches high. The radical leaves are pale-green, large, thick, and fleshy,—those of the stalk long, narrow, and pointed; the flowers are small, pale-blue, washed or stained with red; the seeds are of a light-brown color, somewhat compressed, convex on one side, hollowed on the opposite, and retain their vitality five years,—nearly twenty-two thousand are contained in an ounce.

It is cultivated and used in the same manner as the species before described. It is, however, earlier, milder in flavor, and slower in running to seed. The leaves are sometimes employed early in spring as a substitute for Spinach; but their downy or hairy character renders them less valuable for salad purposes than those of some of the varieties of the Common Corn Salad.

82

Cress/Peppergrass (Lepidium sativum)

The Common Cress of the garden is a hardy annual, and a native of Persia. When in flower, the stem of the plant is smooth and branching, and about fifteen inches high. The leaves are variously divided, and are plain or curled, according to the variety; the flowers are white, very small, and produced in groups, or bunches; seeds small, oblong, rounded, of a reddish-brown color, and of a peculiar, pungent odor,—about fourteen thousand are contained in an ounce, and they retain their germinative properties five years.

Newly Sprouted Cress

Soil and Cultivation—Cress will flourish in any fair garden soil, and is always best when grown early or late in the season. The seed vegetates quickly, and the plants grow rapidly. As they are milder and more tender while young, the seed should be sown in succession, at intervals of about a fortnight; making the first sowing early in April. Rake the surface of the ground fine and smooth, and sow the seed rather thickly, in shallow drills six or eight inches apart. Half an ounce of seed will be sufficient for thirty feet of drill.

To raise Seed—Leave a dozen strong plants of the first sowing uncut. They will ripen their seed in August, and yield a quantity sufficient for the supply of a garden of ordinary size.

Use—The leaves, while young, have a warm, pungent taste; and are eaten as a salad, either separately, or mixed with lettuce or other salad plants. The leaves should be cut or plucked before the plant has run to flower, as they then become acrid and unpalatable. The curled varieties are also used for garnishing.

Broad-Leaved Cress

A coarse variety, with broad, spatulate leaves. It is sometimes grown for feeding poultry, and is also used for soups; but it is less desirable as a salad than most of the other sorts.

Common or Plain-Leaved Cress

This is the variety most generally cultivated. It has plain leaves, and consequently is not so desirable a sort for garnishing. As a salad kind, it is tender and delicate, and considered equal, if not superior, to the Curled varieties.

Curled Cress

(Garnishing Cress)

Leaves larger than those of the common plain variety, of a fine green color, and frilled and curled on the borders in the manner of some kinds of Parsley. It is used as a salad, and is also employed as a garnish.[Pg 343] It is very liable to degenerate by becoming gradually less curled. To keep the variety pure, select only the finest curled plants for seed.

Golden Cress

This variety is of slower growth than the Common Cress. The leaves are of a yellowish-green, flat, oblong, scalloped on the borders, sometimes entire, and of a much thinner texture than any of the varieties of the Common Cress. It is very dwarf; and is consequently short, when cut as a salad-herb for use. It has a mild and delicate flavor. When run to flower, it does not exceed eighteen inches in height.

It deserves more general cultivation, as affording a pleasant addition to the varieties of small salads.

The seeds are of a paler color, or more yellow, than those of the other sorts.

Normandy Curled Cress

A very excellent variety, introduced by Mr. Charles M'Intosh, and described as being hardier than the other kinds, and therefore better adapted for sowing early in spring or late in summer.

The leaves are finely cut and curled, and make not only a good salad, but a beautiful garnish. The seed should be sown thinly, in good soil, in drills six inches apart. In gathering, instead of cutting the plants over, the leaves should be picked off singly. After this operation, fresh leaves are soon put forth.

It is difficult to procure the seed true; the Common Curled being, in general, substituted for it.

83

Cuckoo Flower (Cardamine pratensis)

(Small Water-Cress)

A hardy, perennial plant, introduced from Europe, and naturalized to a limited extent in some of the Northern States. Stem about fifteen inches high, erect, smooth; leaves deeply divided,—the divisions of the radical or root leaves rounded, those of the stalk long, narrow, and pointed; the flowers are comparatively large, white, or rose-colored, and produced in erect, terminal clusters; the seeds are of a brown color, small, oblong, shortened on one side, rounded on the opposite, and retain their vegetating powers four years,—nearly thirty thousand are contained in an ounce.

Soil—It succeeds best in moist, loamy soil; and should have a shady situation.

Propagation and Cultivation—It may be propagated from seeds, or by a division of the roots. The seeds are sown in April or May, in shallow drills a foot asunder. The roots may be divided in spring or autumn.

Use—The leaves have the warm, pungent taste common to the Cress family; and are used in their young state, like Cress, as a salad. Medically, they have the reputation of being highly antiscorbutic and of aiding digestion. There are four varieties:

White Flowering—A variety with white, single flowers.
Purple Flowering—Flowers purple, single. Either of these varieties may be propagated from seeds, or by a division of the roots.
Double Flowering White—Flowers white, double.
Double Flowering Purple—A double variety, with purple blossoms. These varieties are propagated by a division of the roots. Double-flowering plants are rarely produced from seeds.

Cardamine pratensis

84

The Dandelion (Leontodon taraxacum)

(Taraxacum officinale)

The Dandelion, though spontaneously abundant, is not a native of this country. Introduced from Europe, it has become extensively naturalized, abounding in gardens, on lawns, about cultivated lands; and, in May and June, often, of itself alone, constituting no inconsiderable portion of the herbage of rich pastures and mowing-fields.

It is a hardy, perennial plant, with an irregular, branching, brownish root. The leaves are all radical, long, runcinate, or deeply and sharply toothed; the flower-stem is from six to twelve inches and upwards in height, leafless, and produces at its top a large, yellow, solitary blossom; the seeds are small, oblong, of a brownish color, and will keep three years.

Taraxacum officinale)

Soil and Cultivation—Although the Dandelion will thrive in almost any description of soil, it nevertheless produces much the largest, most tender, and best-flavored leaves, as well as the greatest crop of root, when grown in mellow, well-enriched ground. Before sowing, stir the soil, either by the spade or plough, deeply and thoroughly; smooth off the surface fine and even; and sow the seeds in drills half an inch deep, and twelve or fifteen inches apart. If cultivated for spring greens, or for blanching for salad, the seed must be sown in May or June. In July, thin out the young plants to two or three inches apart; cultivate during the season in the usual form of cultivating other garden productions; and, in April and May of the ensuing spring, the plants will be fit for the table.

For very early use, select a portion of the bed equal to the supply required; and, in November, spread it rather thickly over with coarse stable-manure. About the beginning of February, remove the litter, and place boards or planks on four sides, of a square or parallelogram, in the manner of a common hot-bed, providing for a due inclination towards the south. Over these put frames of glass, as usually provided for hot-beds; adding extra protection by covering with straw or other material in intensely cold weather. Thus treated, the plants will be ready for cutting two or three weeks earlier than those in the open ground.

When grown for its roots, the ground must be prepared in the manner before directed; and the seeds should be sown in October, in drills fourteen or fifteen inches asunder. In June following, thin out the young plants to two or three inches apart; keep the ground loose, and free from weeds, during the summer; and, in October, the roots will have attained their full size, and be ready for harvesting, which is usually performed with a common subsoil plough. After being drawn, they are washed entirely clean, sliced, and dried in the shade; when they are ready for the market.

Use—The Dandelion resembles Endive, and affords one of the earliest, as well as one of the best and most healthful, of spring greens. "The French use it bleached, as a salad; and if large, and well bleached, it is better than Endive, much more tender, and of finer flavor." The roots, after being dried as before directed, constitute an article of considerable commercial importance; being extensively employed as a substitute for, or mixed in various proportions with, coffee.

It may be grown for greens at trifling cost; and a bed twelve or fourteen feet square will afford a family an abundant supply.

Under cultivation, and even in its natural state, the leaves of different plants vary in a marked degree from each other, not only in size, and manner of growth, but also in form. Judicious and careful cultivation would give a degree of permanency to these distinctions; and varieties might undoubtedly be produced, well adapted for the various purposes for which the plant is grown, whether for the roots, for blanching, or for greens.

85

Endive (Chicorium endivia)

Endive is a hardy annual, said to be a native of China and Japan. When fully developed, it is from four to six feet in height. The leaves are smooth, and lobed and cut upon the borders more or less deeply, according to the variety; the flowers are usually of a blue color, and rest closely in the axils of the leaves; the seeds are small, long, angular, and of a grayish color; their germinative properties are retained for ten years; nearly twenty-five thousand are contained in an ounce.

Soil—All of the varieties thrive well in any good, mellow garden soil. Where there is a choice of situations, select one in which the plants will be the least exposed to the effects of drought and heat.

Propagation—The plants can be raised only from seed. This may be sown where the plants are to remain; or it may be sown broadcast, or in close drills in a nursery-bed for transplanting. If sown where the plants are to remain, sow thinly in shallow drills a foot apart for the smaller, curled varieties, and fifteen inches for the larger, broad-leaved sorts. Thin out the plants to a foot asunder as soon as they are large enough to handle, and keep the ground about them, as well as between the rows, loose, and free from weeds, by repeated hoeings. If required, the plants taken out in thinning may be reset in rows at the same distances apart.

If sown in a nursery-bed, transplant when the young plants have eight or ten leaves; setting them at the distances before directed. This should be done at morning or evening; and the plants should afterwards be watered and shaded for a few days, until they are well established.

The first sowing may be made as early in spring as the weather will permit; and a sowing may be made a month or six weeks after, for a succession: but as it is for use late in autumn, or during the winter and spring, that Endive is most required, the later sowings are the most important. These are usually made towards the end of July.

Blanching—Before using, the plants must be blanched; which is performed in various ways. The common method is as follows: When the root-leaves have nearly attained their full size, they are taken when entirely dry, gathered together into a conical form, or point, at the top, and tied together with matting, or any other soft, fibrous material; by which means, the large, outer leaves are made to blanch the more tender ones towards the heart of the plant.

After being tied in this manner, the plants are sometimes blanched by earthing, as practised with Celery or Cardoons. This process is recommended for dry and warm seasons: but in cold, wet weather, they are liable to decay at the heart; and blanching-pots, or, in the absence of these, common flower-pots, inverted over the plants, will be found a safe and effectual means of rendering them white, crisp, and mild flavored.

"Some practise setting two narrow boards along each side of the row; bringing them together at the top in the form of a triangle, and afterwards drawing earth over them to keep them steady. Some cover the dwarfish sorts with half-decayed leaves, dry tanner's bark, sand, coal-ashes, and even sawdust; but all of these methods are inferior to the blanch-pot or the tying-up process."

Time required for Blanching—In summer weather, when vegetation is active, the plants will blanch in ten days; but in cool weather, when the plants have nearly attained their growth or are slowly developing, three weeks will be required to perfect the operation.

Harvesting, and Preservation during Winter—

"Before **frost** sets in, they must be tied up in a conical form, as before directed; and all dead or yellow leaves must be taken off. Then take them up with a ball of soil to each, and put them into light earth in a cellar or some warm building. Put only the roots into the earth. Do not suffer the plants to touch each other; and pour a little water round the roots after they are placed in the earth. If they are perfectly dry when tied up, they will keep till spring."[1]

Seed—Two or three vigorous plants, left unblanched, will yield sufficient to supply a garden of ordinary size for years. Half an ounce will sow a seed-bed of forty square feet.

Use—

"The leaves are the parts used, and these only when blanched to diminish their natural bitterness of taste. It is one of the best autumn, winter, and spring salads."[2]

Varieties—The descriptions of many of the varieties have been prepared from an interesting paper read before the London Horticultural Society by Mr. Matthews, clerk of the society's garden.

The different sorts are divided into two classes,—the "Batavian" and the "Curled-leaved."

Batavian Endives

Under the Batavian Endives are included all the varieties with broad leaves, generally rounded at the points, with the margin slightly ragged or torn, but not curled. These are called, by the French, *Scarolles.* As most of the sorts require more room than the Curled-leaved kinds, the rows should be about fourteen inches apart, and the plants thinned out from nine to twelve inches in the rows.

Broad-Leaved Batavian Endive

(Common Yellow Endive, of the Dutch)

Leaves yellowish-green, large, long and broad, thick and fleshy, the edges slightly ragged: when fully grown, they are about ten inches long, and an inch wide at the base; increasing regularly in width towards the end, and measuring five or six inches in diameter at the broadest part. The leaves of the centre of the plant are of the same form, but shorter, and much paler. The plants form but little heart of themselves; but the length of the outer leaves is such, that they tie up well for blanching. In quality, as well as in appearance, it is inferior to the Curled sorts; and its flavor is not so mild and agreeable as that of some of the other kinds of Batavian endives.

Curled Batavian Endive

The leaves of this variety are neither so large nor so broad as those of the Broad-leaved Batavian Endive: they grow flat on the ground, and are curled at their edges. The whole appearance of the plant is very different from the Common Broad-leaved; approaching the Curled endives, in general character. The heart, which forms of itself, is small, and lies close to the ground.

The plants require twelve or fourteen inches' space between the rows, and eight or ten inches in the row.

Large Batavian Endive

(Scarolle grande, of the French)

This differs from the Small Batavian Endive in the size and shape of its leaves, which are broader and more rounded: they are a little darker, but yet pale. The inner ones are turned over like the small variety, though not so regularly; but form a large, well-blanched heart, of good flavor. This and the Small Batavian will blanch perfectly if a mat is laid over them, and do not require to be tied up. Both the Small and the Large sorts are considered hardier than the Curled varieties.

Endive (Scarole type)
Adolphe Millot

Lettuce-Leaved or White Batavian Endive

(Scarolle Blonde)

Leaves broad and large, obtuse, ragged at the edges, of a paler color and thinner texture than either of the other Batavian sorts; the exterior leaves are spreading, fourteen inches long, two inches wide at the base, and, growing regularly broader to the end, measure six or seven inches in diameter at the widest part; the central leaves are short, and the head is less compact than that of the Common Broad-leaved; the seeds are of a paler color than those of the Green Curled Endive.

To blanch it, the leaves must be tied up; and it should be grown for summer use, as it is comparatively tender, and will not endure severe weather. It is best if used while young; for, when fully developed, the leaves are not tender, and, if not well blanched, are liable to have a slightly bitter taste.

Sow in May or June, in rows fifteen inches apart, and thin to a foot in the rows; or transplant, giving the plants the same space.

Small Batavian Endive

(Scarolle Courte, of the French)

Leaves whitish-green, broad, of moderate length, and slightly cut at the edges. The inner leaves are numerous, and turn over like a hood at the end; forming a larger head than any of the other kinds. It is one of the best of the endives, and a valuable addition to our winter salads. It blanches with little trouble; and is mild and sweet, without being bitter.

Curled Endives

Curled endives are those with narrow leaves, more or less divided, and much curled. They are usually full in the heart. The French call them, by way of distinction, *Chicorées*.

Dutch Green Curled Endive

This approaches the Large Green Curled Endive in appearance and growth; but the divisions of the leaves are deeper, the outer leaves are broader, not so much curled, and the inner ones more turned into the heart: the outer leaves are about ten inches long. It blanches well, and is hardy.

Green Curled Endive

(Small Green Curled Endive)

Leaves six or seven inches long, finely cut, and beautifully curled; the outer leaves lying close to the ground, the inner ones thickly set, forming a compact heart. Easily blanched, very hardy, and well adapted for winter use. The leaves are longer, and of a darker-green color, than those of the Green Curled Summer Endive, and will tie up much better for blanching. It is a fortnight later.

Sow in rows a foot or fourteen inches apart, and thin to six or eight inches in the row.

It may be quickly blanched by simply covering the plant with a deep flower-pot saucer. In summer, while the plants are growing vigorously, the process will be completed in about a week: later in the season, two-weeks, or even more, may be necessary.

Green Curled Summer Endive

Leaves not quite so large as those of the Green Curled; finely and deeply cut: the outer ones are five or six inches long, and grow close to the ground; the inner are short, numerous, curled, and form a close, full heart. It is much the smallest of any of the kinds, and is somewhat tender. The outer leaves are so short, that they will not tie up; but blanch well by being covered simply with a flat garden-pan, as directed for the Green Curled.

This variety is distinguished from the last named by its shorter, broader, deeper cut, and less curled leaves: the head is more solid at the centre, and is also much harder. The seeds should be sown early; for, if sown late, the plants are liable to be affected by dampness and wet weather, and to rot at the heart.

Cultivate in rows twelve or fourteen inches apart, and eight or ten inches apart in the rows.

Italian Green Curled Endive

Leaves from ten to twelve inches long, deep-green, narrow, and divided to the mid-rib. They grow erect, and the segments are much cut and curled.

It is a well-marked variety; readily distinguished by the length of the leaf-stalks, and the pinnatifid character of the leaves. It blanches well, and is of good quality.

Large Green Curled Endive

A sub-variety of the Common Green Curled, of stronger growth, and larger hearted. The exterior leaves are ten or twelve inches long, looser and more erect than those of the last named: the inner ones are less numerous, and not so much divided.

It is hardy, blanches quickly, and is not liable to decay at the heart.

Long Italian Green Curled

Leaves long, deeply divided, and more upright in their growth than those of the Large Green Curled; the divisions of the leaves are large, and toothed, or cut, but are not curled; the heart-leaves are few and short.[Pg 354] The variety is quite distinct; and, though not so neat and regular as some others, it is of excellent quality, and recommended for cultivation.

Picpus Fine Curled Endive

Exterior leaves seven or eight inches long, deeply lobed; the lobes divided in the same manner as those of the Common Green Curled. The inside leaves are finely cut, and much curled; and form a kind of head more compact than that of the Green Curled, but comparatively loose-hearted.

It blanches well and quickly, and is a good variety; though neither its foliage nor its general habit presents any very distinctive peculiarities.

Ruffec Curled

(Chicorée Frisée de Ruffec)

This variety attains a remarkable size, much exceeding that of the Common Green Curled. The leaves sometimes measure nearly a foot and a half in length. Quality tender and good.

Staghorn Endive

(Early Fine Curled Rouen)

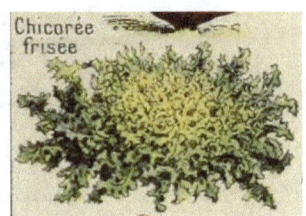

Chicorée frisée (Curled Endive)
Adolphe Millot

A recently introduced variety. The leaves are deep-green, divided into numerous segments, not frilled or curled, but much cut or jagged at the points, the borders having a branched appearance; whence the name. The leaves gradually shorten towards the centre of the plant, are more finely cut, and become closer together; thus forming a moderately firm heart, or head; less compact, however, than that of the Green Curled Summer Endive.

It is well adapted to humid climates, is hardier than the Common Green Curled, and is preferred by market-gardeners for cultivation in autumn and winter.

Triple-Curled Moss Endive

(Winter Moss Endive; Chicorée mousse)

This is a sub-variety of the Staghorn Endive, and comparatively of recent introduction. It is a unique sort, exceedingly well curled; and, when the variety is genuine and the plant well developed, has an appearance not unlike a tuft of moss.

It is liable to degenerate; and, though sometimes classed as a Winter Endive, is less hardy than many other sorts.

It may be grown in rows a foot apart, six inches being allowed between the plants in the rows.

White Curled or Ever-Blanched Endive

Leaves pale yellowish-green, nearly white when young, ten inches long, rather narrow, lobed, cut, and beautifully frilled, or curled, on the borders; the upper surface of the mid-ribs generally tinged with red. The leaves of the centre are not numerous, and much curled: resting upon those of the exterior, they form no head, but leave the heart loose and open.

It is distinguished from all others by its color; both the leaves and the seeds being paler than those of any other sort. Its principal recommendation is signified in the name; but it should be used while young, cut and served in the form of lettuce. It is then tender and of good quality; though the plants yield a small amount of salad, compared with many other sorts. When fully grown, the leaves become tough, and often bitter. As a variety for winter culture, it is of little value.

1. The American Gardener. By William Corbett. Concord, Boston, and New York, 1842.

2. The Book of the Garden. By Charles M'Intosh. 2 vols. Edinburgh and London, 1855.

86

Horse-Radish (Armoracia rusticana)

(Cochlearia armoracia)

Horse-radish is a hardy perennial, introduced from Europe, growing naturally along old roads, and about gardens and waste places in long settled towns. The root is white within and without, long, nearly cylindrical, and from an inch to two inches and a half in diameter; stalk two feet or more in height, smooth and branching; the radical leaves are from fifteen to eighteen inches in length, oval-oblong, and toothed on the margin,—those of the stalk narrow, pointed, smooth, and shining; the flowers are white, and are put forth in June; the seed-pods are globular, but are very rarely formed, the flowers being usually abortive.

There is but one variety.

Propagation and Culture—

"Propagation is always effected by planting portions of the roots, which grow readily. The soil most conducive to it is a deep, rich, light sand, or alluvial deposit, free from stones or other obstructions; as, the longer, thicker, and straighter the roots are, the more they are valued. There is scarcely another culinary vegetable, of equal importance, in which cultivation is, in general, so greatly neglected as in this. It is often found planted in some obscure corner of the garden, where it may have existed for years; and is only visited when needed for the proprietor's table. The operation of hastily extracting a root or two is too often all that is thought of; and the crop is left to fight its way amongst weeds and litter as best it may."[1]

A simple method of cultivation is as follows: Trench the ground eighteen inches or two feet deep, and set the crowns or leading buds of old roots, cut off about three inches in length, in rows a foot apart, and nine inches from each other in the rows; cover six inches deep, and cultivate in the usual manner during the summer. The shoots will soon make their appearance, and the large leaves of the plant completely occupy the surface of the bed. After two seasons' growth, the roots will be fit for use.

Taking the Crop—Its season of use is from October till May; and, whenever the ground is open, the table may be supplied directly from the garden.

For winter use, take up the requisite quantity of roots in November, pack them in moist sand or earth, and store in the cellar, or in any situation out of reach of frost.

Use—The root shredded or grated, with the addition of vinegar, is used as a condiment with meats and fish. It has an agreeable, pungent flavor; and, besides aiding digestion, possesses other important healthful properties.

1. The Book of the Garden. By Charles M'Intosh. 2 vols. Edinburgh and London, 1855.

Cochlearia armoracia (Modern botanical name: Armoracia rusticana)
Flora Batava, plate 303

87

Lettuce (Lactuca sativa)

Lettuce is said to be of Asiatic origin. It is a hardy, annual plant, and, when fully developed, from two to three feet in height, with an erect, branching stem. The flowers are compound, yellow, usually about half an inch in diameter; the seeds are oval, flattened, and either white, brown, or black, according to the variety,—nearly thirty thousand are contained in an ounce, and their vitality is retained five years.

Soil—Lettuce succeeds best in rich and comparatively moist soil; and is also best developed, and most crisp and tender, if grown in cool, moist weather. A poor soil, and a hot, dry exposure, may produce a small, tolerable lettuce early in spring, or late in autumn; but, if sown in such situations during the summer months, it will soon run to seed, and prove nearly, if not entirely, worthless for the table. The richer the soil may be, and the higher its state of cultivation, the larger and finer will be the heads produced; and the more rapidly the plants are grown, the more tender and brittle will be their quality.

Propagation—It is always grown from seeds, which are small and light; half an ounce being sufficient to sow a nursery-bed of nearly a hundred square feet. It is necessary that the ground should be well pulverized and made smooth before it is sown, and the seeds should not be covered more than a fourth of an inch deep.

Cultivation—Some recommend sowing where the plants are to remain, in drills from ten to fifteen inches apart, and thinning the plants to nearly the same distance in the lines; adapting the spaces between the drills, as well as between the plants in the drills, to the habit and size of the variety in cultivation. Others recommend sowing in a small nursery-bed, and transplanting. The process of transplanting unquestionably lessens the liability of the plants to run to seed, and produces the largest and finest heads. The first sowing in the open ground may be made as soon in March or April as the frost leaves the ground; and, if a continued supply is desired, a sowing should afterwards be made, at intervals of about four weeks, until September.

"During spring, the young crops must be protected from frost, and in summer from drought by copious manure-waterings and frequent stirring of the ground between the plants. In the growing season, every stimulant should be applied; for much of the excellence of the crop depends on the quickness of its growth."

Forcing—Lettuce is now served at table the year round; not, of course, of equal excellence at all seasons. Sowings are consequently required for each month: those intended for the spring supply being made from December to February; about twelve weeks being required for its full development, when reared in the winter months. The seed is sown rather thinly, broadcast, in a hot-bed; and, when the plants have made two or three leaves, they are pricked out to three or four inches apart in another portion of the bed,—thus affording them more space for growth, and opportunity to acquire strength and hardiness. When two or three inches high, they are finally transplanted into yet another part of the bed, at distances corresponding with the size of the variety, varying from ten to fourteen inches in each direction. As the plants increase in size, the quantity of air should be increased; and water should be given, whenever the surface of the bed becomes dry. In severe cold or in cloudy weather, and almost always at night, straw matting (made thick and heavy for the purpose), woollen carpeting, or a similar substitute, should be extended over the glass, for the retention of heat.

Some practise transplanting directly from the nursery-bed to where the plants are to remain; but the finest Lettuce is generally obtained by the treatment above described.

"Lettuces are sometimes required for cutting young, or when about two inches high. These are termed, by the French, *Laitues à couper.* The small, early sorts (such as the Hardy Hammersmith and Black-seeded Gotte) are preferred for this purpose; but any sort that is green or pale-green, and not brown or otherwise colored, will do. They should be sown in the open ground about once a week, or every ten days, from April, throughout the season. In winter, they are best raised on heat. They should be sown rather thickly in drills six inches apart."[1]

To save Seed—

"This should be done from plants raised from early sowings. The finest specimens should be selected; avoiding, however, those that show a disposition to run quickly to seed. Those that heart readily, and yet are slow to run up, are to be preferred. Care should be taken that no two different varieties be allowed to seed near each other, in order that the sorts may be kept true. The seed which ripens first on the plant is the best: therefore it should be secured, rather than wait for the general ripening. The branchlets which first ripen their seed should be cut off, and laid on a cloth in the sun; or, when the forward portion of the seed is as near maturity as will safely bear without shaking off, the plants should be carefully pulled up, and placed upright against a south wall, with a cloth under them to perfect their ripening. The seed should in no case be depended on without trial. Plants from seeds two years old heart more readily than those from one-year-old seed."—*Thomp.*

Use—Lettuce is well known as one of the best of all salad plants. It is eaten raw in French salads, with cream, oil, vinegar, salt, and hard-boiled eggs. It is also eaten by many with sugar and vinegar; and some prefer it with vinegar alone. It is excellent when stewed, and forms an important ingredient in most vegetable soups. It is eaten at almost all meals by the French; by the English after dinner, if not served as adjuncts to dishes during the repast; and by many even at supper. In lobster and chicken salads, it is indispensable; and some of the varieties furnish a beautiful garnish for either fish, flesh, or fowl.

In a raw state, Lettuce is emollient, cooling, and in some degree laxative and aperient, easy of digestion, but containing no nourishment.

Varieties—These are exceedingly numerous. Some are of English origin; many are French and German; but comparatively few are American. The number of kinds grown to any considerable extent in this country is quite limited. Cultivators generally select such as appear to be best adapted to the soil and climate of their particular locality; and, by judicious management, endeavor to give vigor and hardiness to the plants, and to increase the size, compactness, and crispy quality of the head. Some of the varieties have thus been brought to a remarkable degree of perfection; the plants producing heads with as much certainty, and nearly as well proportioned and solid, as those of the Common Cabbage. They are generally divided into two classes; viz., Cabbage lettuces and Cos lettuces.

Cabbage Lettuces

Brown Dutch
(Black-Seeded)

Head of medium size, rather long and loose; the leaves, which coil or roll back a little on the borders about the top of the head are yellowish-green, washed or stained with brownish-red,— the surplus leaves are large, round, waved, green, washed with bronze-red, and coarsely, but not prominently, blistered; diameter twelve to fourteen inches; weight about eight ounces.

This Lettuce cabbages readily, forms a good-sized head, is tender, of good quality, hardy, and tolerably early. It does not, however, retain its head well in dry and warm weather; and, as it is little affected by cold, seems best adapted to winter or very early culture. It resembles the Yellow-seeded Brown Dutch, but is not so early, and the head is looser and larger.

Brown Silesian or Marseilles Cabbage

(Brown Batavian)

Head green, tinted with brown, remarkably large,—not compactly, but regularly, formed; ribs and nerves of the leaves large and prominent; the leaves disconnected with the head are large, bronze-green, coarsely blistered, and frilled and curled on the margin. The diameter of a well-grown plant is about eighteen inches, and its weight twenty-eight ounces. The seeds are white.

This Lettuce, though somewhat hard, is brittle and mild flavored, but is better when cooked than when served in its crude state as a salad. It is a hardy, late sort; succeeds well in winter, and retains its head a long period; but is rarely employed for forcing, on account of its size,—one of the plants occupying, in a frame or hot-bed, the space of two plants of average dimensions.

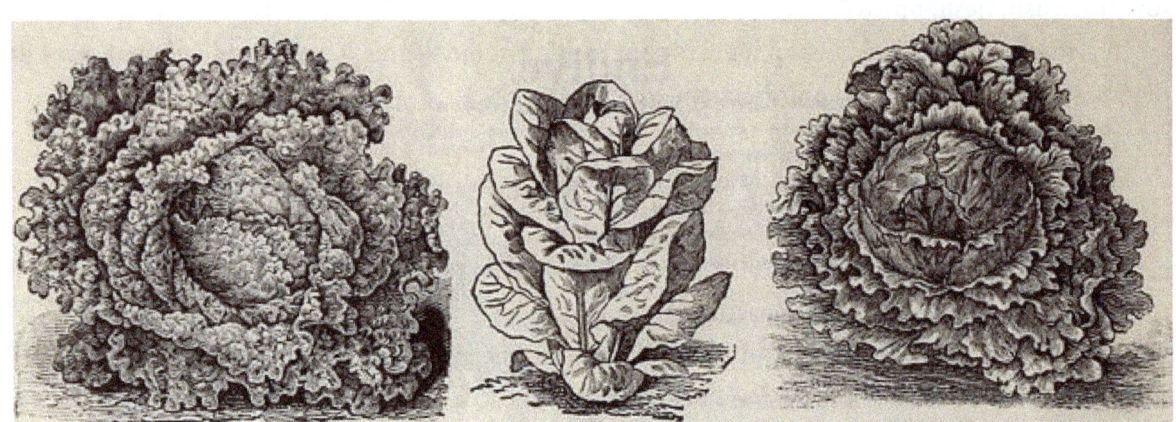

Lettuce Varieties
Annual catalogue of seeds and bulbs, 1896

Brown Winter Cabbage

(Large Brown Winter)

Head of medium size, green, washed or stained with brownish-red, regularly formed, and moderately compact; the exterior leaves are round and short, much wrinkled, and coarsely blistered. When grown in winter or in cool weather, the plants measure fourteen inches in diameter, and weigh from fourteen to sixteen ounces. The seeds are white.

Hardy, and well adapted for winter culture. The heads are not so firm as those of some varieties; but they are well retained, blanch white and tender, and are of excellent flavor.

Early or Summer Cape

(Royal Cape)

Head roundish, usually well formed, and moderately close and firm; the outer leaves are large, loose, golden-green, undulated, and coarsely blistered; the interior leaves are more finely blistered, and nearly of the same color as those of the outside; head, when divided, yellowish to the centre; the plants, when fully grown, measure nearly a foot in diameter, and weigh from six to ten ounces.

The variety is not well adapted for forcing or for early culture in the open ground. As a summer Lettuce, it is one of the best; enduring the heat well, and not running soon to seed. Though not so crisp and brittle as some of the winter or spring grown varieties, it is comparatively well flavored and of good quality. It is similar to the Summer or Royal Cabbage.

Early Simpson

(Hâtive de Simpson)

Head large, pale-green, a little irregular in its form, and only of medium solidity; the outside leaves are large and broad, plaited, and much blistered; diameter fourteen or fifteen inches; weight twelve or fourteen ounces; seeds white.

This Lettuce is brittle, and of excellent flavor; but its head is not compactly formed. Its season is near that of the Versailles Cabbage; but it runs quicker to seed. It is said to be an American variety, and is much grown in the vicinity of New York City for marketing.

Early White Spring or Black-Seeded Gotte

A small spring Cabbage Lettuce, growing close to the ground. Its heart is hard and firm, and measures about four inches in diameter when stripped of its outer leaves; color pale-green; the leaves are thin, nearly round, rugose, and waved on the margin.

This Lettuce comes early into use, and, besides, is of excellent flavor; but its chief merit is, that it remains longer than almost any other sort before running to seed, and even sometimes bursts before the flower-stem is formed. It is one of the smallest of the Cabbage lettuces, and somewhat resembles the Tennis-ball; from which, however, it differs in the leaves being more curled and of a lighter-green color, and by not running to seed so soon by three weeks or a month.

The variety has black seeds; and this fact should be particularly attended to in obtaining it from seedsmen, as the White-seeded Gotte lettuces run much sooner to flower. Various other Gotte lettuces are described by authors. "All are of great merit, but are little cultivated in the United States. Where small, hard, compact, and delicate sorts are required, this class should be selected."

Endive-Leaved

(Laitue Chicorée)

This variety forms no head. The leaves are finely frilled and curled, and spread regularly from a common centre in the form of a rosette. A well-developed plant resembles Curled Endive. It appears to be nearly identical with the Green Curled Lettuce.

The seeds are black, and smaller than those of any other variety.

English Endive-Like Curled-Leaved

Like the Common Green Curled Lettuce, this variety forms no head. The plant has the form of a rosette, and the foliage a silvery-gray appearance. The leaves are short, undulated on the border, but not frilled and curled like the Common variety; nerves purplish; the heart of the plant is large and full; seeds black.

This Lettuce is hardy, tender, and well flavored, and equal, if not superior, to the Common Green Curled, both in respect to quality and its adaptation to winter culture.

Green Curled

(Curled; Endive-Leaved; Boston Curled)

The Green Curled strongly resembles, if it is not identical with, the Endive-leaved. When well grown, the plant measures about ten inches in diameter, and is one of the most beautiful of all the lettuces. The exterior leaves are finely frilled and curled, and of a rich, golden-green color; the central leaves are smaller, but frilled and curled like those of the exterior. When in perfection, the plants have the form of a rosette, and make an excellent garnish. The seeds are white.

It is hardy, well adapted for forcing, and is extensively grown in the vicinity of Boston, Mass., for early marketing. As respects its value for the table, it cannot be considered equal to many of the Cabbage varieties, as it is deficient in crispness, and tenderness of texture,—qualities essential in all salad plants. Its recommendations are its hardiness, its adaptation to early culture and forcing, and particularly its beautiful appearance.

Market-gardeners and cultivators make three sub-varieties, which are known as "Single-curled," "Double-curled," and "Triple-curled;" the difference consisting in the finer frilling, or curling, of the last named. A well-grown plant resembles some varieties of Endive; whence the term "Endive-leaved."

Green Winter Cabbage

(Hardy Winter Cabbage; Morine)

Head pale-green, of medium size, round and regular, firm and solid; leaves of the head much wrinkled, and coarsely blistered; the outside leaves are broad and large, glossy-green, wrinkled and blistered like those composing the head. Winter-grown plants will measure in their full diameter about twelve inches, and weigh from fourteen to sixteen ounces. Seeds white.

The Green Winter Cabbage Lettuce is tender, and of excellent flavor, particularly if cultivated in cool weather. It is hardy, forms its head promptly and uniformly, is slow in the development of its flowers, and must be classed as one of the best of the hardy, winter varieties.

Hammersmith Hardy

(Hardy Green Hammersmith; Early Frame; Early Dwarf Dutch; Green Dutch)

A popular, old variety, with a comparatively small, dark-green head. The leaves are much wrinkled, concave, thick, and fleshy; the seeds are white. It is considered the hardiest sort in cultivation, and is one of the best for growing in winter or for forcing. When raised in spring, late in autumn, or in cool, moist weather, the plants attain a diameter of nearly ten inches, and weigh from six to eight ounces; but summer-grown specimens are much smaller, rarely measuring more than six or seven inches in diameter, or weighing above three or four ounces. In warm, dry weather, it soon runs to seed.

Ice Cabbage

This variety belongs to the division of the Silesian or Batavian lettuces, and must not be confounded with the White Cos. The leaves are of a light shining green, blistered on the surface, much undulated, and slightly jagged on the edges, nearly erect, eight inches long, and five or six inches broad; the outer leaves spread a little at the top, but grow close at the heart. It blanches without tying up, and becomes white, crisp, and tender.

The Ice Cabbage Lettuce comes into use with the White Silesian, from which it differs, as it also does from any other of its class, in being much more curled, having a lucid, sparkling surface (whence probably its name), and not turning in so much at the heart. It lasts as long in crop as the White Silesian.

Imperial Head

(Turkey Cabbage; Union)

A large and excellent variety, but inferior to the Versailles or the Ice Cabbage. Head large, regular, a little oblong, of a dull, pale-green color, and not compactly formed; the outside leaves are large, rounded, undulated or waved on the borders, thin in texture, and of a soiled or tarnished light-green colour; diameter fourteen inches; weight twelve to fifteen ounces; seeds white.

This is a crisp and tender lettuce, though sometimes slightly bitter. It is not early, and soon shoots up to seed; but is quite hardy, and well adapted for winter cultivation.

Top and middle: Cabbage Headed Varieties: Boston Forcing, Tennisball and White-Seeded. Bottom: Trianon Cos/Romaine Lettuce
Farquhar's 1910 garden annual

The Imperial Head, or Imperial Cabbage Lettuce, with white seeds, was at one period more generally cultivated in small gardens than any other variety; and though some of the recently introduced sorts excel it, not only in size, but in tender consistency and flavor, the Imperial is still extensively cultivated and much esteemed.

With the exception of the color of its seeds, it resembles the Turkey Cabbage.

India

(Large India)

Head large, moderately compact; leaves large, with coarse and hard mid-ribs and veins. Its recommendation is its remarkable adaptedness to summer culture; as it withstands heat and drought, and retains its head to a remarkable degree before running to seed. For the table, it is inferior to many other sorts; although the large ribs and veins of the leaves are comparatively brittle, and of tender texture.

Large Brown Cabbage or Mogul.

(Grosse Brune Paresseuse; Large Gray Cabbage; Mammoth)

Head remarkably large, round, regularly formed, grayish-green, tinted or washed with reddish-brown at the top: the leaves not composing the head are large, plaited, coarsely blistered, of a grayish-green color, stained here and there with spots of pale-brown. The diameter of a well-grown plant is about fourteen inches, and its weight nearly a pound; seeds black.

The Large Brown Cabbage Lettuce is crisp and tender, but is sometimes slightly bitter. Its season is near that of the Versailles; but it is slower in forming its head, and sooner runs to flower. It is hardy, good for forcing and well adapted for cultivation during winter. In summer, the heads are comparatively small, and loosely formed.

Large Red Cabbage

(Rouge Charteuse)

Head green, washed with red, of medium size, regularly but loosely formed; the exterior leaves are large, undulated, blistered, and stained with brownish-red, like those of the head; diameter thirteen or fourteen inches; weight twelve ounces; seeds black.

Its season is near that of the Large Brown Cabbage. When grown in warm weather, the head is small, and the plant soon runs to seed: in winter, the head is much larger, more solid, and longer retained. It resembles the Brown Dutch, but differs in the deeper color of the leaves.

Large Winter Cabbage or Madeira

(Laitue Passion)

Head of medium size, regular in form, not compact, green, washed with red at the top: the leaves not composing the head are broad and large, a little undulated or waved on the border, plaited or folded at the base, thin in texture, somewhat blistered, and stained with spots of clear brown. When grown in winter, or in cool, moist weather, the plants will measure about a foot in diameter, and weigh nearly a pound. Seeds white.

It is quite brittle, though not remarkable for tenderness of texture; hardy; succeeds well when grown in cold weather; and remains long in head before shooting up to seed. Season, the same with that of the Green Winter Cabbage.

Malta or Ice Cabbage

(Ice Cos; Drumhead; White Cabbage; De Malte)

In its general character, this variety resembles the White Silesian. The head is remarkably large, somewhat flattened, compact, pale-green without, and white at the centre; the outer leaves are large and broad, glossy-green, and coarsely blistered; the mid-ribs and nerves are large and prominent. The extreme diameter of a full-grown plant is about sixteen inches, and the weight from twenty to twenty-four ounces. The seeds are white.

The variety heads readily, blanches naturally, and is crisp, tender, and well flavored. It is hardy, but not early; and remains long in head without running to seed.

It is extensively cultivated in England; and in some localities succeeds better, and is of finer quality, than the White Silesian or Marseilles Cabbage. The name is derived from the glazed or polished surface of the leaves.

Neapolitan

(Naples Cabbage)

Plant dwarfish; head of large size, round, regularly formed, solid,—when in perfection, resembling a well-developed cabbage; the exterior leaves are broad and large, green, frilled on the margin, and coarsely blistered. If well grown, the plants will measure sixteen inches in diameter, and weigh from twenty to twenty-four ounces. Seeds white.

The Neapolitan Lettuce blanches naturally, is well flavored, and so slow in the development of its flower-stalk, that the heads are sometimes artificially divided at the top to facilitate its growth, and to secure the seeds, a supply of which is always obtained with difficulty; as, aside from the tardiness of the plant in flowering, the yield is never abundant.

It is not so good for forcing as many others, and must be classed as a summer rather than as a winter variety.

Palatine

(Brown Cabbage)

A variety of medium size, with a round, somewhat depressed head, stained with red about the top. The foliage is yellowish-green, strongly marked or clouded with brownish-red. Extreme diameter of the plant ten or eleven inches; weight about twelve ounces. The seeds are black.

It is remarkably crisp and tender; of excellent flavor; yields a large quantity of salad in proportion to its size; flourishes well at all seasons, even during winter; and must be classed as one of the best, and recommended for general cultivation.

Spotted Cabbage (Black-Seeded)

(Sanguine à Graine Noire)

The heads of this variety are of medium size, round and regular in their form, and comparatively solid; the sides are brownish-red, but at the crowns the color is changed to clear, bright-red; the outer leaves are short, broad, and round, and strongly marked or clouded with brownish-red, like those composing the head. If grown in winter or in cool weather, the plants attain a diameter of about twelve inches, and will weigh twelve ounces.

It retains its head longer than almost any other variety; and, though sometimes slightly bitter, is considered superior to the White-seeded. Compared with the last-named, the head is not so well formed, the foliage is deeper colored, and it is not so well adapted for forcing or for cultivation during winter.

Spotted Cabbage (White-Seeded)

(Sanguine à Graine Blanche)

Head yellowish-green, spotted and clouded with brownish-red, of medium size, round and regular. The surplus leaves are small and numerous, round, prominently blistered, copper-green, streaked and variegated with brownish-red. Summer-grown plants will measure ten inches in diameter, and weigh about eight ounces. Winter-grown plants, or those grown in cool and moist weather, will give an increase of the diameter, and weigh nearly a pound.

It is a brittle, well-flavored lettuce, hardy, and well adapted for growing in frames during winter. When grown in the summer months, the head is seldom well formed, and the plants soon run to seed.

Stone Tennis-Ball.

(Gotte Lente à Monter)

Plant quite small, with a uniformly green, regular, solid head; all of the leaves to the heart being strongly wrinkled and coarsely blistered. The exterior leaves are comparatively few and small, green, undulated, and prominently blistered. Summer-grown plants measure six or seven inches in diameter, and weigh about three ounces. When grown early or late in the season, or under the influence of cool and moist weather, the plants attain a larger size; often measuring nine or ten inches in diameter, and weighing eight ounces. The seeds are black.

The Stone Tennis-ball hearts well, is of excellent quality, and, in proportion to its size, yields a large quantity of salad. It retains its head a long period, even in warm weather, without shooting up to seed; and, as most of the leaves of the plant are embraced in the head, it occupies but a small space of ground in cultivation. Hardy and early.

Summer Cabbage

(Large White Cabbage; Royal Cabbage; Summer Blond; Sugar Cabbage)

Foliage pale yellowish-green; head of medium size, round, somewhat flattened, firm and close; the leaves composing it are wrinkled and blistered,—those of the outside being frequently torn and broken on the margins about the crown. The entire diameter of a well-grown plant is about twelve inches, and the weight from ten to twelve ounces. The seeds are white.

It is one of the best sorts for summer cultivation, as it not only forms its head readily in warm and dry weather, but remains long in head before running to flower. For forcing, or for sowing early in the season, some other varieties would succeed better. Though sometimes slightly bitter, it is crisp, tender in texture, appears to be adapted to our climate, and is recommended for cultivation.

Tennis-Ball

(Green Ball; Button; Capuchin; Hardy Hammersmith)

One of the oldest and most esteemed of the Cabbage lettuces. The head is below medium size, dark-green, remarkably solid if grown in cool weather, but often loose and open-hearted if cultivated during the summer months; the surplus leaves are few in number, deep-green, slightly curled, and broadly, but not prominently, blistered; the seeds of the genuine variety are black.

The Tennis-ball Lettuce is remarkable for its extreme hardiness. Winter-grown plants, or those raised in cool, moist weather, will measure about ten inches in diameter, and weigh eight ounces; whilst those raised under opposite conditions rarely exceed seven or eight inches in diameter, or weigh more than four or five ounces.

It is slow in running to seed, and the head blanches white and tender. "It requires little room in frames in winter, and yields a great return in spring, as almost the whole plant is eatable." A large Cabbage Lettuce, tinted with brown about the head, is erroneously known in some localities as the "Tennis-ball."

Turkey Cabbage

Similar to the Imperial Head; the principal if not the only difference consisting in the color of the seeds, which are black.

Versailles

(Swedish; Blond Versailles; Sugar-lettuce)

Head pale yellowish-green, large, long, and compactly formed; the exterior leaves are large, numerous, wrinkled, and coarsely blistered. When in its greatest perfection, the extreme diameter of the whole plant is about fourteen inches, and its weight twelve or fourteen ounces. The seeds are white.

This variety forms its head quickly and uniformly; cabbages white and crisp; is slow in shooting up to seed; flourishes in almost every description of soil, and at all seasons, except, perhaps, in extreme cold; and, though sometimes slightly bitter to the taste, is crisp, tender, and of good quality.

With the exception of its paler color, it resembles the Neapolitan. It is one of the best of all varieties for summer cultivation.

Victoria or Red-Bordered

An excellent early and hardy variety. The head is of medium size, tinted or washed with red at the top, round and regular in form, and comparatively solid; leaves large, yellowish-green, wrinkled, and blistered. If grown in summer, the plants measure eight or nine inches in diameter, and weigh four ounces. In cool weather, the plants attain a diameter of twelve inches, and weigh from ten to twelve ounces; seeds white.

The Victoria Lettuce is larger than the Tennis-ball, heads freely, and is crisp and well flavored. When sown in summer, it soon runs to flower; but, in cool weather, the heads are well retained.

White Gotte (Black-Seeded)

A small, low-growing, yellowish-green Cabbage Lettuce, with a comparatively loose head. The plants rarely measure more than six inches in their full diameter, or weigh above four ounces.

It is one of the earliest of all the lettuces, crisp, of good flavor, and well adapted for forcing or for frame culture. Besides the distinction in the color of the seeds, it differs from the White-seeded White Gotte in its smaller and more loosely formed heads.

White Gotte (White-Seeded)

(White Tennis-Ball)

This variety has a small, long, firm, and close head; and is uniformly of a yellowish-green color. The outer leaves are small, light greenish-yellow, waved on the borders, and prominently blistered. The plant is of small dimensions; rarely measuring more than six or seven inches in diameter, or weighing above three ounces. The variety is early, crisp, and well flavored, but soon runs to seed, and is much better adapted for growing in winter, or for forcing, than for cultivation in the summer months.

White Silesian, or White Batavian

(Drumhead Cabbage; Large Drumhead; Spanish)

One of the largest of the Cabbage lettuces. Head golden-green, tinted with brownish-red about the top, regularly but not compactly formed. The outer leaves are large and broad, yellowish-green, bordered with brown, wrinkled, and coarsely blistered. When well grown, the entire diameter of the plant is about eighteen inches, and its weight twenty ounces. The seeds are white.

This variety appears to be adapted to all seasons. It is hardy, retains its head well, withstands heat and drought, blanches white and crisp, and is of excellent flavor. It succeeds well in frames; but, on account of its large size, is not a profitable sort for forcing.

A variety, known as the "Tennis-ball" in some localities, is very similar to this; and the "Boston Cabbage" of New England, if not identical, seems to be but an improved form of the White Silesian.

White Stone Cabbage

(Large Golden Summer Cabbage)

Head of medium size, yellowish-green, stained with brownish-red, firm and solid. When fully developed, the entire diameter of the plant is about fourteen inches, and its weight sixteen ounces. The seeds are white.

This lettuce is brittle, of tender texture and good quality, though it is sometimes slightly bitter. It is hardy, heads readily, is slow in running to flower, succeeds well in warm and dry weather, and is also well adapted for frame-culture or for forcing.

Yellow-Seeded Brown Dutch

(White Dutch; American Brown Dutch)

Head of medium size, yellowish-green, variegated with red, rounded at the top, and tapering to a point at the base; compact; seeds yellow.

A half-early sort, of good quality, hardy, and well adapted for winter culture, or for sowing early in spring. It somewhat resembles the Black-seeded Brown Dutch: but, apart from the difference in the color of the seeds, its foliage is more blistered, and more colored with red; and the plant produces numerous sprouts, or shoots, about the base of the head.

Cos Lettuces

These are quite distinct from the Cabbage lettuces before described. The heads are long, erect, largest at the top, and taper towards the root,—the exterior leaves clasping or coving over and around the head in the manner of a hood, or cowl. As a class, they are remarkable for hardiness and vigor; but the midribs and nerves of the leaves are comparatively coarse and hard, and most of the kinds will be found inferior to the Cabbage lettuces in crispness and flavor. They are ill adapted for cultivation in dry and hot weather; and attain their greatest perfection only when grown in spring or autumn, or in cool and humid seasons.

Varieties—

Alphange or Florence Cos (Black-Seeded)

In the form of the head, and in its general character, this variety resembles the White-seeded. Both of the sorts are remarkable for size, for hardiness and healthy habit, for the length of time they remain in head before running to seed, and for the brittle and tender character of the ribs and nerves of the leaves.

Besides the difference in the color of the seeds, the head of this variety is smaller, and the foliage paler, than that of the White-seeded.

'Half-Century' Lettuce
Henry G. Gilbert Nursery and Seed Trade Catalog Collection, 1904

Alphange or Florence Cos (White-Seeded)

(Magnum Bonum Cos)

Head large, long, not compact, and forming well only when the exterior leaves are tied loosely together. The midribs and nerves of the leaves are large, but brittle, and of tender texture.

It is ten or twelve days later than the Green Paris Cos, retains its head well, is hardy and of healthy habit, but is deficient in flavor, and inferior to either of the Paris sorts.

Artichoke-Leaved

This variety forms no head; and, in its foliage and general habit, is quite distinct from all of the Cos varieties. The leaves are numerous, twelve or fourteen inches long; of a lively-green color, often stained with brownish-red; erect, narrow, pointed, and toothed on the margin, like those of the Artichoke. Before blanching, the leaves are slightly bitter; but mild, crisp, and tender, with no savor of bitterness, after being blanched. The seeds are black.

The plant grows uprightly, groups its leaves together, and thus blanches the interior parts spontaneously; but a much larger portion will be fit for use, if the leaves are collected, and tied loosely about the tips in the manner of treating Cos lettuces.

It is remarkably hardy, slow in running to flower, and the seeds may be sown till August. Late in the season, it is mild and pleasant, and furnishes a tender salad when most of the Cos lettuces become bitter and strong-flavored.

Bath Green Cos

This variety has much merit as a hardy, winter, green sort; and is nearly related to the Brown Cos, but is less brown on the outer leaves: but, while that has white seeds, the seeds of this variety are black. Hence there are found, upon the catalogues of seedsmen, Black-seeded Bath, or Brown Cos; and White-seeded Bath, or Brown Cos; the latter seeming to be the hardiest, while the former appears to be the best.

Brown Cos

Bath Cos. Sutton's Berkshire Brown Cos. Wood's Improved Bath Cos. Bearfield Cos. White-seeded Brown Cos.

This is one of the oldest of the Cos lettuces, and considered the hardiest of the class. The head is of large size, pointed, not compact, and requires to be tied in order to obtain it in its greatest perfection; the leaves are of a copper-green color, stiff and firm, toothed and blistered; the seeds are white.

The Brown Cos blanches white and tender, and is exceedingly crisp and well flavored; but the dark-brownish color of the exterior leaves is deemed an objection, and it is often displaced by really inferior varieties. In weight and measurement, it differs little from the Green Paris Cos. Extensively cultivated and much esteemed in England.

Gray Paris Cos

Head of the form of an inverted cone; green, with a grayish tone about the top; compact, and forming well without tying. The exterior leaves are numerous, deep-green, erect, firm, and prominently blistered. The full diameter of the plant is nearly twelve inches, and its weight about twenty ounces; the seeds are white.

The Gray Paris Cos is brittle, and of tender texture; but is considered inferior to the other Paris Cos sorts, and is but little cultivated.

Green Paris Cos

(Kensington Cos; Sutton's Superb Green Cos; Wellington; Ady's Fine Large)

Head inversely conical, compact; leaves deep-green, erect, firm, hooded or cowl-formed towards the ends, and serrated on the margin; the ribs and nerves are large and prominent. When fully grown, the entire diameter of the plant is fifteen or sixteen inches, and its weight twenty-four ounces; the seeds are white.

It is considered one of the best of the Cos lettuces; and, though not so hardy as the Brown Cos, is a good variety for forcing, and furnishes a tender, well-flavored head during summer. Whether for spring, summer, or autumn, it is an excellent sort. It attains a large size, is of a fine green color, and, "from the manner in which the outer leaves cove over the interior ones, blanches well without having to be tied together."

It has a tender, brittle leaf; is some days earlier than the White Paris; and is the principal variety employed by the market-gardeners of Paris for cultivating under glass.

Green Winter Cos

Head elongated, somewhat of the form of the preceding variety; deep-green, and not forming well, unless the exterior leaves are tied together at the tips; the outer leaves are large, erect, concave, toothed on the margin, and prominently blistered; the seeds are black.

It blanches well; but the ribs and nerves of the leaves are comparatively coarse and hard. Well adapted to winter culture; but, as a summer lettuce, of little value.

Monstrous Brown Cos

(Two-Headed)

Head of remarkable size, long, loose, and open; leaves large, equalling in size those of the Alphange or Florence Cos; green, washed with brown; pointed; seeds white.

The plant sends out numerous side-shoots, or suckers; and sometimes produces several distinct heads: these, however, are generally loosely formed, and not of the fine, tender quality of the Paris varieties.

Oak-Leaved Cos

(Romaine à Feuille de Chêne)

The Oak-leaved Lettuce produces no head, but forms a loose and open heart at the centre of the plant. The leaves are numerous, bronze-green, and deeply cut, or lobed, on the margin, in the form of the leaves of some species of the oak; the seeds are black.

The plants put forth fresh sprouts after having been cut; but the quality is inferior, and the variety is rarely cultivated.

Red Winter Cos

Foliage deep-brown, smooth, and glossy,—gathered at the centre of the plant into a loose heart, rather than head; seeds black.

The hardiness of this lettuce is its principal merit. It is little affected by severe weather; and, as a sort for winter culture, is desirable. When grown in summer, it is of poor quality.

Spotted Cos (Black-Seeded)

(Red-Spotted; Bloody; Aleppo; Panachée à Graine Noire)

This variety is similar to the White-seeded, and, like it, forms no head: the leaves are green, much stained or clouded with brownish-red, erect, firm, rounded at the ends, concave or spoon-shaped, and grouped at the centre into a long and comparatively close heart.

It is crisp and well flavoured, but attains its greatest perfection only when the outer leaves are tied loosely together about the top of the plant.

Spotted Cos (White-Seeded)

Like the preceding, this variety forms no head; but the interior leaves are formed into an erect, oblong, close heart, which, by tying the exterior leaves together, becomes white, crisp, and of excellent flavor.

Though late, it is hardy, remains long in head before running to seed, and is well worthy of cultivation.

Waite's White Cos

An excellent variety, apparently intermediate between the Green Paris and White Paris; not of quite so deep a green as the former, yet deeper than the latter. With regard to its comparative excellence, it is considered fully equal to the Paris Cos varieties; as it is grown as easily, and is equally crisp and tender. Size and weight nearly the same.

White Brunoy Cos (Black-Seeded)

Leaves of large size, yellowish-green, pointed, slightly undulated, entire on the borders, and often revoluted like those of the White-seeded. It rarely produces a head; or, if so, it is loose and open. Its greatest perfection is obtained by collecting the exterior leaves about the top of the plant, and tying them loosely together.

The variety is not considered superior to the White-seeded, though both of the sorts are inferior to the Paris Cos or Florence sorts.

White Brunoy Cos (White-Seeded)

The heads of this variety are long and loose, and rarely form well unless the exterior leaves are tied loosely together. It somewhat resembles the Alphange in the form and character of its foliage, though the head is longer and larger.

The plant attains a remarkable size, is hardy, and of good quality; but soon runs to seed, and appears to be a winter rather than a summer lettuce.

White Paris Cos

(London White Cos; Sutton's Superb White Cos)

The head of this variety has the form of the Green Paris, and blanches well without tying; the outside leaves are erect, yellowish-green, and rather numerous. The extreme diameter of the entire plant, when well grown, is about fourteen inches, and its weight nearly twenty-four ounces. The seeds are white.

This is the sort most generally grown by the London market-gardeners, millions of it being produced annually within a few miles of London alone; and it has been adopted almost exclusively, by the gardeners of Paris, for cultivation in the open air. Next to the Green Paris Cos, this is the best, the largest, and the longest in running to seed, of all the summer lettuces. It is tender, brittle, and mild flavored, less hardy and a few days later than the Green Paris Cos.

Endive-Leaved Lettuce

(Lactuca intybacea)

The leaves of this species have the form of those of some of the varieties of Endive; whence the name. They are small, pale-green, broad towards the ends, cut and irregularly lobed on the borders. While young, the plants have the appearance of Green Curled Endive.

As it runs to flower much earlier than the Spinach Lettuce, it is less esteemed than that variety. The seeds should be sown thickly, in shallow drills ten or twelve inches apart; and the plants should be cut for use when they are three or four inches high.

Perennial Lettuce

(Lactuca perennis)

This species is a native of Europe; and, in habit and duration, is distinct from all others. The leaves are about ten inches long, of a glaucous or sea green color, thick and fleshy, deeply cut or divided on the margin, and spread regularly from the centre of the plant in the form of a rosette. When fully developed, the plant is two feet and a half high; separating into numerous branches, which terminate in large purple flowers.

The seeds, which are of a brownish-black color, are sown in drills fifteen inches apart; and the plants should be thinned to six inches apart in the drills.

The leaves are eaten as salad; but, when so used, they should be blanched, either by earthing up or by tying the plant together. They are also sometimes eaten boiled as Spinach or Endive.

Lactuca perennis
Andrea Moro

Spinach Lettuce

(Oak-Leaved Lettuce; Lactuca quercina)

The leaves of this species are six inches long, pale yellowish-green, lyrate, with obtuse and entire divisions: when fully developed, they somewhat resemble those of the oak, as implied by the name. The plants form no heart, or head; and are never cultivated singly like the Cabbage or Cos lettuces. The leaves are produced in moderate abundance, and are crisp and well flavored.

The seeds should be sown, like those of the Endive-leaved, thickly, in drills; and, when the lower leaves are four or five inches long, they may be cut for use. If not taken off too closely, the plants will afford a second cutting. The seeds are sown early with other spring salads.

Lactuta quercina
Dietrich, A. (1843)

1. The Gardener's Assistant. By Robert Thompson.

88

Madras Radish (Raphanus caudatus)

(Rat-Tailed Radish; Serpent Radish; Podding Radish)

The roots of the Madras Radish are sometimes eaten while they are quite young and small; but they soon become fibrous, strong flavored, and unfit for use. The plant is generally cultivated for its pods, which sometimes measure ten or twelve inches in length: these are solid, crisp, and tender, and, while young, are used for pickling and for salad; being much superior for these purposes to those of the Common Radish.

When cultivated for its pods, the seeds should be sown in drills two feet apart, and the plants thinned to nine inches in the drills.

The podding radish, or rattail radish,
Nature and Art Vol.I, 1866

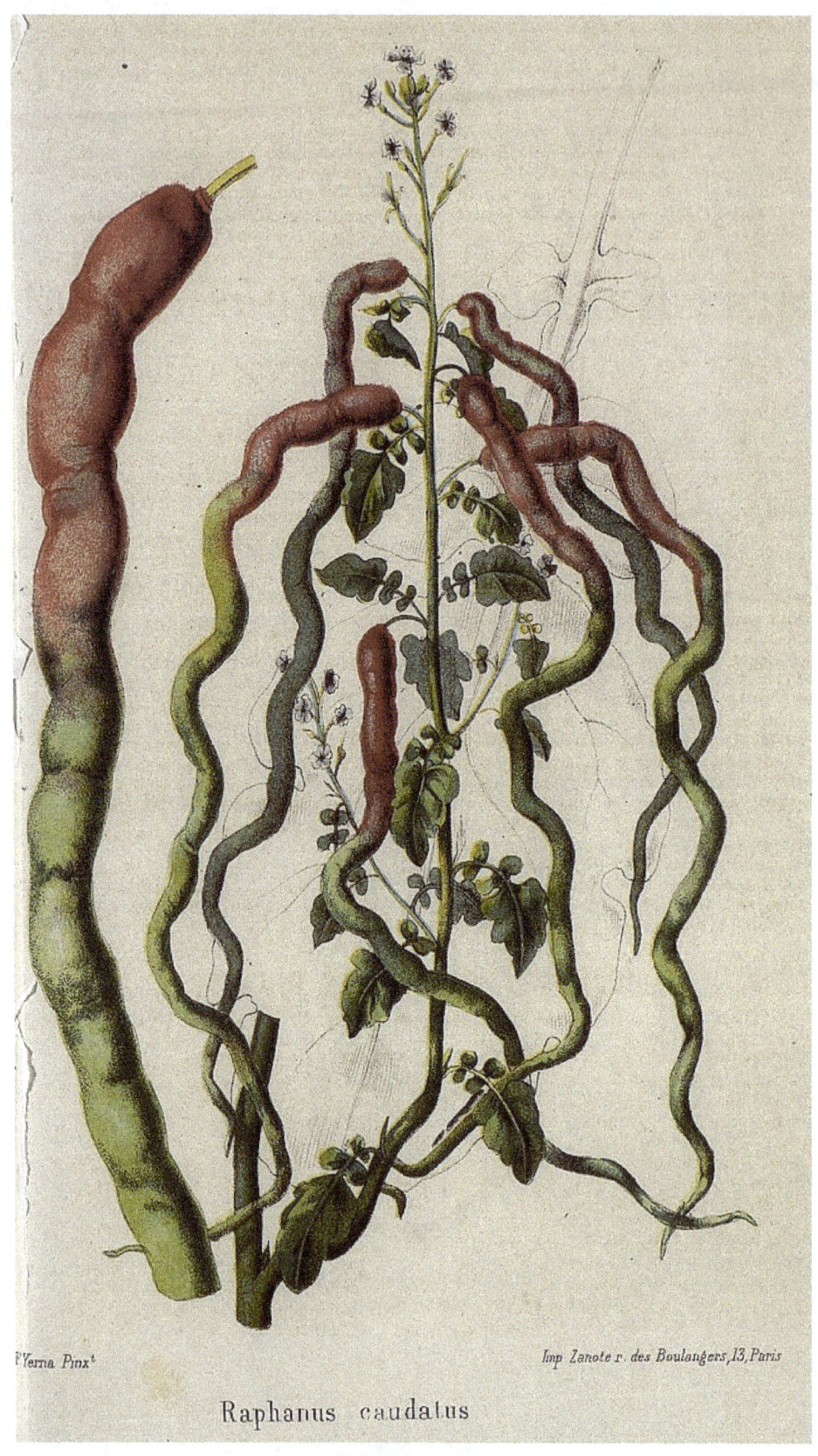

Raphanus caudatus
La Revue horticole 1866

89

Marrow, Curled-Leaved (Malva crispa)

An annual plant, introduced from Europe, and occasionally found growing spontaneously in the vicinity of gardens where it has been once cultivated. The stem is frequently more than six feet in height; the leaves are nearly five inches in diameter, smooth, and of a rich green color, lobed, and beautifully frilled or curled on the borders; flowers axillary, white, and small; the seeds are somewhat kidney-shaped, of a yellowish-brown color, and retain their powers of germination five years.

Cultivation—The seeds are sown the last of April or beginning of May, and covered about an inch deep. The plants require much space, and should be grown at least eighteen inches asunder. The best method is to drop a few seeds where the plants are to grow; or to rake in a few seeds sown broadcast, and transplant.

Malva crispa
Hortus Eystettensis, 1640

90

Mustard (Sinapis nigra/alba)

Black Mustard
(Brown Mustard; Red Mustard; Sinapis nigra)

Black Mustard is a hardy, annual plant, introduced from Europe. In some localities, it grows naturally in great abundance; and is regarded as a troublesome weed, though its seeds furnish the common table mustard. Its stem is four or five feet in height, round, smooth, and branching; the leaves are lobed and toothed on the margin,—the radical or lower ones rough, those of the upper portion of the stalk smooth; the flowers are numerous, rather large, bright-yellow; the pods are erect, somewhat four-sided, and are set closely against the sides of the stalk; the seeds are small, round, brownish-black, and retain their germinative powers many years; nearly eighteen thousand are contained in an ounce.

Propagation and Cultivation—It is raised from seeds, about four quarts of which will be required for sowing an acre. It is sometimes grown in the vegetable garden, but is generally cultivated in fields for its seeds, which, as before remarked, furnish the common table mustard. The sowing is usually made from the middle of April to the middle of May. After making the surface of the ground fine and smooth, sow broadcast, or thinly in shallow drills fourteen or fifteen inches apart; cultivate during the season in the usual manner; and, in August, the crop will be ready for harvesting. Cut the stalks at the ground before the pods shed their seeds; and spread in a dry, light, and airy situation, till they are sufficiently dried for threshing.

When grown for salad in the vegetable garden, it should be sown, and cut for use, as directed for White Mustard.

> "If the seed is covered to the depth of three inches or more, it will lie dormant, and retain its powers of vegetation for ages: from which circumstance, together with the liability of the seed to become shaken out in the harvesting of the crop, such lands as are once employed for the growing of Mustard cannot be fairly cleaned of it for a considerable length of time, and only by judicious fallowing or fallow-cropping, with repeated hoeing and weeding."[1]

Use—Besides the use of the flour of the seeds as a condiment, the seed-leaves are used as salad, in the manner of those of the White species; and the young plants, cut to the ground, are used as spring greens, either boiled alone, or mixed with Spinach.

Chinese Or Pekin Mustard

(Sinapis Pekinensis)

A hardy annual, introduced from China. Stem four feet high, with remarkably large leaves; the flowers, which are produced in loose, terminal spikes, are yellow and showy; the seeds are small, and retain their vitality five years.

Cultivation—The seeds are sown in April or May, in shallow drills ten or twelve inches apart. If cultivated for its seeds, the drills should be eighteen inches or two feet apart, and the plants thinned to six or eight inches in the drills.

Use—The leaves are employed in salads, in the manner of Cress; and they are also sometimes boiled and served as Spinach.

Cabbage-Leaved Mustard

(Moutarde à Feuilles de Chou; Sinapis sp.)

A hardy, annual, Chinese plant, similar in habit to the species last described. Stem from three to four feet high; leaves large, roundish, lobed, and wrinkled; flowers yellow; the seeds are small, reddish-brown or black, and retain their powers of germination a long period.

Cultivation and Use—This species is cultivated in the same manner, and is used for the same purpose, as the Chinese Mustard.

Curled Mustard.

(West-India Cress)

A comparatively small species. Stem two feet and a half high; flowers bright-yellow; seeds small, blackish-brown,—scarcely distinguishable from those of the Black Mustard. The leaves are of medium size, greenish-yellow, broadest near the ends, deeply and finely cut on the borders, and beautifully frilled, or curled: they make an excellent garnish; and, when used as salad, have a pleasant, cress-like flavor.

Cut-Leaved Mustard

(Moutarde Lacinée)

In its general character, this species resembles the Chinese or Pekin Mustard: the leaves, however, are much smaller, and divided quite to the mid-rib.

When young, the leaves make an excellent small salad; having the warm, pleasant flavor of Cress.

White Mustard.
(Sinapis alba)

White Mustard is a hardy annual, introduced from Europe, and occasionally found growing spontaneously in the vicinity of fields and gardens where it has been once cultivated. The stem is three feet and upwards in height; the leaves are large, deeply lobed, and of a rich, deep-green color; the flowers are large, yellow, produced in loose, terminal spikes; the seeds are yellow, much larger than those of the preceding species, and retain their vitality five years,—seventy-five hundred are contained in an ounce.

Propagation—White Mustard is always raised from seeds; about four quarts of which will be necessary for seeding an acre. When grown for salad, an ounce will sow forty feet of drill.

Soil and Cultivation—It succeeds best in rich, loamy soil; which, previously to sowing, should be thoroughly pulverized. When cultivated in the vegetable garden for salad or greens, the first sowing may be made as early in the season as the frost will admit. Sow the seeds thickly, in drills eight or ten inches apart; and cover half an inch deep with fine mould. Remove all weeds as they make their appearance; and, in continued dry weather, water freely.

The plants should be cut for use while in the seed-leaf; as, when much developed, they become strong, rank, and ill-flavored.

For a succession, a small sowing may be made every week until September.

In field culture, the seeds are sometimes sown broadcast; but the more common method is to sow in drills fifteen or eighteen inches apart. When the crop is ready for harvesting, the plants are cut to the ground, stored and threshed, as directed for Black Mustard.

Use—The plants, before the development of the rough leaves, are used as salad: when more advanced, they are boiled and eaten as Spinach. The flour of the seeds furnishes a table mustard of good quality; though the seeds of the Black species possess greater piquancy, and are generally employed for the purpose. The seeds of both species are much used in medicine, and are considered equally efficacious.

1. The Agriculturist's Manual. By Peter Lawson and Son. Edinburgh, 1836.

91

Nasturtium (Tropæolum, sp. et var)

(Indian Cress; Capucine, of the French)

This plant is a native of Peru; and, though generally treated as an annual, is a tender perennial. When cultivated for its flowers or seeds, it should be planted in poor, light soil; but when foliage and luxuriant growth are desired, for the covering of arbors, trellises, and the like, the soil can hardly be made too rich.

The planting should be made in April or May. As the seeds are quite large, they should be covered two inches deep. When planted in drills, they are made three feet apart, and the young plants thinned to six inches apart in the drills. The growing crop may be supported by staking or bushing, as practised with pease; or the taller-growing sorts may be shortened in, which will induce a strong, stocky habit of growth.

While the plants are young, they will require some attention, in order that they may be properly attached to the stakes or trellises provided for their support; after which, little care need be bestowed, beyond the ordinary stirring of the soil, and keeping the ground free from weeds.

Use—The unexpanded flower-buds, and the seeds while young and succulent, have a warm, aromatic taste, and are pickled and used as capers. The young shoots are eaten as salad; and the flowers, which are large and richly colored, are used for garnishing. Few ornamental plants are better known or more generally cultivated than the Nasturtium.

The species and varieties are as follow:

Tall Nasturtium

(Tropæolum majus)

Stem from six to eight feet high, succulent; leaves alternate, smooth, rounded,—the leaf-stems being attached to the disc, or under-surface; flowers large, on long stems, yellow,—the two upper petals streaked and marked with purple; the seeds are large, somewhat triangular, convex on one of the sides, of a drab or pale-brown colour, and retain their germinative properties five years,—from a hundred and eighty to two hundred are contained in an ounce.

Dark-Flowering

A variety of the preceding; differing only in the brown color of the flowers. Cultivation and uses the same.

Variegated

Also a sub-variety of the Tall Nasturtium, with orange-yellow flowers; each of the petals being stained or spotted with purple.

Other varieties occur, differing in color, but equally useful for the purposes before described.

Small Nasturtium
(Dwarf Capucine; Tropæolum minus)

Much smaller, in all respects, than the common Dwarf variety of *Tropæolum majus*; the stem rarely measuring more than two feet in length, or rising above a foot in height. The flowers are yellow; the lower petals with a blotch of scarlet at their base, and the upper ones delicately striped with the same color.

It yields abundantly; and, though the pods are comparatively small, they are generally preferred to those of the Tall Nasturtium for pickling.

'Indian cress' (Nasturtium)
Johannes Evert Akkeringa

92

Picridium (Picridium vulgare)

(Garden Picridium)

A hardy, annual plant, from the south of Europe. Stem eighteen inches high; leaves six to eight inches long, irregular in form, but generally broad at the ends, and heart-shaped and clasping at the base; flowers yellow, compound, produced in clusters; the seeds are long, slightly curved, four-sided, brown or blackish-brown, and retain their vitality five years.

Sowing and Cultivation—The seeds should be sown in April or May, in drills a foot apart, and half an inch in depth. As the plants, when allowed to run to seed, produce but little foliage, it is necessary, in order to secure a continued supply of fresh leaves, to cut or nip off the flowering-shoot as it makes its appearance. Under proper management, the leaves grow rapidly, and are produced in great abundance.

Use—The leaves have a pleasant, agreeable flavor; and, while young and tender, are mixed in salads.

Picridum vulgare
(Flore coloriée de poche du littoral méditerranéen de Gênes à Barcelone y compris la Corse)

93

Purslain (Portulaca)

Purslain is a hardy, annual plant. Most of the cultivated kinds are but improved forms of the Common Purslain (*P. oleracea*), introduced into this country from Europe, and so troublesome as a weed in most vegetable gardens.

Stem usually about a foot in length, succulent and tender; leaves fleshy, broad and round at the ends, and tapering to the stalk; flowers yellow, resting closely in the axils of the leaves; the seeds are black, exceedingly small, and retain their germinating powers ten years.

Soil, Propagation, and Culture—Purslain thrives well in all soils,—dry, wet, or intermediate; and is propagated by seeds sown in shallow drills at any time from April to July.

Use—The plants may be cut for use when they have made a growth of four or five inches. They are mixed in salads, eaten boiled as Spinach, or pickled.

The species and varieties are as follow:

Common Purslain

(Portulaca oleracea)

Abundant in gardens, cultivated fields, and waste grounds. The Green and the Golden Purslain are improved sub-varieties. The Common Purslain is used in all the forms in which the cultivated sorts are used; and, though some of the latter are considered more succulent, the difference in quality will scarcely repay the cost of cultivation, where the present variety would be the ceaseless competitor for the supremacy.

Golden Purslain

(Pourpier Doré; Vil. P. oleracea var. aurea)

Similar to the Green Purslain, but differing in the paler or yellowish color of the stalks and leaves.

Green Purslain

(Pourpier Vert)

Leaves an inch and three-fourths in length, and upwards of an inch in width, deep-green.

Large-Leaved Golden Purslain

Leaves pale yellowish-green, larger than those of the preceding sorts. The plant is a strong grower, and the leaves attain a remarkable size; but the stalks are often comparatively tough and hard, and, for salad purposes, much inferior to those of the Green or Golden varieties.

Portulaca oleracea
Francisco Manuel Blanco

94

Rape (Brassica napus)

This plant is generally cultivated for its seeds, like Mustard. It is, however, sometimes grown for salad; the seeds being sown in April, and, for a succession, once in three or four weeks till August or September. Sow thickly, in drills ten or twelve inches apart, and cover half an inch deep. The soil should be rich and moist, in order to induce a rapid growth, and thus to give a tender, succulent character to the young leaves; these being the parts eaten. They are served like Lettuce, or boiled and treated as Coleworts or Spinach. For mixing with Cress or Lettuce, the plants are cut to the ground before the development of the second leaves.

The species are as follow:

Annual Rough-Leaved Summer Rape

(Turnip Rape; Brassica rapa)

Root fusiform, small, hard, and woody; radical leaves lyrate, vivid green, and without any appearance of the glaucous bloom for which the biennial sorts are so distinguished; the stem-leaves are slightly glaucous, smooth, or nearly so,—the lower ones cut on the borders, the upper entire; the seeds are small, and similar to those of the common field turnip, of which it seems to be either a variety, or the source from which the latter has been derived.

Common or Winter Rape

(Cole-seed; Brassica napus)

Biennial; root long, tapering, hard, and woody, like that of the species before described. The leaves are smooth, thick, and fleshy, and of much the same form as those of the Annual Rough-leaved Summer Rape; this species, however, being readily distinguished, when young, by its uniformly smooth leaves. The seeds, also, are larger than those of the last-named species; but this is not to be relied upon as a distinguishing characteristic, as the size of the seeds, in this as in most other plants, is liable to be materially altered by the soil as well as by the previous culture of the seed-stock.

The seeds are sown in summer, and the crop ripens the following year. It is not adapted to the climate of the Northern States.

In England, the foregoing species are extensively cultivated both for forage and for seed; the latter being used to a limited extent for feeding birds, but chiefly for the production of rape-seed oil.

German Rape

(Annual or Early Rape; Smooth-Leaved Summer Rape; Brassica præcox)

The German Rape somewhat resembles the Common or Winter. It differs in being of annual duration; in its more deeply divided leaves, more erect pods, and smaller seeds.

It would unquestionably succeed well in almost any part of the Northern or Middle States, and might prove as remunerative a crop as corn or wheat. The seeds should be sown in May; and the plants should be treated and the crop harvested, in all respects, as Mustard. It is sometimes sown broadcast, but generally in drills. When sown broadcast, eight or ten pounds of seed will be required for an acre; if in drills, three or four pounds will be sufficient. The yield varies from twenty to forty bushels per acre.

Summer Rape

(Colza; Wild Navew; Brassica campestris)

A biennial plant, with a tapering, hard, and fibrous root. The radical leaves are lyrate and roughish when young; those of the stem clasping, or heart-shaped, at base, and of an oblong form,—all somewhat fleshy, of a dark-green colour, with a glaucous bloom. The seeds are larger than those of the Ruta-baga, or Swedish Turnip, but in other respects not distinguishable.

This species is sometimes termed *Brassica campestris olifer*, or Oil-rape, from its being considered the best sort of rape for cultivating for oil; and to distinguish it from the *Campestris Ruta-baga*, or Swedish Turnip, which is only a variety of this species.

It is not sufficiently hardy for cultivation in the Northern States.

95

Rocket (Eruca vesicaria)

(Garden Rocket; Arugula; Roquette, of the French; Brassica eruca)
A hardy, annual plant, from the south of Europe. Stem about two feet high; leaves long, lobed or lyrate, smooth and glossy, succulent and tender; flowers pale citron-yellow, with blackish-purple veins, very fragrant, having the odor of orange-blossoms; the seeds are small, roundish, brown, or reddish-brown, and retain their vitality two years,—fifteen thousand are contained in an ounce.

Sowing and Cultivation—The seed is sown thinly, in shallow drills a foot asunder. The first sowing may be made as early in spring as the frost will permit; afterwards, for a succession, a few seeds may be sown at intervals of three or four weeks. In poor soil and dry seasons, the leaves are liable to be tough and acrid: the seeds should, therefore, be sown in rich loam, and the plants thoroughly watered in dry weather; as, the more rapid and vigorous the growth, the more succulent and mild-flavored will be the foliage.

Use—The leaves, while young and tender, are eaten as salad.

Eruca versicaria flowers, Spain
Javier Martin

96

Samphire (Crithmum maritimum)

(Sea-Fennel; Parsley-Pert; St. Peter's Herb)

This is a half-hardy, perennial plant, common to rocky localities on the seacoast of Great Britain. Stalk from a foot to two feet in height, tender and succulent; leaves half an inch long, somewhat linear, glaucous-green, fleshy; flowers in terminal umbels,—small, white, or yellowish-white; the seeds are oblong, yellowish, and, though somewhat larger, resemble those of Fennel,—they retain their germinative power but one year.

The plant blossoms in July and August, and the seeds ripen in September and October.

Cultivation—

> "It is rather difficult to cultivate in gardens; and the produce is never so good as that obtained from the places where it naturally grows. It may be propagated either by dividing the plant, or by sowing the seed in April or in autumn, soon after it is ripe. The latter period is preferable; for, if kept till spring, the seed does not germinate so well.
>
> "It succeeds best in a light, sandy, or gravelly soil, kept constantly moist, and sprinkled occasionally with a little sea-salt or barilla, or watered with a solution of these substances, in order to supply the plant with soda, which is a necessary element of its food. It will grow still better if planted or sown among stones at the foot of walls, with a south or east aspect. This, and an occasional watering, with a solution of sea-salt, will give conditions nearly the same as those under which the plant naturally grows. As it is rather delicate, and liable to be injured by frost, it should be protected by dry litter or leaves during the winter. Towards the end of summer, the leaves may be cut for use."[1]

*Use—*The leaves have a warm, pleasant, aromatic flavour; and, when pickled in vinegar, are used in salads and as a seasoning.

Golden Samphire

(Inula crithmifolia)

A hardy perennial, growing, like the preceding, naturally, on the marshes and seacoast of Great Britain. The stalk is a foot and a half in height, erect, with clusters of small, fleshy leaves; flowers yellow, in small, umbel-like clusters.

Propagation and Cultivation—It may be propagated by seeds, or by a division of the roots. It thrives best in a shady situation, and requires frequent watering. If salt be occasionally dissolved in the water, it will promote the growth of the plants, and render the branches and foliage more succulent and tender.

Use—The fleshy leaves and the young branches are pickled in vinegar, and added to salads as a relish. The plant, however, has none of the pleasant aromatic flavor of the true Samphire, though often sold under the name, and used as a substitute.

1. The Gardener's Assistant. By Robert Thompson.

Crithmum maritimum

97

Scurvy Grass (Cochlearia officinalis)

This is a hardy, annual, maritime plant, common to the seacoast of France and Great Britain. The root-leaves spread regularly from a common centre, are heart-shaped, fleshy, smooth, and glossy,—those of the stem sessile, oblong, and toothed on the margin; the stalks are numerous, and from six inches to a foot in height; the flowers are small, white, and produced in compact groups, or clusters; the seeds are small, oval, a little angular, and retain their vitality three years.

Soil, Sowing, and Cultivation—It succeeds best in moist, sandy soil; and flourishes in shady situations. Sow the seeds in August, soon after they ripen, in shallow drills eight or ten inches apart; and, while the plants are young, thin them to five or six inches apart in the rows. The plants taken up in thinning may be transplanted, and new beds formed if occasion require. The growing crop should be kept free from weeds, and liberally watered in dry weather. In the following spring, the leaves will be fit for the table. Those plants not cut for use will flower in June, and the seeds will ripen in July. The seeds seldom vegetate well if sown late in spring, or during warm, dry weather.

Cochlearia officinalis
Köhler's Medizinal-Pflanzen

Use—The radical leaves are used as a salad, and are sometimes mixed with Cress. When bruised, they emit an unpleasant odor; and have an acrid, bitter taste when eaten. The plant is more generally used for medicinal purposes than as an esculent.

98

Snail Trefoil (Medicago orbicularis)

From the south of Europe. It is a hardy, annual plant, with reclining steins, compound or winged leaves, and yellow flowers. The pods, or seed-vessels, are smooth, and coiled in a singular and remarkably regular manner. As they approach maturity, they gradually change to a dark-brown color; and, seen from a short distance, have the appearance of snails feeding on the plant.

The seeds are large, flat, somewhat kidney-shaped, of a yellowish-brown color, and retain their powers of germination five years. They are usually sold in the pods, but should be taken out before planting.

Sowing and Culture—It is propagated by seeds, which should be sown in April or May where the plants are to remain. Sow in drills fifteen inches apart. The plants should be thinned out where they are too close, and kept clean from weeds; which is all the culture they require. They will blossom in July, and the seeds will ripen in autumn.

Medicago Species
Jean Henri Jaume Saint-Hilaire

Use—Though entirely inoffensive, no part of the plant is used for food. The pods resemble some species of snails in a remarkable degree, and are placed on dishes of salad for the purpose of exciting curiosity, or for pleasantly surprising the guests at table.

99

Sweet Cicely (Myrrhis odorata)

(Sweet-Scented Chervil; Scandix odorata)

A hardy perennial. When fully grown, the stalk is three feet or more in height; the leaves are large, and many times divided; the stems and nerves downy; the flowers are white, fragrant, and terminate the stalks in flat, spreading bunches, or umbels; the seeds are large, brown, and retain their vitality but one year.

Sowing and Culture—It is usually grown from seeds; and is of easy cultivation, as it thrives in almost any soil or situation. When allowed to scatter its seeds after ripening in the autumn, the plants will spring up spontaneously in great numbers in the following April or May, and may then be transplanted where they are to remain; or the seed may be sown in October, in beds, making the rows fifteen or eighteen inches apart, and thinning the plants to a foot apart in the rows. When practicable, the seed should be sown in the autumn; as it seldom vegetates well, unless subjected to the action of the winter. After the plants have become established, they will require only ordinary treatment, and yield abundantly.

Myrrhis odorata

Use—In this country, it is sometimes cultivated with other aromatic plants; but its use in soups, or as a seasoner or garnish, is very limited.

> "In England, the leaves were formerly put into salads; but the strong flavour of aniseed, which the whole plant possesses, renders them disagreeable to most persons. It is now not cultivated in Britain; but the leaves and roots are still used in France: the former for the same purposes as those of Chervil; the latter in soups, to which they are said to communicate an agreeable taste."[1]

1. The Gardener's Assistant. By Robert Thompson.

100

Tarragon (Artemesia dracunculus)

A hardy, perennial plant, said to be a native of Siberia. Stalk herbaceous, about three feet in height; the leaves are long, narrow, pointed, smooth, and highly aromatic; the flowers are small, somewhat globular, greenish, and generally infertile. There is but one variety.

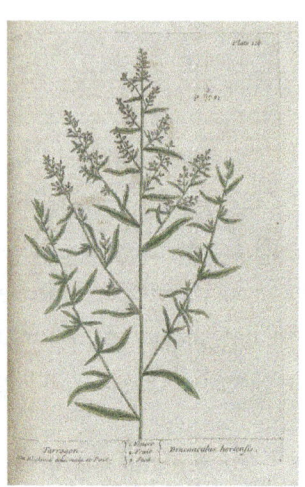

Artemisia dracunculus

Soil, Planting, and Culture—As the plants seldom produce seed, Tarragon is usually propagated by dividing the roots. Select a warm and comparatively dry situation; stir the ground deeply and thoroughly; and, in April, set the roots in rows fifteen inches apart, ten or twelve inches apart in the rows, and cover two or three inches deep. They will soon send up vigorous shoots, which may be cut for use the first season.

It is sometimes increased by cuttings, set three or four inches deep in moist earth. If seeds can be obtained, they should be sown in April or May, in a nursery-bed or in a common frame. Sow in shallow drills six or eight inches apart; and, when the plants are three or four inches high, set them out as directed for the roots. They will early become strong and stocky, and may be used in August or September. The plants are more healthy, yield more abundantly, and are of finer quality, when not allowed to run to flower.

Use—

"Tarragon is cultivated for its leaves and the points of its young shoots; both of which are used as ingredients in salads, soups, stews, pickles, and other compounds. Tarragon vinegar, so much esteemed as a fish-sauce, is made by infusion of the leaves in common vinegar. It is also added to most salads to correct their coldness. Three or four plants will be sufficient for a family."[1]

1. The Book of the Garden. By Charles M'Intosh. 2 vols. Edinburgh and London, 1855.

101

Valeriana (Valeriana cornucopiæ)

This is an annual plant, with a smooth, branching stem about fifteen inches high. The leaves are oblong, stemless, thick, and fleshy, and of a bright, glossy-green color; the flowers are numerous, large, rose-colored, showy, and ornamental; the seeds are oblong, yellowish, somewhat vesiculous, and retain their vitality five years,—twenty-two hundred are contained in an ounce.

Soil and Culture—It succeeds best in a light, warm soil. Prepare a bed four feet and a half wide, spade it thoroughly over, rake the surface smooth and fine, and sow the seed in drills fourteen inches apart. The first sowing should be made the last of April, or early in May; and afterwards, for a succession, sow a row or two every fortnight till July.

Use—It is used as a salad, and is said to be superior to the Common Fetticus, or Corn-salad.

When in blossom, the plant presents a beautiful appearance, and well deserves a place in the flower-garden.

Valeriana officinalis
Flora Batava, plate 22

102

Water-Cress (Nasturtium officinale)

(Sisymbrium nasturtium)

Water-cress is a hardy, aquatic perennial; and is found growing naturally, in considerable abundance, about ponds, and in ditches and small running streams. When in blossom, the plant is about two feet in height, or length; the leaves are winged, with five or six pairs of rounded leaflets, and, in deep water, are often immersed, or float upon the surface; the flowers are small, white, four-petaled, and terminate the stalks in loose spikes; the seeds are very small, reddish-brown, and retain their powers of germination five years,—nearly a hundred and twenty thousand are contained in an ounce.

Planting and Culture—Water-cress is of the best quality when grown in running streams and gravelly soil." The roots may be planted in spring, in situations where the water is from four to eight inches deep. After they are established, the plants will rapidly increase, both from the natural distribution of the seeds and the spreading of the roots, and soon entirely cover the surface of the water with foliage. It may be grown with trifling cost in any small collection of water, and can be easily introduced by dropping a few plants about the borders at the time of the ripening of the seeds. In many localities, it is found growing in spontaneous abundance; and one of the best and most healthful of salads may be obtained for the mere labor of gathering.

Varieties—There are three described varieties,—the Green-leaved, the Small Brown-leaved, and the Large Brown-leaved. These differ slightly, if at all, in flavor; though the Brown-leaved is generally preferred: having a fine appearance, and a small proportion of stalk to the leaves, it is most salable in the market. The variations in foliage and habit do not appear to be caused by the quantity or quality of the water in which the plants are grown, as the three kinds are found growing together.

"The Green-leaved is the easiest of cultivation, and the Small Brown-leaved is the hardiest. The Large Brown-leaved is the best, and is the only one which can be well grown in situations where shallow water is not to be obtained."[1]

Gathering and Use—

"The shoots are *cut* for market, not *broken* off, as is the usual mode of gathering Cress in its natural state, and which is found to be very injurious to the plants in the beds. After they have been cut about three times, they begin to stock; and then, the oftener they are cut, the better. In summer, it is necessary to keep them very closely cut; and in water of a proper depth, and with a good soil, each bed supplies a gathering once a week."

It is extensively employed as an early spring salad; and, on account of its warm and pleasant taste, is by many persons preferred to all other salad plants.

1. The Transactions of the London Horticultural Society. Commenced 1815, and continued at intervals to the present time.

Woman and her two young sons pick wild watercress

103

Winter-Cress (Barbarea præcox)

(Barbarea; American Winter-Cress; Belle-Isle Cress)

Stems from twelve to fifteen inches high; leaves lyrate, the terminal lobe round; flowers small, in erect, loose, terminal spikes, or groups; the seeds are small, wrinkled, of a grayish color, and retain their vitality three years. Introduced from Europe, and naturalized in the Northern States.

Common Winter-Cress, or Yellow Rocket.

(Barbarea vulgaris)

This species somewhat resembles the foregoing; and, like it, grows naturally in moist, shady situations. It is distinguished by its longer, more erect, and more slender pods.

Barbarea praecox
Flora Batava, Volume 15

Soil and Cultivation—Both of the species are hardy, perennial plants; and are raised from seeds, which should be sown in April or May, in shallow drills a foot apart. For a succession, a few seeds may be sown at intervals of three or four weeks till August. For winter use, sow, and subsequently cultivate, as Winter Spinach.

Use—As soon as the plants have made sufficient growth, they may be cut for use. The outer leaves should be first gathered, and the flower-stalks cut or nipped off as they make their appearance, in order to render the plants strong and stocky, and to promote the growth of the leaves; these being the parts of the plants used. They are served as Cress, which they resemble in flavor.

104

Wood-Sorrel (Oxalis acetocella)

Wood-sorrel is a hardy, perennial plant; growing naturally in woods, in cool and shaded situations. The leaves are radical, inversely heart-shaped, and produced three together at the extremity of quite a long stem, or petiole; the flower-stalk is entirely leafless, and supports a solitary bell-shaped flower, the petals of which are white, finely lined or striped with purple; the seed-vessels are of an oblong form, five-angled, and, when ripe, burst open by the touch, in the manner of those of the *Impatiens noli me tangere*, or Common Balsam, of the flower-garden; the seeds are quite small, and of a reddish-brown color.

The flowers are produced in May and June, and the seeds ripen in July.

Propagation and Culture—It may be propagated either by seeds or by dividing the roots. The soil should be rich and moist; and the seeds may be sown in April or May, in shallow drills ten or twelve inches apart; or the roots may be divided in spring or autumn, and set in rows the same distance asunder.

Wood Sorrel (Oxalis acetosella)
William Catto

Use—The leaves possess a pleasant, acid taste; and are mixed with salads, to which they impart an agreeable, refreshing flavor. The plant is considered one of the most valuable of all vegetables cultivated for their acid properties.

105

Worms (Astragalus hamosus)

A hardy, annual plant, indigenous to the south of Europe. Stem ten or twelve inches long, recumbent; leaves pinnate, with ten or twelve pairs of quite small leaflets; flowers yellow, produced five or six together at the extremity of quite a long stem, or peduncle; the seed-pods are about two inches long, nearly a fourth of an inch thick, peculiarly bent or curved, and contain ten or twelve brown seeds.

There is but one species or variety cultivated.

Sowing and Culture—The plants may be started by sowing the seeds in a hot-bed in March, or the seeds may be sown in the open ground in May. They are cultivated in rows fourteen inches apart, and ten or twelve inches apart in the rows; and are also grown in groups, or hills, three or four together. The plants blossom in July, and the pods attain their growth in August and September.

Use—The pods, in their green state, much resemble some descriptions of worms; and, like Caterpillars (*Scorpiurus*) and Snails (*Medicago*), are sometimes placed on dishes of salad to excite curiosity, or for pleasantly surprising the guests at table. Though inoffensive, they are seldom eaten.

Astragalus hamosus
Florae Graecae prodromus - v.2 - 1760

Part Eight: Oleraceaous Plants

Angelica; Anise; Balm; Basil; Borage; Caraway; Clary; Coriander; Costmary; Cumin; Dill; Fennel; Lavender; Lovage; Marigold; Marjoram; Nigella; Parsley; Peppermint; Rosemary; Sage; Savory; Spearmint; Tansy; Thyme.

Bouquet de lavande d'Auvergne-Rhône-Alpes, France

106

Angelica (Angelica archangelica)

Angelica is a native of Hungary and Germany, and is also indigenous to Great Britain. It is a hardy, biennial plant, with a cylindrical, hollow, herbaceous stem four or five feet high. The radical leaves are from two to three feet long, compound, or divided in threes, purplish-red at the base; flowers small, pale-yellow, in large, terminal, spherical umbels; the seeds are of a yellowish color, oblong, flattened on one side, convex on the opposite, ribbed, thin, and membraneous on the borders, and retain their germinative power but a single season,—nearly six thousand are contained in an ounce.

Garden Angelica
Franz Eugen Köhler, 1887

Soil and Culture—The plants thrive best in damp, and even wet, localities; but may be grown in any good, well-enriched soil. As the seeds soon lose their vitality, they should be sown in August, immediately after ripening. Make a small bed, sow the seeds in drills ten inches apart, and cover three-fourths of an inch deep. In this seed-bed allow the young plants to remain until the following spring, when they should be set out two feet asunder in each direction. The stalks will be fit for use in May and June of the following year. If the flower-stem is removed as it makes its appearance, the plants will put forth fresh sprouts from the sides of the root, and survive three years; but when allowed to blossom, and to perfect their seeds, the plants soon after perish.

Use—Angelica was formerly used, after being blanched, as a salad, like Celery. In the vicinity of London, it is raised to a considerable extent for confectioners,—the tender leaf-stalks and flowering-shoots serving as a basis for sweetmeat. The seeds are sometimes employed for flavoring liquors.

107

Anise (Pimpinella anisum)

This is an annual plant, originally from Egypt. Though but little cultivated in this country, neither our soil nor climate is unsuitable; and it might be successfully, if not profitably, grown in the Middle and warmer parts of the Northern States. Large quantities of the seeds are raised on the Island of Malta and in some parts of Spain, and thence exported to England and America for the purpose of distillation or expression.

The stem is from a foot and a half to two feet high, and separates into numerous slender branches; the leaves are twice pinnate,—those of the upper part of the stalk divided into three or four narrow segments; the flowers are small, yellowish-white, produced in large, loose umbels, at the extremities of the branches; the seeds are of a grayish-green color, oblong, slightly bent or curved, convex and ribbed on one side, concave on the opposite, and terminate in a small bunch, or knob,—nearly nine thousand are contained in an ounce, and they retain their vitality three years.

Culture—Anise is raised from seeds sown annually, and thrives best in light, rich, comparatively dry soil, and in a warm, sunny situation. As early in spring as the appearance of settled warm weather, lay out a bed four feet and a half wide, and as long as may be desired; spread on a thin dressing of well-digested compost, and spade it thoroughly in with the soil; then rake the surface fine and even, and sow the seed thinly in drills twelve inches apart and an inch deep, allowing an ounce of seed for a hundred and fifty linear feet. When the plants are an inch high, thin them to five or six inches apart; and, as they increase in size, keep the ground between the rows loose, and the spaces between the plants free from weeds. Towards the close of the season, the seed will be ripened sufficiently for harvesting; when the plants should be pulled up, and spread in a sunny place until dry. The seed should then be threshed from the heads, riddled and winnowed, and again exposed to the sun, or spread in a dry, airy room, to evaporate any remaining moisture; when they will be ready for use or the market.

In field culture, the grower should follow substantially the same method, with the exception of laying out the ground; omitting, in this particular, its division into beds. After the land has been well prepared, the seed can be sown with great facility by a common sowing-machine, adjusted as when employed for sowing carrots. At the time of harvesting, the plants may be cut near the surface of the ground, or even mowed; thereby avoiding much of the inconvenience arising from the soil that adheres to the roots when the plants are pulled up.

There are no varieties.

Use—The seeds and leaves are used both in medicine and cookery. The green leaves are employed in salads, and for seasoning and garnishing, like Fennel. The seeds have a fragrant odor, a pleasant, warm taste, and are highly carminative. Large quantities are used for distillation and in flavouring liquors, and also for expressing for their essential oil.

Anise
Franz Eugen Köhler, 1887

108

Balm (Melissa officinalis)

A hardy, perennial plant, from the south of Europe. The stalk is four-sided, branching, and from two to three feet high; leaves opposite, in pairs, ovate, toothed on the borders; the flowers are small, nearly white, produced in spikes, or clusters, at or near the top of the plant.

Foliage of Melissa officinalis

Soil, Propagation, and Culture—Any warm, mellow, garden soil is suited to its growth. It is generally propagated by dividing the roots, which may be done either in spring or in autumn. After thoroughly stirring the soil, set the roots in rows fifteen inches apart, and a foot apart in the rows. Under good management, the plants will soon completely cover the surface of the ground, and the bed will not need renewal for many years.

Gathering—If required for drying, the plants should be cut as they come into flower, separating the stems at the surface of the ground. They should not be exposed to the sun in drying, but placed in an airy, shady place, and allowed to dry gradually. The leaves, in their green state, may be taken directly from the plants as they are required for use.

Use—The plant has a pleasant, lemon-like odor; an agreeable, aromatic taste; and, in flavoring certain dishes, is used as a substitute for lemon-thyme. It is beneficial in hemorrhage, and other diseases of the lungs; and, in the form of tea, constitutes a cooling and grateful diluent in fevers. A mixture of balm and honey, or sugar, is sometimes applied to the interior of beehives, just previous to receiving the swarm, for the purpose of "attaching the colony to its new settlement."

109

Basil (Ocimum basilicum)

There are two species of Basil cultivated in gardens; viz., the Common Sweet Basil (*O. basilicum*) and the Small Bush Basil (*O. minimum*). Of the Common Sweet Basil, there are three varieties; and of the Bush Basil, two varieties. They are all annuals, and are grown from seeds, which are black, small, oblong, and retain their vitality from six to ten years.

Common Sweet Basil

(Large Sweet Basil; Ocimum basilicum)

Stem from a foot to a foot and a half in height; leaves comparatively large, green, ovate, sharply pointed; flowers white, in whorls at the extremities of the stems and branches. The whole plant, when bruised, is highly aromatic; having the odor and flavor of cloves.

Flowering Common Green Basil

The seeds of the Common Sweet Basil, and also those of the two following varieties, may be sown in a hot-bed in March, and the plants set out in May in rows a foot apart, and five or six inches apart in the rows; or the seeds may be sown in the open ground the last of April or early in May, and the plants thinned while young, as directed for transplanting. In removing the plants from the hot-bed, retain as much of the earth about the roots as possible; water freely as soon as transplanted, and also in dry weather; and they will soon yield an abundance of tender stems and leaves.

Varieties—

Purple Basil

(Basilic Grand Violet)

Leaves and flowers purple. When grown in sunny situations, the leaf-stems and young branches are also purple. In other respects, the variety is similar to the Common Sweet Basil. Its properties and uses are the same.

Lettuce-Leaved Basil

The leaves of this variety are large, pale-green, wrinkled and blistered like those of some kinds of Lettuce: whence the name. It resembles the foregoing varieties in taste and odor, and is used for the same purposes.

Bush Basil

(Ocimum minimum)

The Bush Basils are small, low-growing, branching plants; and are propagated and cultivated like the Common Sweet Basil.

Green Bush Basil

(Basilic Fin Vert)

Stem about eight inches high; leaves small, green, oval; flowers white, produced in whorls about the upper portion of the principal stalk and towards the extremities of the branches.

Purple Bush Basil

(Basilic Fin Violet)

Leaves purple. In other respects, similar to the Green Bush Basil.

Use—The leaves and young branches have a strong, clove-like taste and odor, and are used in highly seasoned soups and meats. They are also sometimes added to salads. For winter use, the stalks are cut while in flower, dried, powdered, and preserved, like other pot-herbs.

Large-leafed purple basil (Ocimum basilicum)

110

Borage (Borago officinalis)

Borage is generally classed as a hardy annual, though it is sometimes biennial. Stem two feet high; the leaves are oval, alternate, and, in common with the stalk and branches, thickly set with stiff, bristly hairs; the flowers are large and showy,—they are red, white, or blue, and often measure more than an inch in diameter; the seeds are large, oblong, slightly curved, and retain their germinative property three years.

Soil and Cultivation—Borage thrives best in light, dry soil. The seeds are sown in April or May, in drills ten or twelve inches apart, and half an inch deep. They should be sown quite thinly, or so as to secure a plant for every six or eight inches; to which distance they should be thinned. When a continued supply is required, a second sowing should be made in July. The plants seed abundantly; and, when once introduced into the garden, spring up spontaneously.

Use—The plant is rarely cultivated and little used in this country. It is sometimes employed as a pot-herb, and the young shoots are occasionally mixed in salads. They are also sometimes boiled and used as Spinach. The flowers make a beautiful garnish, and it is well worthy cultivation as an ornamental plant. "The stalks and foliage contain a large proportion of nitre; and, when dried, burn like match-paper."

Varieties—There are several varieties, differing slightly, except in the color of the flowers; the Red-flowering, White-flowering, and Blue-flowering being the principal. A variety, with variegated foliage, is described by some authors. Miller states that "they generally retain their distinctions from seeds."

FIELD AND GARDEN VEGETABLES OF THE LATE NINETEENTH CENTURY | 395

Borage officinalis
Lydia Penrose

111

Caraway (Carum carvi)

The Common Caraway is a hardy, biennial plant; a native of various parts of Europe; and, to a considerable extent, naturalized in this country. The root is long and tapering, of a yellowish-white color, and about three-fourths of an inch in diameter near the crown or at its broadest part; the flesh of the root is white, fine-grained, with a flavor not unlike that of the carrot; the flower-stalks are put forth the second season, and are about two feet and a half in height, with numerous spreading branches; the leaves are finely cut, or divided, and of a deep-green color; the flowers are small, white, and produced in umbels at the ends of the branches; the seeds, which ripen quite early in the season, are of an oblong form, somewhat curved, furrowed, slightly tapering towards the extremities, of a clear olive-brown color, and pleasant, aromatic flavor and odor,—nearly eight thousand five hundred seeds are contained in an ounce, and they retain their vitality three years.

Soil and Cultivation—Caraway is one of the hardiest of plants, and succeeds well in almost any soil or situation. In the coldest parts of the United States, and even in the Canadas, it is naturalized to such an extent about fields and mowing lands, as to be obtained in great abundance for the mere labor of cutting up the plants as the ripening of the seeds takes place.

When cultivated, the sowing may be made in April or May: but, if sown just after ripening, the seeds not only vegetate with greater certainty, but the plants often flower the ensuing season; thus saving a summer's growth. Sow in drills twelve or fifteen inches apart, and cover half an inch deep. When the plants are well up, thin to six or eight inches apart, and keep the ground loose, and free from weeds. The seeds will ripen in the July of the year after sowing. For other methods of culture, see *Coriander*.

Use—It is principally cultivated for its seeds, which constitute an article of some commercial importance; a large proportion, however, of the consumption in this country being supplied by importation from Europe. They are extensively employed by confectioners, and also for distillation. They are also mixed in cake, and, by the Dutch, introduced into cheese.

It is sometimes cultivated for its young leaves, which are used in soups and salads; or as a pot-herb, like Parsley. The roots are boiled in the manner of the Carrot or Parsnip, and by some preferred to these vegetables; the flavor being considered pleasant and delicate.

There are no described varieties.

Carum carvi
Franz Eugen Köhle

112

Clary Sage (Salvia sclarea)

Clary is a hardy, biennial plant. It is indigenous to the south of Europe, and has been cultivated in gardens for upwards of three centuries. The radical leaves are large, rough, wrinkled, oblong-heart-shaped, and toothed on the margin; stalk two feet high, four-sided, clammy to the touch; flowers pale-blue, in loose, terminal spikes; seeds round, brownish, and, like others of the family, produced four together,—they retain their vitality two years.

Sowing and Culture—It is generally grown from seeds, which are sown annually in April or May, in drills fifteen or eighteen inches apart, and half or three-fourths of an inch deep. When the young plants are two or three inches high, thin them to ten or twelve inches apart, and treat the growing crop in the usual form during summer. The leaves will be in perfection in the ensuing autumn, winter, and spring; and the plants will blossom, and produce their seeds, in the following summer.

Salvia sclarea
Jacob Sturm

Use—The leaves are used for flavoring soups, to which they impart a strong, peculiar flavor, agreeable to some, but unpleasant to most persons. It has some of the properties of Common Sage, and is occasionally used as a substitute.

The plant is seldom employed in American cookery, and is little cultivated.

113

Coriander (Coriandrum sativum)

(Cilantro)

A hardy annual, supposed to have been introduced from the south of Europe, but now naturalized in almost all temperate climates where it has once been cultivated.

Stem about two feet in height, generally erect, but, as the seeds approach maturity, often acquiring a drooping habit; stem-leaves more finely cut or divided than those proceeding directly from the root, and all possessed of a strong and somewhat disagreeable odor. The generic name is derived from *Koris* (a bug), with reference to the peculiar smell of its foliage. Flowers white, produced on the top of the plant, at the extremities of the branches, in flat, spreading umbels, or bunches; seeds globular, about an eighth of an inch in diameter, of a yellowish-brown color, with a warm, pleasant, aromatic taste,—they become quite light and hollow by age, and are often affected by insects in the manner of seed-pease. Though they will sometimes vegetate when kept for a longer period, they are not considered good when more than two years old.

Fresh Coriander

Propagation and Cultivation—Like all annuals, it is propagated from seed, which should be sown in April or May, in good, rich, mellow soil well pulverized. Sow in drills made fourteen or sixteen inches asunder and about three-fourths of an inch in depth, and thin to nine inches in the rows. It soon runs to flower and seed, and will be ready for harvesting in July or August.

In the south of England, Coriander is generally cultivated in connection with Caraway; eighteen pounds of Caraway seed being mixed with fifteen pounds of Coriander for an acre. The Coriander, being an annual, yields its crop the first season. After being cut, it is left on the field to dry, and the seeds afterwards beaten out on cloths; the facility with which these are detached not admitting of the usual method of harvesting.

An unquestionably preferable mode of cultivation would be to sow them both in drills alternately, by which means the Caraway would be more easily hoed and cleaned after the removal of the Coriander.

Use—It is generally cultivated for its seeds, which are used to a considerable extent by druggists, confectioners, and distillers. In the garden, it is sometimes sown for its leaves, which are used as Chervil in soups and salads; but, when so required, a sowing should be made at intervals of three or four weeks.

There are no varieties.

Coriander cultivation

114

Costmary/Alecost (Tanacetum balsamita)

Costmary is a hardy, perennial plant, with a hard, creeping root, and an erect, branching stem two or three feet high. The radical leaves, which are produced on long footstalks, are oval, serrated, and of a greyish colour,—those of the stalk are sessile, smaller than the radical ones, but similar in form; the flowers are deep-yellow, in erect, terminal, spreading corymbs; the seeds are small, slightly curved, and of a greyish-white colour.

Hoary-Leaved Costmary

A variety with deeply divided and hoary leaves, less fragrant than the preceding.

Propagation and Cultivation—Costmary may be cultivated in almost any description of soil or situation. It is sometimes grown from seeds, but is generally propagated by dividing the roots, which increase rapidly, and soon entirely occupy the ground. They are taken up for planting out either in spring or autumn, and should be set two feet apart in each direction. By occasionally thinning out the plants as they become too thick, a bed may be continued many years.

Costmary
Flora Batava, plate 185

Use—The plant has a soft, agreeable odor, and is sometimes used as a pot-herb for flavoring soups. The leaves are used in salads, and also for flavoring ale or beer: hence the name "Alecost."

115

Cumin (Cuminum cyminum)

Cumin is a native of Egypt. It is a tender, annual plant, from nine to twelve inches high. The leaves are deep-green, and divided into long, linear segments, not unlike those of Fennel; the flowers are white or pale-blue, and produced in small umbels at the extremities of the branches; the seeds are long, furrowed, of a pale-brownish color, and somewhat resemble those of Anise,—about seven thousand are contained in an ounce, and they retain their power of germination three years.

Close-up of cumin seeds

Soil and Cultivation—Cumin requires a light, warm-loamy soil. The seed should be sown about the beginning of May, in drills fourteen inches apart and half an inch deep. When the plants are well up, they should be thinned to three or four inches apart in the lines. The treatment of the growing crop, and the usual method of harvesting, are the same as directed for Anise or Coriander.

The seed is sometimes sown broadcast; the soil being first finely pulverized, and raked smooth and even. This may be successfully practised upon land naturally light and warm, if free from weeds.

Though a native of a warm climate, Cumin may be successfully grown throughout the Middle States, and in the warmer portions of the Northern and Eastern.

Use—The plant is cultivated for its seeds, which are carminative, and used as those of Caraway and Coriander. They are sometimes employed for flavoring spirits.

The plant is rarely grown, and the seeds are but little used, in the United States. There are no varieties.

116

Dill (Anethum graveolens)

Dill is a hardy, biennial plant. There is but one species cultivated, and there are no varieties. The stem is erect and slender, and the leaves are finely divided; the flowers are produced in June and July of the second year, and the seeds ripen in August. The plant resembles Fennel in its general character, though smaller and less vigorous.

Propagation and Cultivation—Dill flourishes best in light soil, and is propagated from seeds sown annually. As these retain their vitality but a single year, and, even when kept through the winter, vegetate slowly, they are frequently sown late in summer, or early in autumn, immediately after ripening. The drills are made a foot apart, and the seeds covered half an inch deep. The young plants should be thinned to six inches apart in the rows; and the leaves may be gathered for use from July till winter, and in the following spring till the plants have run to flower.

Dill plant with young seeds

Use—The whole plant is strongly aromatic. Its leaves are used to give flavor to pickles, particularly cucumbers; and occasionally are added to soups and sauces: the seeds are also employed for flavoring pickles. All parts of the plant are used in medical preparations.

Dill Pickles Rag, 1906

117

Fennel (Fœniculum)

Three species of Fennel are cultivated, differing not only in habit, but, to some extent, in their properties. The stems vary in height from two to four feet, and are smooth and branching; the flowers are yellow, in terminal umbels; the seeds are oval, ribbed, or furrowed, generally of a light, yellowish-brown color, and retain their vitality from three to five years.

Soil, Sowing, and Culture—A light, dry soil is best adapted to the growth of Fennel; though it will thrive well in any good garden loam. It is generally raised from seeds, which may be sown in August, just after they ripen, or in April and May. They are generally sown in drills fifteen or eighteen inches apart, and about three-fourths of an inch deep,—the young plants being afterwards thinned to twelve or fifteen inches apart in the drills; or a few seeds may be scattered broadcast on a small seed-bed, raked in, and the seedlings, when two or three inches high, transplanted to rows, as before directed.

Fennel is sometimes propagated by a division of the roots and by offsets. This may be performed either in spring, summer, or autumn. Set the roots, or shoots, fifteen inches apart in each direction; and they will soon become stocky plants, and afford an abundance of leaves for use. When cultivated for its foliage, the flowering-shoots should be cut off as they may make their appearance, to encourage the production of fresh shoots, and to give size and succulency to the leaves.

The species and their peculiar uses are as follow:

Common or Bitter Fennel

(Fœniculum vulgare)

A perennial species, with deep, strong, fleshy roots; stem three or four feet high, with finely divided leaves. The flowers are put forth in July, and the seeds ripen in August: the latter are about one-sixth of an inch long, of a greenish-brown color, and, in common with the leaves, of a decidedly bitter taste.

Soil, Sowing, and Culture — This species may be grown in almost any soil or situation. Sow the seeds soon after ripening, or early in spring. The plants require no other care than to be kept free from weeds.

Use — The young leaves are used for flavoring soups and sauces, and are sometimes mixed in salads. The seeds are carminative, and the roots and leaves have reputed medicinal properties.

Dark-Green Leaved

A variety with deep-green foliage. Its uses, and modes of culture, are the same as those of the foregoing species.

Florence Or Italian Fennel

(Finochio; Sweet Azorian Fennel; Fœniculum dulce)

Quite distinct from the Common Fennel, and generally cultivated as an annual. The stem, which is about eighteen inches high, expands near the surface of the ground; and, when divided horizontally, presents an oval form, measuring four or five inches in one direction, and two inches in the opposite. The flowers are produced in umbels, as in the other species. The seeds are slender, yellow, somewhat curved, sweet and pleasant to the taste, and of an agreeable, anise-like odor.

Sowing and Culture — The plant should be grown in well-enriched, mellow soil. Sow the seeds in April or May, thinly, in shallow drills from eighteen inches to two feet apart. Half an ounce of seeds will be sufficient for fifty feet of drill; or, by transplanting when they spring up too thickly, will furnish seedlings for a hundred feet.

The plants should be eight or ten inches apart; and, when the stems have attained a sufficient size, they should be earthed up for blanching, in the manner of Celery. Two or three weeks will be required to perfect this; and, if properly treated, the stems will be found white, crisp, tender, and excellent.

Plants from the first sowing will be ready for use in July and August. For a succession, a few seeds may be sown in June, or early in July.

Use — The blanched portion of the stem is mixed in soups, and also used as a salad. It is served like Celery, with various condiments; and possesses a sweet, pleasant, aromatic taste.

It is a popular vegetable in some parts of Europe, but is rarely cultivated in this country.

Sweet Fennel

(Malta Fennel; Fœniculum officinale)

By some writers, this has been described as a variety of the Common Fennel; but its distinctive character appears to be permanent under all conditions of soil and culture. The leaves are long and narrow, and, compared with those of the last named, less abundant, and not so pointed. The stem is also shorter, and the seeds are longer, more slender, and lighter coloured.

Sowing and Cultivation—It is propagated and cultivated as the Common Fennel.

Use—It is used in all the forms of the last named. The seeds have a sweet, pleasant, anise-like taste and odor, are strongly carminative, and yield an essential oil by distillation.

Fennel grown as a vegetable

118

Lavender (Lavendula spica)

Lavender is a hardy, low-growing, shrubby plant, originally from the south of Europe. There are three varieties; and they may be propagated from seeds by dividing the roots, or by slips, or cuttings.

The seeds are sown in April or May. Make the surface of the soil light and friable, and sow the seeds in very shallow drills six inches apart. When the seedlings are two or three inches high, transplant them in rows two feet apart, and a foot apart in the rows.

The slips, or cuttings, are set in April, two-thirds of the length in the soil, and in rows as directed for transplanting seedlings. Shade them for a few days, until they have taken root; after which, little care will be required beyond the ordinary form of cultivation.

The roots may be divided either in spring or autumn. Though Lavender grows most luxuriantly in rich soil, the plants are more highly aromatic, and less liable to injury from severe weather, when grown in light, warm, and gravelly situations.

Lavender Flower

Use—Lavender is sometimes used as a pot-herb,

"...but is more esteemed for the distilled water which bears its name, and which, together with the oil, is obtained in the greatest proportion from the flower-spikes which have been gathered in dry weather, and just before the flowers are fully expanded. The oil of lavender is obtained in the ratio of an ounce to sixty ounces of dried flowers."[1]

"In the neighbourhood of Mitcham, in Surrey, England, upwards of two hundred acres are occupied with Lavender alone."[2]

Varieties—

Broad-Leaved Lavender
(Spike Lavender)

Compared with the Common Lavender, the branches of this variety are shorter, more sturdy, and thicker set with leaves; the latter being short and broad.

The Broad-leaved Lavender rarely blossoms; but, when this occurs, the leaves of the flower-stalk are differently formed from those of the lower part of the plant, and somewhat resemble those of the Common variety. The stalks are taller, the spikes lower and looser, and the flowers smaller, than those of the last named.

Common or Blue-Flowering Lavender
(Narrow-Leaved Blue-Flowering)

A shrubby, thickly-branched plant, from a foot to upwards of three feet high, according to the depth and quality of the soil in which it is cultivated. The leaves are opposite, long, and narrow; flowers blue or purple, in spikes.

The whole plant is remarkably aromatic; but the flowers have this property in a greater degree than the foliage or branches. The plants are in perfection in July and August, and are cut for drying or distillation, close to the stem, as the blossoms on the lower part of the spikes begin to change to a brown color.

Narrow-Leaved White-flowering

A sub-variety of the Common Lavender, with white flowers. It is of smaller growth and less hardy than the last named, though not so generally cultivated. Its properties and uses are the same.

---

1. The Elements of Practical Agriculture. By David Low. London, 1843.
2. The Gardener's Assistant. By Robert Thompson.

119

Lovage (Ligusticum levisticum)

Lovage is a hardy, perennial plant, with a hollow, channelled, branching stem six or seven feet high. The leaves are winged, smooth, deep, glossy-green, and somewhat resemble those of Celery; the flowers are yellow, and produced in large umbels at the extremities of the branches; the seeds are oblong, striated, of a pale, yellowish-brown color, and retain their germinative powers but one year.

Ligusticum levisticum, 1804

Soil, Propagation, and Culture—Lovage requires a deep, rich, moist soil; and is propagated either by seeds or dividing: the roots. The seeds should be sown in August, or immediately after ripening; as, when sown in spring, they seldom vegetate well. When the young plants have made a growth of two or three inches, they should be transplanted three feet apart in each direction; and, when well established, will require little care, and continue for many years.

The roots may be divided in spring or autumn; and should be set three feet apart, as directed for seedling plants; covering the crowns three inches deep.

Use—Lovage was formerly cultivated as an esculent; but its use as such has long been discontinued. It is now cultivated for its medicinal properties; both the seeds and roots being used. The latter are large, fleshy, dark-brown without, yellowish within, and of a peculiar, warm, aromatic taste. They are sliced and dried, and in this state are used to some extent by confectioners. The seeds are similar to the roots in taste and odor, but have greater pungency. In appearance and flavor, the plant is not unlike Celery.

There are no varieties.

120

Calendula (Calendula officinalis)

***(Marigold)*[1]**

This hardy annual is a native of France and the south of Europe. Aside from its value for culinary purposes, its large, deep, orange-yellow flowers are showy and attractive; and it is frequently cultivated as an ornamental plant. The stem is about a foot in height; the leaves are thick and fleshy, rounded at the ends, and taper to the stalk; the flowers are an inch and a half or two inches in diameter, yellow,—differing, however, in depth of color, and single or double according to the variety; the seeds are large, light-brown, much curved and contorted, and very irregular both in their size and form.

Sowing and Cultivation—The plant is of easy culture. The seeds are sown in autumn, just after ripening; or in April, May, or June. Make the drills a foot apart; cover the seed three-fourths of an inch deep; and, when the plants are an inch or two inches high, thin them to eight or ten inches apart. Plants from the first sowing will blossom early in July, and continue in bloom until destroyed by frost.

Calendula flowers
Elias Verhulst

Gathering—The flowers are gathered when fully expanded, divested of their calyxes, and spread in a light, airy, shaded situation until they are thoroughly dried. They are gathered as they come to perfection; for, when the plants are allowed to ripen their seeds, they become much less productive.

To raise Seed—Leave one or two of the finest plants, without cutting the flowers; and, when the heads of seed begin to change from a green to a brownish color, cut them off, spread them a short time as directed for drying the flowers, and pack away for use.

Use—The flowers are used in various parts of Europe for flavouring soups and stews, and are much esteemed. Though often grown as an ornamental plant, the flowers are but little used in this country for culinary purposes. The varieties are as follow:

Common Orange-Flowered

Flowers single, deep orange-yellow, high-flavored. It is considered the best variety for cultivation.

Lemon-Flowered

This differs from the foregoing in the paler color of the flowers, which are also less aromatic. The plants are not distinguishable from those of the Common Orange-flowered.

Double Orange-Flowering

Of the same color with the first named, but with fine, large, double ornamental flowers. The petals are flat, and rest in an imbricated manner, one on the other, as in some varieties of the Anemone. It is more productive, but less aromatic, than the Single-flowering.

Double Lemon-Flowering

A variety of the second-named sort, with double flowers like those of the preceding.

To raise good seeds of either of the double-flowering kinds, all plants producing single flowers must be removed as soon as their character is known. When the single and double-flowering plants are suffered to grow together, the latter rapidly deteriorate, and often ultimately become single-flowering.

Childing, or Proliferous Marigold

This variety produces numerous small flowers from the margin of the calyx of the large central flowers. It is quite ornamental, but of little value as an esculent.

---

1. The author called this plant 'marigold', though giving it the Latin name Calendula officinalis. At the time, 'marigold' was the name for the plant now known as known as 'Calendula', not the 'marigold' of the Tagetes genus.

121

Oregano and Marjoram (Origanum vulgare)

Oregano[1]

(Origanum vulgare)

A perennial species, with a shrubby, four-sided stem, a foot and a half high; leaves oval, opposite,—at the union of the leaves with the stalk, there are produced several smaller leaves, which, in size and form, resemble those of the Common Sweet Marjoram; the flowers are pale-red, or flesh-coloured, and produced in rounded, terminal spikes; the plants blossom in July and August, and the seeds ripen in September.

Oregano vulgare in flower

Propagation and Culture—It may be grown from seeds, but is generally propagated by dividing the roots, either in spring or autumn. Set them in a dry and warm situation, in rows fifteen inches apart, and ten or twelve inches from plant to plant in the rows.

The seeds may be sown in a seed-bed in April or May, and the seedlings transplanted to rows as directed for setting the roots; or they may be sown in drills fifteen inches apart, afterwards thinning out the young plants to ten inches apart in the drills.

There is a variety with white flowers, and another with variegated foliage.

Use—The young shoots, cut at the time of flowering and dried in the shade, are used as Sweet Marjoram for seasoning soups and meats. The whole plant is highly aromatic.

Sweet Marjoram

(Knotted Marjoram; Origanum majorana)

Sweet Marjoram is a native of Portugal. Though a biennial, it is always treated as an annual; not being sufficiently hardy to withstand the winters of the Middle or Northern States in the open ground. The plant is of low growth, with a branching stem, and oval or rounded leaves. The flowers, which appear in July and August, are of a purplish color, and produced in compact clusters, or heads, resembling knots: whence the term "Knotted Marjoram" of many localities. The seeds are brown, exceedingly small, and retain their germinative properties three years.

Oregano majorana in flower

Sowing and Cultivation—Sweet Marjoram is raised from seeds sown annually in April, May, or June. Its propagation, however, is generally attended with more or less difficulty, arising from the exceeding minuteness of the seeds, and the liability of the young seedlings to be destroyed by the sun before they become established. The seeds are sown in drills ten or twelve inches apart, and very thinly covered with finely pulverized loam. Coarse light matting is often placed over the bed immediately after sowing, to facilitate vegetation; and, if allowed to remain until the plants are well up, will often preserve a crop which would otherwise be destroyed.

The seeds are sometimes sown in a hot-bed, and the plants set out in May or June, in rows twelve inches apart, and six inches apart in the rows.

Gathering—The plants, when in flower or fully developed, are cut to the ground; and, for winter use, are dried and preserved as other pot-herbs.

Use—Sweet Marjoram is highly aromatic, and is much used, both in the green state and when dried, for flavoring broths, soups, and stuffings.

Pot Marjoram

(Origanum onites)

A perennial species, from Sicily. Stem a foot or more in height, branching; leaves oval, comparatively smooth; the flowers are small, of a purplish colour, and produced in spikes.

Propagation and Cultivation—The species is propagated, and the crop in all respects should be treated, as directed for Common Marjoram. The properties and uses of the plant are also the same. Both, however, are much inferior to the Sweet Marjoram last described.

Origanum onites

Winter Sweet Marjoram

(Origanum heracleoticum)

A half-hardy perennial, from the south of Europe. Stem eighteen inches high, purplish; the leaves are opposite, oval, rounded at the ends, and resemble those of Sweet Marjoram; the flowers are white, and are put forth in July and August, in spikelets about two inches in length; the seeds ripen in September.

Propagation and Culture—It may be grown from seeds, but is generally propagated by dividing the roots either in the spring or fall, and planting the divisions ten inches apart, in rows eighteen inches asunder. It succeeds best in dry localities, and requires no other attention than to have the soil kept loose, and free from weeds.

There is a variety with variegated leaves, but differing in no other respect from the foregoing.

Use—The leaves and young branches are used in soups, and stuffing for meats; and should be cut when just coming into flower, and dried in the shade.

1. The identification 'oregano' has bee added by the authors to help modern readers distinguish it from other varieties known as 'marjoram.'

122

Nigella (Nigella sativa)

Nigella Sativa
Köhler's Medizinal-Pflanzen

(Black Cumin; Quatre Epices, of the French)

A hardy, annual plant from the East Indies. Stem twelve to eighteen inches high, with alternate, sessile, finely divided leaves; the flowers are large, white, variegated with blue; the seeds, which are produced in a roundish capsule, are somewhat triangular, wrinkled, of a yellowish colour, and pungent, aromatic taste,—about thirteen thousand are contained in an ounce, and they retain their vitality three years. There is a species cultivated, the seeds of which are black.

Soil and Cultivation—It is always raised from seed, and thrives best in light, warm soil. The seed may be sown from the middle of April to the middle of May. Pulverize the soil well, make the surface smooth and even, and sow in drills twelve or fourteen inches apart and about half an inch deep. When the plants are two inches high, thin them to five or six inches apart in the rows. During the summer, cultivate in the usual manner, keeping the soil loose, and watering occasionally if the weather be dry; and in August or September, or when the seed ripens, cut off the plants at the roots, spread them in an airy situation, and, when sufficiently dried, thresh out; after which, spread the seed a short time to evaporate any remaining moisture, and they will be ready for use.

Use—The seeds have a warm, aromatic taste; and are employed in French cookery, under the name of *quatre épices*, or "four spices."

123

Parsley (Apium petroselinum)

Parsley is a hardy, biennial plant from Sardinia. The leaves of the first year are all radical, compound, rich, deep-green, smooth, and shining. When fully developed, the plant measures three or four feet in height; the flowers are small, white, in terminal umbels; the seeds are ovoid, somewhat three-sided, slightly curved, of a greyish-brown colour and aromatic taste,—seven thousand are contained in an ounce, and they retain their vitality three years.

Soil and Propagation—Parsley succeeds best in rich, mellow soil, and is propagated from seeds sown annually; an ounce of seed being allowed to a hundred and fifty feet of drill.

Sowing—As the seed vegetates slowly,—sometimes remaining in the earth four or five weeks before the plants appear,—the sowing should be made as early in spring as the ground is in working condition. Lay out the bed of a size corresponding to the supply required, spade it deeply and thoroughly, level the surface (making it fine and smooth), and sow the seed in drills fourteen inches apart, and half an inch deep. When the plants are two or three inches high, thin them to eight or ten inches apart; being careful, in the thinning, to leave only the best and finest curled plants.

According to Lindley, the finest curled kinds will rapidly degenerate and become plain, if left to themselves; while, on the other hand, really excellent sorts may be considerably improved by careful cultivation.

The best curled Parsley is obtained by repeated transplantings. When the seedlings are two inches high, they are set in rows ten inches apart, and six inches apart in the rows. In about four weeks, they should be again transplanted to where they are to remain, in rows eighteen inches apart, and fourteen inches apart in the rows. When thus treated, the plants become remarkably close, of a regular, rosette-like form, and often entirely cover the surface of the ground. When grown for competition or for exhibition, this process of transplanting is thrice and often four times repeated.

Seed—In autumn, select two or three of the finest curled and most symmetrical plants; allow them to remain unplucked; give a slight protection during winter; and, in the following summer, they will yield abundantly. Much care is requisite in keeping the varieties true. This is especially the case with the curled sorts. The seed-growers, who value their stock and character, select the best and finest curled plants, and allow no others to flower and seed. When the object is to improve a variety, but few seeds are saved from a plant; and, in some cases, but few seeds from a head.

Use—The leaves of the curled varieties afford one of the most beautiful of garnishes: they are also used for flavoring soups and stews. The seeds are aromatic, and are sometimes used as a substitute for the leaves; though the flavor is much less agreeable.

Varieties—

Dwarf Curled Parsley

(Curled Parsley; Sutton's Dwarf Curled; Usher's Dwarf Curled)

A fine, dwarfish, curled variety, long cultivated in England. In some gardens, it is grown in such perfection as to resemble a tuft of finely curled, green moss.

It is hardy, and slow in running to seed, but liable to degenerate, as it constantly tends to increase in size and to become less curled.

From the Dwarf Curled Parsley, by judicious cultivation and a careful selection of plants for seed, have originated many excellent sorts of stronger growth, yet retaining its finely curled and beautiful leaves.

Mitchell's Matchless Winter

A fine, curled sort, larger than the Dwarf Curled; and, on account of its remarkable hardiness, recommended as one of the best for winter culture.

Myatt's Triple-Curled

(Myatt's Garnishing; Myatt's Extra Fine Curled; Windsor Curled)

The leaves of this variety are large and spreading, bright-green above, paler beneath. When true, the foliage is nearly as finely curled as that of the Dwarf, though the plant is much larger and stronger in its habit.

Plain Parsley

(Common Parsley)

The leaves of this sort are plain, or not curled; and the plant produces them in greater quantity than the curled sorts. It is also somewhat hardier.

For many years, it was the principal variety grown in the gardens of this country; but has now given place to the curled sorts, which, if not of better flavor, are generally preferred, on account of their superior excellence for garnishing.

Rendle's Treble Garnishing

A variety of the Dwarf Curled, of larger size; the leaves being as finely curled and equally beautiful.

Hamburg or Large-Rooted Parsley

(Turnip-Rooted Parsley; Petroselinum crispum var. tuberosum)

A variety of the Common Plain Parsley, with stronger foliage. Though the leaves are sometimes used in the manner of those of the Common Parsley, it is generally cultivated for its fusiform, fleshy roots.

To obtain these of good size and quality, the soil should not be too rich, but deeply and thoroughly trenched. Sow the seeds in April or May, in drills a foot or fourteen inches apart, and three-fourths of an inch deep; and, when the seedlings are two or three inches high, thin them to six or eight inches apart in the rows. Cultivate during the season as carrots or parsnips; and, in October, the roots will have attained their growth, and be suitable for use. Take them up before the ground closes, cut off the tops within an inch or two of the crowns, pack in earth or sand, and store in the cellar for winter.

Root parsley

To raise Seeds—Reset a few roots in April, two feet apart; or leave a few plants in the open ground during the winter. They will blossom in June and July, and ripen their seeds in August.

Use—The roots are eaten, boiled as carrots or parsnips. In connection with the leaves, they are also mixed in soups and stews, to which they impart a pleasant, aromatic taste and odor.

Naples or Celery-Leaved Parsley

(Neapolitan Parsley; Celery Parsley)

This variety somewhat resembles Celery; and, by writers on gardening, is described as a hybrid between some of the kinds of Celery and the Large-rooted or Hamburg Parsley. With the exception of their larger size, the leaves are similar to those of the Common Plain Parsley.

Use—The leaves are sometimes employed for garnishing; but are generally blanched, and served as Celery.

Sowing and Cultivation—The plants are started in a hot-bed in March, or the seeds may be sown in a seed-bed in the open ground in May. When the seedlings are four or five inches high, transplant to trenches two feet apart and six or eight inches deep, setting the plants a foot apart in the trenches; afterwards gather the earth gradually about the stems, in the process of cultivation; and, when they are sufficiently grown and blanched, harvest and preserve as Celery.

To raise Seeds—Leave two or three plants unblanched. They should be eighteen inches asunder, and may remain in the open ground during winter. They will flower, and yield a plentiful supply of seeds, the following summer.

124

Peppermint (Mentha piperita)

Peppermint is a hardy, perennial plant, introduced from Europe, and growing naturally in considerable abundance along the banks of small streams, and in rich, wet localities. Where once established, it spreads rapidly, and will remain a long period.

Stem smooth, erect, four-sided, and from two to three feet in height; leaves opposite, ovate, pointed, toothed on the margin; flowers purplish, or violet-blue, in terminal spikes; the seeds are small, brown, or blackish-brown, and retain their vitality four years.

Propagation and Culture—It may be grown from seeds; but this method of propagation is rarely practised, as it is more readily increased by dividing the roots.

The agreeable odor, and peculiar, warm, pleasant flavor, of the leaves are well known. The plant, however, is little used as a pot-herb, but is principally cultivated for distillation. For the latter purpose, the ground is ploughed about the middle of May, and furrowed in one direction, as for drill-planting of potatoes; making the furrows about eighteen inches apart. The best roots for setting are those of a year's growth; and an acre of these will be required to plant ten acres anew. These are distributed along the furrows in a continuous line, and covered sometimes with the foot as the planter drops the roots, and sometimes by drawing the earth over them with a hoe. In about four weeks, the plants will be well established, and require hoeing and weeding; which is usually performed three times during the season, the cultivation being finished early in August.

"The cutting and distilling commence about the 25th of August, except in very dry seasons, when it stands two or three weeks longer, and continues until the 1st of October; during which period the plant is in full inflorescence, and the lower leaves begin to grow sear. It is raked together in small heaps; when it is suffered to wilt ten or twelve hours, if convenient.

"The next year, little is done to the mint-field but to cut and distil its product. During this (the second) year, a few weeds make their appearance, but not to the injury of the crop; though the most careful of the mint-growers go through their fields, and destroy them as much as possible. The second crop is not so productive as the first.

"The third year, little labor is required other than to harvest and distil the mint. The stem is coarser than before, and the leaves still less abundant. The weeds this year abound, and are not removed or destroyed; half or more of the product of the field often being weeds.

"The fourth year, the field is ploughed up early in the spring; and this 'renewing' is sometimes done every third year.

"The fifth year, without any further attention, produces a crop equal to the second; after which, the field is pastured and reclaimed for other crops.

"The first year produces the best quality of oil, the highest yield per acre, and the greatest amount to the quantity of herbage."[1]

1. *F. Stearns.*

Fresh peppermint leaves used as tea

125

Rosemary (Salvia rosmarinus)

(Rosmarinus officinalis)

Rosemary is a half-hardy, shrubby plant, from three to six feet in height. The leaves vary in form and color in the different varieties; the flowers are small, generally blue, and produced in axillary clusters; the seeds are brown, or blackish-brown, and retain their vitality four years.

Propagation and Cultivation—Like most aromatic plants, Rosemary requires a light, dry soil; and, as it is not perfectly hardy, should have a sheltered situation. The Common Green-leaved and the Narrow-leaved are best propagated by seeds; but the variegated sorts are propagated only by cuttings or by dividing the roots. The seeds are sown in April, in a small nursery-bed; and the seedlings, when two or three inches high, transplanted in rows two feet apart, and eighteen inches apart in the rows.

When propagated by cuttings, they should be taken off in May or June, six inches long, and set two-thirds of the length in the earth, in a moist, shady situation: when well rooted, transplant as directed for seedlings. The roots may be divided in spring or autumn.

Use—It is sometimes employed, like other pot-herbs, for flavoring meats and soups. It is used in the manufacture of "eau de Cologne," and its flowers and calyxes form a principal ingredient in the distillation of "Hungary Water." Infusions of the leaves are made in some drinks, and the young stems are used as a garnish.

There are four varieties, as follow:

Common or Green-Leaved

Leaves narrow, rounded at the ends,—the upper and under surface green; the flowers are comparatively large, and deep-colored.

The plant is of spreading habit; and, in all its parts, is more strongly aromatic than the Narrow-leaved. It is decidedly the best sort for cultivation.

Gold-Striped

A variety of the Common or Green-leaved, with foliage striped, or variegated with yellow.

This and the Silver-leaved are generally cultivated as ornamental plants. The Gold-striped is much the hardier sort, and will succeed in any locality where the Common Green-leaved is cultivated.

Narrow-Leaved

The plants of this variety are smaller and less branched than those of the Common or Green-leaved, and are also less fragrant; the leaves are hoary beneath, and the flowers are smaller and of a paler color.

It is used in all the forms of the Common or Green-leaved, but is less esteemed.

Silver-Striped

This is a sub-variety of the Common or Green-leaved, and the most tender of all the sorts. It is principally cultivated for its variegated foliage; the leaves being striped, or variegated with white.

Like the Gold-striped, it can only be propagated by slips or by dividing the roots, and must be well protected during winter.

Close-up of flowering rosemary

126

Sage (Salvia officinalis)

Sage is a low-growing, hardy, evergreen shrub, originally from the south of Europe. Stem from a foot and a half to two feet high,—the leaves varying in form and color in the different species and varieties; the flowers are produced in spikes, and are white, blue, red, purple, or variegated; the seeds are round, of a blackish-brown color, and retain their power of germination three years,—nearly seven thousand are contained in an ounce.

Soil and Propagation—Sage thrives best in light, rich, loamy soil. Though easily grown from slips, or cuttings, it is, in this country, more generally propagated from seeds. These may be sown on a gentle hot-bed in March, and the[Pg 439] plants set in the open ground in June, in rows eighteen inches apart, and a foot asunder in the rows; or the seeds may be sown in April, where the plants are to remain, thinly, in drills eighteen inches apart, and three-fourths of an inch deep. When the plants are two inches high, thin them to a foot apart in the rows; and, if needed, form fresh rows by resetting the plants taken up in thinning.

If grown from cuttings, those from the present year's growth succeed best. These should be set in June. Cut them four or five inches in length, remove the lower leaves, and set them two-thirds of their length in the earth. Water freely, and shade or protect with hand-glasses. By the last of July, or first of August, they will have taken root, and may be removed to the place where they are to remain.

It may also be propagated by dividing the roots in spring or autumn, in the manner of other hardy shrubs.

Gathering and Use—Sage should be gathered for drying before the development of the flowering-shoots; and, when cultivated for its leaves, these shoots should be cut out as they make their appearance. When thus treated, the product is largely increased; the leaves being put forth in much greater numbers, and of larger size.

It is sometimes treated as an annual; the seeds being sown in April, in drills fourteen inches apart, and the plants cut to the ground when they have made sufficient growth for use.

The leaves are employed, both in a green and dried state, for seasoning stuffings, meats, stews, and soups. Sage is also used for flavoring cheese; and, in the form of a decoction, is sometimes employed for medical purposes.

Species and Varieties—

Broad-Leaved Green Sage

(Balsamic Sage)

Stems shrubby, less erect and more downy than those of the succeeding species; the leaves are comparatively large, broad, heart-shaped, woolly, toothed on the margin, and produced on long footstalks,—those of the flower-stalks are oblong, sessile, and nearly entire on the borders; the flowers are small, pale-blue, and much less abundant than those of the Common Sage.

It is rarely employed in cookery, but for medical purposes is considered more efficacious than any other species or variety.

Common or Red-Leaved

(Purple-Top; Red-Top)

This is the Common Sage of the garden; and with the Green-leaved, which is but a sub-variety, the most esteemed for culinary purposes. The young stalks, the leaf-stems, and the ribs and nerves of the leaves, are purple: the young leaves are also sometimes tinged with the same color, but generally change by age to clear green.

The Red-leaved is generally regarded as possessing a higher flavor than the Green-leaved, and is preferred for cultivation; though the difference, if any really exists, is quite unimportant. The productiveness of the varieties is nearly the same. The leaves of the Green Sage are larger than those of the Red; but the latter produces them in greater numbers.

Green-Leaved

(Green-Top)

A variety of the preceding; the young shoots, the leaf-stalks, and the ribs and nerves of the leaves, being green.

There appears to be little permanency in the characters by which the varieties are distinguished. Both possess like properties, and are equally worthy of cultivation. From seeds of either of the sorts, plants answering to the description of the Red-leaved and Green-leaved would probably be produced, with almost every intermediate shade of colour.

Narrow-Leaved Green Sage

(Sage of Virtue)

Leaves narrow, hoary, toothed towards the base; the spikes of flowers are long, and nearly leafless; flowers deep-blue; the seeds are similar to those of the Red-leaved, and produced four together in an open calyx. Compared with the Common Red-leaved or Green-leaved, the leaves are much narrower, the spikes longer and less leafy, and the flowers smaller and of a deeper color.

The variety is mild flavored, and the most esteemed of all the sorts for use in a crude state; as it is also one of the best for decoctions.

> "At one period, the Dutch carried on a profitable trade with the Chinese by procuring the leaves of this species from the south of France, drying them in imitation of tea, and shipping the article to China, where, for each pound of sage, four pounds of tea were received in exchange."[1]

Variegated-Leaved Green Sage

A sub-variety of the Green-leaved, with variegated foliage. It is not reproduced from seeds, and must be propagated by slips or by dividing the roots.

Variegated-Leaved Red Sage

This is but an accidental variety of the Common Red-leaved Sage, differing only in its variegated foliage. It can be propagated only by cuttings or by a division of the roots.

1. The Book of the Garden. By Charles M'Intosh. 2 vols. Edinburgh and London, 1855.

Salvia officinalis
Carl Stupper

127

Savory (Satureja)

The cultivated species are as follow:

Headed Savory

(Satureja capitata)

A perennial plant, with a rigid, angular, branching stem a foot and a half high. The leaves are firm, pointed, and, when bruised, emit a strong, pleasant, mint-like odor; the flowers are white, and are produced in terminal, globular heads; the seeds are quite small, of a deep-brownish color, and retain their vitality three years.

It may be propagated from seeds or by dividing the roots; the latter method, however, being generally practised. The young shoots are used in all the forms of Summer Savory.

Satureja hortensis
Jacob Sturm

Shrubby Savory

(Satureja viminea)

A shrub-like, perennial species, cultivated in the same manner as the Winter Savory. The plant has the pleasant, mint-like odor of the species first described, but is little used either in cookery or medicine.

Summer Savory

(Satureja hortensis)

An annual species, from the south of Europe. Stem twelve or fifteen inches high, erect, rather slender, and producing its branches in pairs; the leaves are opposite, narrow, rigid, with a pleasant odour, and warm, aromatic taste; the flowers are pale-pink, or flesh-coloured, and are produced at the base of the leaves, towards the upper part of the plant, each stem supporting two flowers; the seeds are quite small, deep-brown, and retain their vitality two or three years.

Propagation and Cultivation—Summer Savory is always raised from seeds, sown annually in April or May. It thrives best in light, mellow soil; and the seed should be sown in shallow drills fourteen or fifteen inches apart. When the plants are two or three inches high, thin them to five or six inches apart in the rows, and cultivate in the usual manner during the summer.

When the plants have commenced flowering, they should be cut to the ground, tied in small bunches, and dried in an airy, shady situation.

For early use, the seeds are sometimes sown in a hot-bed on a gentle heat, and the seedlings afterwards transplanted to the open ground in rows, as directed for sowing.

Use—The aromatic tops of the plant are used, green or dried, in stuffing meats and fowl. They are also mixed in salads, and sometimes boiled with pease and beans. It is sold in considerable quantities at all seasons of the year, in a dried and pulverized state, packed in hermetically-sealed bottles or boxes.

Winter Savory

(Saturjea montana)

A hardy, evergreen shrub, with a low, branching stem about a foot in height. The leaves are opposite, narrow, and rigid, like those of the preceding species; the flowers resemble those of the Summer Savory, but are larger and of a paler color; the seeds, which ripen in autumn, are small, dark-brown, and retain their vitality three years.

Propagation and Culture—

> "It may be raised from seed sown in April or May; but is generally propagated by dividing the plants in April, or by cuttings of the young shoots taken off in April or May. The cuttings should be planted two-thirds of their length deep, on a shady border, and, if necessary, watered until they take root. When well established, they may be planted out a foot apart, in rows fifteen inches asunder. Some may also be planted as an edging.
>
> "The plants should be trimmed every year in autumn, and the ground between the rows occasionally stirred; but, in doing this, care must be taken not to injure the roots. Fresh plantations should be made before the plants grow old and cease to produce a sufficient supply of leaves."

Use—It is used for the same purposes as Summer Savory. The leaves and tender parts of the young branches are mixed in salads: they are also boiled with pease and beans; and, when dried and powdered, are used in stuffings for meats and fowl.

128

Spearmint (Mentha spicata)

(Green Mint; Mentha viridis)

A hardy, perennial plant, introduced from Europe, and generally cultivated in gardens, but growing naturally in considerable abundance about springs of water, and in rich, wet localities. The stem is erect, four-sided, smooth, and two feet or more in height; the leaves are opposite, in pairs, stemless, toothed on the margin, and sharply pointed; the flowers are purple, and are produced in August, in long, slender, terminal spikes; the seeds are small, oblong, of a brown color, and retain their vitality five years,—they are generally few in number, most of the flowers being abortive.

Soil, Propagation, and Culture—It may be grown from seed, but is best propagated by a division of the roots, which are long and creeping, and readily establish themselves wherever they are planted. Spearmint thrives best in rich, moist soil; but may be grown in any good garden loam. The roots may be set either in the autumn or spring.

Where large quantities are required for marketing in the green state, or when grown for distillation, lay out the land in beds three or four feet in width, and make the drills two or three inches deep and a foot apart. Having divided the roots into convenient pieces, spread them thinly along the drills, and earth them over to a level with the surface of the bed. Thus treated, the plants will soon make their appearance; and may be gathered for use in August and September. Just before severe weather, give the beds a slight dressing of rich soil; and, the ensuing season, the plants will entirely occupy the surface of the ground.

Use—Mint is sometimes mixed in salads, and is used for flavoring soups of all descriptions. It is often boiled with green pease; and, with the addition of sugar and vinegar, forms a much-esteemed relish for roasted lamb. It has also much reputed efficacy as a medicinal plant.

Curled-Leaved Spearmint

A variety with curled foliage. It is a good sort for garnishing; but, for general use, is inferior to the Common or Plain-leaved species before described. Propagated by dividing the roots.

129

Tansy (Tanacetum vulgare)

Tansy is a hardy, perennial, herbaceous plant, naturalized from Europe, and abundant by roadsides and in waste places. Its stem is from two to three feet high; the leaves are finely cut and divided, twice-toothed on the margin, and of a rich, deep-green colour; flowers in corymbs, deep-yellow, and produced in great abundance; the seeds are small, of a brownish colour, and retain their vitality three years.

Soil and Cultivation—Tansy may be grown in almost any soil or situation, and is propagated from seeds or by dividing the roots; the latter method being generally practised. In doing this, it is only necessary to take a few established plants, divide them into small pieces or collections of roots, and set them six inches apart, in rows a foot asunder, or in hills two feet apart in each direction. They will soon become established; and, if not disturbed, will completely occupy the ground. In most places, when once introduced, it is liable to become troublesome, as the roots not only spread rapidly, but are very tenacious of life, and eradicated with difficulty.

When cultivated for its leaves, the flowering-shoots should be cut off as they make their appearance. It is but little used, and a plant or two will afford an abundant supply.

Use—The leaves have a strong, peculiar, aromatic odor, and a bitter taste. They were formerly employed to give color and flavor to various dishes, but are now rarely used in culinary preparations. The plant possesses the tonic and stomachic properties common to bitter herbs.

There are three cultivated varieties, as follow:

Curled-Leaved Tansy

(Double Tansy; Tanacetum vulgare, var. crispum)

This differs from the Common Tansy in the frilled or curled character of the leaves, which have some resemblance to the leaves of the finer kinds of Curled Cress or Parsley. They are of a rich green color, and are sometimes employed for garnishing. In the habit of the plant, color of the flowers, odor and flavor of the leaves, the variety differs little, if at all, from the Common Tansy. It is more beautiful than the last-named; and, in all respects, much more worthy of cultivation.

Propagated only by dividing the roots.

Large-Leaved Tansy

Leaves larger than those of any other variety, but much less fragrant. It is of little value, and rarely cultivated.

Variegated-Leaved

A variety with variegated foliage. Aside from the peculiar color of the leaves, the plant differs in no respect from the Common Tansy: it grows to the same height, the flowers are of the same color, and the leaves have the same taste and odor.

It must be propagated by dividing the roots; the variegated character of the foliage not being reproduced from seeds.

Tanacetum vulgare

130

Thyme (Thymus)

Two species of Thyme are cultivated for culinary purposes,—the Common Garden Thyme (*T. vulgaris*) and the Lemon or Evergreen Thyme (*T. citriodorus*).

They are hardy, perennial plants, of a shrubby character, and comparatively low growth. They are propagated from seeds and by dividing the roots; but the finest plants are produced from seeds.

Of the Common Garden Thyme, there are three varieties:

Broad-Leaved

The Broad-Leaved Thyme is more cultivated in this country than any other species or variety. The stem is ten or twelve inches high, shrubby, of a brownish-red colour, and much branched; the leaves are small, narrow, green above, and whitish beneath; flowers purple, in terminal spikes; the seeds are black, and exceedingly small,—two hundred and thirty thousand being contained in an ounce; they retain their vitality two years.

Wild Thyme
William Catto

Propagation and Cultivation—When propagated by seeds, they are sown in April or May, thinly, in shallow drills ten or twelve inches apart. When the plants are up, they should be carefully cleared of weeds, and thinned to eight or ten inches apart, that they may have space for development. They may be cut for use as soon as they have made sufficient growth; but, for drying, the stalks are gathered as they come into flower.

If propagated by dividing the roots, the old plants should be taken up in April, and divided into as many parts as the roots and tops will admit. They are then transplanted about ten inches apart, in beds of rich, light earth; and, if the weather be dry, watered till they are well established. They may be cut for use in August and September.

Use—The leaves have an agreeable, aromatic odor; and are used for flavoring soups, stuffings, and sauces.

Narrow-Leaved

The stalks of this variety are shorter than those of the Broad-leaved; the leaves also are longer, narrower, and more sharply pointed; and the flowers are larger.

It is propagated, cultivated, and used as the Broad-leaved.

Variegated-Leaved

A sub-variety of the Broad-leaved, with variegated foliage. It is generally cultivated as an ornamental plant; and is propagated only by dividing the roots, as directed for the Broad-leaved.

Lemon Thyme

(Thymus citriodorus)

A low, evergreen shrub, with a somewhat trailing stem, rarely rising more than six or eight inches high. It is readily distinguished from the Common or Broad-leaved by the soft, pleasant, lemon-like odor of the young shoots and leaves.

It is used for flavoring various dishes, and by some is preferred to the Broad-leaved.

The species is propagated from seeds by dividing the roots, and by layers and cuttings. Seedling plants, however, are said to vary in fragrance; and, when a choice stock can be obtained, it is better to propagate by dividing the plants.

Part Nine: Leguminous Plants

Common Garden-bean; Asparagus-bean; Lima Bean; Scarlet-runner; Sieva; Chick-pea; Chickling Vetch; English Bean; Lentil; Lupine; Pea; Peanut; Vetch, or Tare; Winged Pea.

Various types of common and lima beans

131

The Common Garden Bean (Phaseolus vulgaris)

(French Bean; Kidney-Bean; Haricot, of the French)

The Common Garden-bean of the United States is identical with the French or Kidney Bean of England and France, and is quite distinct from the English or Garden Bean of French and English catalogues.

The American Garden-bean is a tender, annual plant from the East Indies, with a dwarfish or climbing stem and trifoliate leaves. The flowers are variable in colour, and produced in loose clusters; the seeds are produced in long, flattened, or cylindrical, bivalved pods, and vary, in a remarkable degree, in their size, form, and colour,—their germinative powers are retained three or four years.

As catalogued by seedsmen, the varieties are divided in two classes,—the Dwarfs, and the Pole or Running Sorts.

Dwarf/Bush Beans

The plants of this class vary from a foot to two feet in height. They require no stakes or poles for their support; and are grown in hills or drills, as may suit the taste or convenience of the cultivator.

All of the varieties are comparatively tender, and should not be planted before settled, mild weather. They succeed best in warm, light soil; but will flourish in almost any soil or situation, except such as are shaded or very wet.

When planted in drills, they are made about two inches deep, and from fourteen to twenty inches apart. The seeds are planted from three to six inches apart; the distance in the drills, as well as the space between the drills, being regulated by the habit of the variety cultivated.

If planted in hills, they should be three feet apart in one direction, and about two feet in the opposite. If the variety under cultivation is large and vigorous, four or five plants may be allowed to a hill; if of an opposite character, allow twice this number.

To raise Seed—Leave a row or a few hills entirely unplucked. Seed is of little value when saved at the end of the season from a few scattered pods accidentally left to ripen on plants that have been plucked from time to time for the table.

Bagnolet

A half-dwarf, French variety. Plant strong and vigorous, with remarkably large, deep-green foliage; flowers bright lilac; the pods are straight, seven inches long, half an inch wide, streaked and spotted with purple when sufficiently grown for shelling in their green state, nankeen-yellow when fully ripe, and contain six seeds, which are nearly straight, rounded at the ends, a little flattened on the sides, three-fourths of an inch long, a fourth of an inch thick, and of a violet-black colour, variegated or marbled with drab.

About sixteen hundred beans are contained in a quart; and, as the plants are vigorous growers, this amount of seed will be sufficient for three hundred feet of drill, or for nearly three hundred hills. If planted in drills, they should be made twenty inches apart, and two plants allowed to a linear foot.

The variety is not early, and requires the entire season for its full perfection. When sown as soon as the weather is suitable, the plant will blossom in about seven weeks. In sixty days, pods may be plucked for use; and the crop will be ready for harvesting in fifteen weeks from the time of planting. For its green pods, the seeds may be planted until the middle of July.

1. Haricot de Bagnolet; 2. Haricot blanc géant sans parchemin; 3. Haricot d'Alger, (beurre) noir nain; 4. Haricot d'Alger, noir à rames, haricot beurre noir.
Vilmorin-Andrieux, 1891

The Bagnolet is of little value as a shelled-bean, either green or ripe. As a string-bean, it is deservedly considered one of the best. The pods are produced in great abundance; and are not only tender, succulent, and well flavoured, but remain long on the plants before they become tough, and unfit for use. If the pods are plucked as they attain a suitable size, new pods will rapidly succeed, and the plants will afford a continued supply for several weeks.

Black-Eyed China

Plant fifteen inches high, less strong and vigorous than that of the Common Red-eyed China; the flowers are white; the pods are comparatively short, usually about five inches long, green and straight while young, straw-yellow when sufficiently advanced for shelling, yellow, thick, hard, and parchment-like when ripe, and contain five or six seeds,—these are white, spotted and marked about the eye with black, of an oblong form, usually rounded, but sometimes shortened at the ends, slightly compressed on the sides, and measure half an inch in length, and three-eighths of an inch in thickness.

A quart contains fifteen hundred beans, and will plant a drill, or row, of two hundred feet, or a hundred and fifty hills.

The variety is early. When sown at the commencement of the season, the plants will blossom in six weeks, produce pods for the table in seven weeks, pods for shelling in ten weeks, and ripen in eighty-seven days. It yields well, ripens off at once, and, on account of the thick, parchment-like character of the pods, suffers much less from wet and unfavourable seasons than many other sorts.

As a string-bean, it is of fair quality, good when shelled in the green state, and farinaceous and mild flavoured when ripe.

Blue Pod

A half-dwarf variety, growing from two to three feet high, with a branching stem, deep-green foliage, and white flowers. The pods are five inches long, pale-green while young, light-yellow as the season of maturity approaches, cream-white when fully ripe, and contain five or six seeds.

Its season is intermediate. If sown early, the plants will blossom in seven weeks, afford pods for stringing in eight weeks, green beans in ten or eleven weeks, and ripen their seeds in ninety-seven days. It is a week earlier than the White Marrow, and ten days in advance of the Pea-bean. Plantings may be made as late as the last week in June, which will yield pods for the table in seven weeks, and ripen the middle of September, or in about twelve weeks.

The ripe seed is white, oblong, flattened, rounded on the back, often squarely or angularly shortened at the ends, half an inch long, and a fourth of an inch thick: twenty-seven hundred will measure a quart.

It is a field rather than a garden variety; though the green pods are tender and well flavoured. If planted in drills two feet apart, five pecks of seed will be required for an acre; or four pecks for the same quantity of ground, if the rows are two feet and a half apart. If planted in hills, six or eight seeds should be put in each; and, if the hills are three feet apart, twelve quarts of seed will plant an acre.

The Blue Pod is the earliest of the field varieties; more prolific, more generally cultivated, and more abundant in the market, than either the Pea-bean or the White Marrow. It is, however, much less esteemed; and, even in its greatest perfection, is almost invariably sold at a lower price.

On account of its precocity, it is well suited for planting in fields of corn, when the crop may have been partially destroyed by birds or insects, and the season has too far advanced to admit of a replanting of corn. In field-culture, Blue-pod beans are planted till the 25th of June.

Seed annual 1903

Canada Yellow

(Round American Kidney)

The plants of this variety are from fourteen to sixteen inches high, and of medium strength and vigor; flowers lilac-purple; the pods are five inches long, nearly straight, green while young, yellow at maturity, and contain from four to six seeds.

Season intermediate. If sown early, the plants will blossom in six or seven weeks, supply the table with pods in eight weeks, green shelled-beans in ten weeks, and ripen off in ninety days. When planted after settled warm weather, the variety grows rapidly, and ripens quickly; blossoming in less than six weeks, and ripening in seventy days, from the time of planting. For green shelled-beans, the seeds may be planted till the middle of July.

The ripe seeds are of an ovoid or rounded form, and measure half an inch in length and three-eighths of an inch in thickness. They are of a yellowish-drab colour, with a narrow, reddish-brown line about the eye; the drab changing, by age, to dull nankeen-yellow. About seventeen hundred are contained in a quart; and this amount of seeds will plant two hundred and fifty feet of drill, or a hundred and seventy-five hills.

The variety is quite productive, and excellent as a shelled bean, green or dry. The young pods are not so tender as those of many other sorts, and are but little used.

Chilian

Plant sixteen or eighteen inches high, sturdy and vigorous; foliage large, deep-green, wrinkled; flowers pale-lilac; the pods are five inches and a half long, slightly curved, pale-green while young, yellowish-white when ripe, and contain five seeds.

If planted early in the season, the variety will blossom in seven weeks, yield pods for the table in about eight weeks, and ripen in a hundred days, from the time of planting.

The ripe seeds are of a clear, bright pink, or rose colour; gradually becoming duller and darker from the time of harvesting. They are kidney-shaped, a little flattened, and of large size; generally measuring three-fourths of an inch long, and three-eighths of an inch thick. Twelve hundred and fifty are contained in a quart, and will be sufficient for planting a row or drill of two hundred feet, or for a hundred and twenty-five hills.

The variety is healthy, and moderately productive; not much esteemed for its young pods, but is worthy of cultivation for the large size and good quality of the beans; which, either in the green or ripe state, are quite farinaceous and mild flavoured.

Crescent-Eyed

Height fourteen or fifteen inches; flowers white,—the upper petals slightly stained with red; the pods are five inches and a half long, pale-green and somewhat curved when young, yellowish-white when fully ripe, and contain five seeds.

Season intermediate. If planted early, the variety will blossom in seven weeks, yield pods for stringing in eight weeks, supply the table with green beans in eleven weeks, and ripen in about ninety days. When planted and grown under the influence of summer weather, pods may be plucked for the table in fifty days, and the crop will ripen in about twelve weeks.

The beans, when ripe, are white, with a large, rose-red patch about the eye; the coloured portion of the surface being striped and marked with brownish-red. The fine rose-red changes by age to a brownish-red, and the red streaks and markings become relatively duller and darker: they are somewhat kidney-shaped, and measure three-fourths of an inch in length and three-eighths of an inch in thickness. A quart contains nearly thirteen hundred seeds, and will plant a hundred and fifty hills, or a row of two hundred feet.

The variety yields well, and the green pods are tender and well flavoured. It is, however, generally cultivated for its seeds, which are of large size and excellent quality, whether used in a green or ripe state.

Dun-coloured

Plant of vigorous, branching habit, sixteen inches in height, with broad, deep-green foliage and purplish-white flowers; the pods are five inches and a half long, half an inch broad, green and nearly straight while young, yellow and slender when fully ripe, and contain five or six beans.

The ripe seeds are dun-coloured or dark-drab, usually with a greenish line encircling the eye, kidney-shaped, five-eighths of an inch long, and about a fourth of an inch thick. A quart contains about seventeen hundred beans, and will plant a row of two hundred and twenty-five feet, or a hundred and seventy-five hills.

It is one of the earliest of the dwarf varieties; blossoming in about six weeks, producing young pods in seven weeks, and ripening in eighty-five days, from the time of planting. When sown after settled warm weather, pods may be gathered for use in six weeks; and, for these, plantings may be made until the 1st of August.

As a shelled-bean, green or dry, it is of little value, and hardly worthy of cultivation. As an early string-bean, it is one of the best. The pods are not only succulent and tender, but suitable for use very early in the season. It is also quite prolific; and, if planted at intervals of two weeks till the last of July, will supply the table to the last of September.

The variety has long been cultivated in England and other parts of Europe, and is much esteemed for its hardiness and productiveness.

Dwarf Cranberry

Plant vigorous; and, if the variety is pure, strictly a Dwarf, growing about sixteen inches high. As generally found in gardens, the plants send out slender runners, eighteen inches or two feet in length. The flowers are pale-purple; the pods are five inches long, sickle-shaped, pale-green in their young state, nearly white when ripe, and contain five or six seeds.

The ripe seeds are smaller than those of the running variety, but of the same form and colour: sixteen hundred are contained in a quart, and will plant nearly two hundred feet of drill, or a hundred and seventy-five hills.

The genuine Dwarf Cranberry is not one of the earliest varieties, but rather an intermediate sort. If sown as soon as the weather will admit, the plants will blossom in seven or eight weeks, and the young pods may be gathered for use in nine weeks. In favorable seasons, the crop is perfected in about ninety days. If planted in June, the variety will ripen in ten weeks.

It is hardy and productive; and the young pods are not only succulent and tender, but are suitable for use at a more advanced stage of growth than those of most varieties. The beans, in their green state, are farinaceous and well flavoured, but, after ripening, are little used; the colour being objectionable.

A variety with a brownish-red, oval, flattened seed, half an inch in length, is extensively known and cultivated as the Dwarf Cranberry. It is ten or twelve days earlier, the plants are smaller and less productive, the young pods less tender and succulent, and the seeds (green or ripe) less farinaceous, than those of the true variety. With the exception of its earlier maturity, it is comparatively not worthy of cultivation.

Dwarf Horticultural

(Variegated Dwarf Prague)

Stem about sixteen inches high; plant of vigorous, branching habit; flowers purple; pods five inches long, green while young, but changing to yellow, marbled and streaked with brilliant rose-red, when sufficiently advanced for shelling in their green state. At maturity, the clear, pale-yellow is changed to brownish-white, and the bright-red variegations are either entirely obliterated, or changed to dull, dead purple. If well formed, the pods contain five (rarely six) seeds.

It is a medium or half-early sort; and, if planted as soon as the weather becomes favorable, will blossom in seven weeks, produce pods for the table in about eight weeks, and ripen in ninety-five or a hundred days. Planted and grown in summer weather, the variety will produce green pods in seven weeks, and ripen in ninety days.

The ripe seeds resemble those of the running variety in form and colour; but they are smaller, a little more slender, and usually flattened slightly at the sides. When pure, they are egg-shaped; and a much compressed or a longer and more slender form is indicative of degeneracy. Fourteen hundred beans are contained in a quart; and this quantity of seed will be sufficient for planting a row of a hundred and seventy-five feet, or a hundred and forty hills.

The Dwarf Horticultural Bean is quite productive, and the young pods are tender and of good quality. It is, however, not so generally cultivated for its young pods as for its seeds, which are much esteemed for their mild flavour and farinaceous quality. For shelling in the green state, it is one of the best of the Dwarfs, and deserves cultivation.

Dwarf Sabre

(Dwarf Case-knife; Dwarf Cimeter)

A half-dwarf, French variety, two and a half to three feet high. As the running shoots are quite slender, and usually decay before the crop matures, it is always cultivated as other Dwarf sorts. Foliage large, wrinkled, and blistered; the flowers are white; the pods are very large, seven to eight inches long, and an inch in width, often irregular and distorted, green while young, paler as the season of maturity approaches, brownish-white when ripe, and contain seven or eight seeds.

The ripe bean is white, kidney-shaped, flattened, often twisted or contorted, three-fourths of an inch in length, and three-eighths of an inch in width: about twelve hundred are contained in a quart. As the variety is a vigorous grower, and occupies much space, this quantity of seed will plant a row of two hundred feet, or two hundred and twenty-five hills.

Season intermediate. The plants blossom in seven weeks, produce young pods in about eight weeks, pods for shelling in their green state in eleven or twelve weeks, and ripen in ninety-seven days, from the time of sowing. If cultivated for its green pods, the seeds may be planted to the middle of July.

The Dwarf Sabre is one of the most productive of all varieties; yielding its long, broad pods in great profusion. From the spreading, recumbent character of the plants, the pods often rest or lie upon the surface of the ground; and, being unusually thin and delicate, the crop often suffers to a considerable extent from the effects of rain and dampness in unfavorable seasons.[Pg 460]

The young pods are remarkable for their tender and succulent character; and the beans, both in a green and dried state, are mild and well flavoured. It is hardy, productive, of good quality, and recommended for cultivation.

Dwarf Soissons

A half-dwarf, French bean, similar in habit to the Dwarf Sabre. While young, the plants produce slender runners, two feet or more in length; but, as they are generally of short duration, the variety is cultivated as a Common Dwarf. The flowers are white; pods six inches long, pale-green at first, cream-yellow when sufficiently advanced for shelling, dull cream-white when fully ripe, and contain five, and sometimes six, beans.

The variety is comparatively early. Plants, from seeds sown in spring, will blossom in six weeks, produce pods for use in seven weeks, and ripen in ninety days. If planted and grown in the summer months, the crop will be ready for harvesting in eleven weeks; and sowings for the ripe seeds may be made till the beginning of July.

Seeds white, kidney-shaped, flattened, often bent or distorted, five-eighths of an inch long, three-eighths of an inch wide, and a fourth of an inch thick: fifteen hundred are contained in a quart, and will plant a drill two hundred and twenty-five feet in length, or about two hundred hills.

The variety is productive, and the young pods are of fair quality; the seeds are excellent, whether used green or ripe; the skin is thin; and they are much esteemed for their peculiar whiteness, and delicacy of flavour.

Early China

(China; Red-Eyed China)

Plant fifteen inches high, with yellowish-green, wrinkled foliage, and white flowers; the pods are five inches long, green and straight while young, yellowish-green as they approach maturity, yellow when fully ripe, and contain five (rarely six) beans.

The ripe seeds are white, coloured and spotted about the eye with purplish-red, oblong, nearly cylindrical at the centre, rounded at the ends, six-tenths of an inch long, and three-eighths of an inch thick: sixteen hundred and fifty measure a quart, and will plant two hundred feet of drill, or two hundred hills.

If planted early in the season, the variety will blossom in six weeks, afford young pods for use in seven weeks, green beans in ten weeks, and ripen in eighty-five days. When planted and grown in summer, the crop will ripen in eleven weeks; and plants from seeds sown as late as the first of August will generally afford an abundant supply of tender pods from the middle to the close of September.

The Early China is very generally disseminated, and is one of the most popular of the Dwarf varieties. It is hardy and productive; but the young pods, though succulent and tender, are inferior to those of some other varieties. The seeds, green or ripe, are thin-skinned, mealy, and mild flavoured.

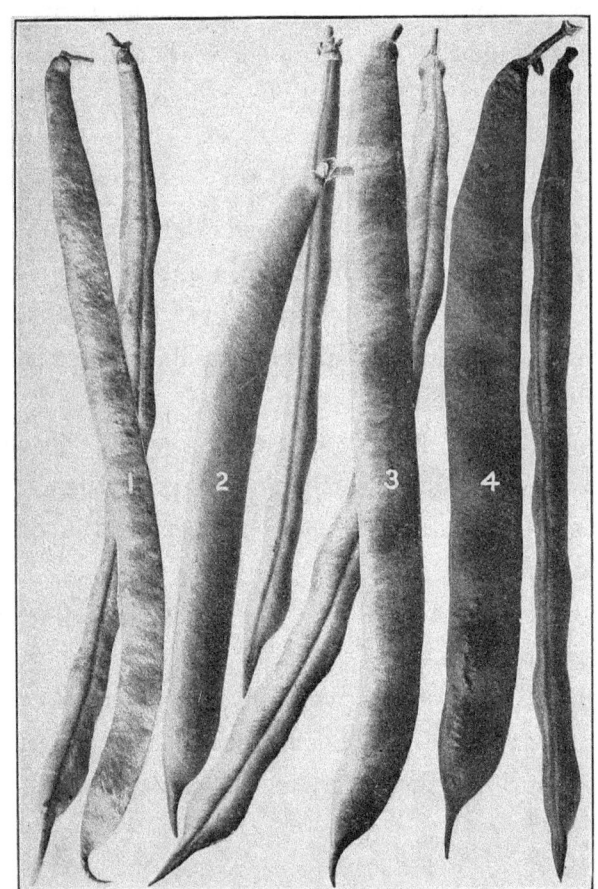

Bush Varieties (Snap Pods). 1.—Thorburn's Prolific Market. 2.—China Red Eye. 3.—Best of All (late type). 4.—Blue Pod Butter.
Bureau of Plant Industry, U. S. Dept. of Agriculture, 1907

Early Rachel

A low-growing, branching variety, twelve to fifteen inches high; flowers white; the pods are five inches and a half long, green while young, becoming paler or greenish-yellow as they approach maturity, cream-white when ripe, and contain five seeds.

Planted early in the season, the variety will blossom in about seven weeks; and, in eight weeks, the young pods will be fit for use. Pods for shelling may be plucked in ten weeks, and the crop will ripen in eighty days. For the green pods, the seeds may be planted till the middle or twentieth of July.

The ripe seed is yellowish-brown, white at one of the ends, kidney-shaped, often abruptly shortened, five-eighths of an inch long, and a fourth of an inch thick: nearly two thousand are contained in a quart.

The Early Rachel is hardy, and moderately productive, and, as an early string-bean, may be desirable; but as a shell-bean, green or dry, it is of little value. In common with many other early sorts cultivated as string-beans, the pods, though crisp and tender at first, soon become too tough and parchment-like for use. In general, the pods of the later sorts remain crisp and tender a much longer period than those of the earlier descriptions.

Early Valentine
(Valentine)

Plant about sixteen inches high, with small, yellowish-green leaves and white flowers; the pods are comparatively short, usually four and a half or five inches long, sickle-shaped, almost cylindrical, green while young, yellow when ripe, and contain five seeds.

The variety is productive, and quite early, though not one of the earliest. When sown at the commencement of the season, the plants will blossom in six weeks, produce pods for use in about seven weeks, and ripen in thirteen weeks, or ninety days, from the time of planting. If planted after the beginning of summer weather, pods may be gathered for the table in fifty days, and the beans will ripen in eleven weeks.

The beans, when ripe, are of a pale-pink colour, marbled or variegated with rose-red, becoming duller and browner by age, oblong, nearly straight, sometimes distorted and irregular as if pressed out of their natural shape, often more or less shortened at the ends, five-eighths of an inch long, three-eighths of an inch wide, and about the same in thickness. A quart will contain eighteen or nineteen hundred seeds; which will be sufficient for a hundred and seventy-five hills, or for a drill, or row, of two hundred or two hundred and twenty-five feet.

The Early Valentine is generally cultivated for its tender and very fleshy pods, which remain long on the plants without becoming hard and tough. They make an excellent, brittle pickle; and, when cooked, are equal to those of any other Dwarf variety. The shelled-beans, either in their green or ripe state, are little esteemed.

The variety has long been grown in England and other parts of Europe, and is common to gardens in almost every section of the United States.

Golden Cranberry

(Canadian; Round American Kidney)

Height about sixteen inches; flowers purple; the pods are five inches and a half long, five-eighths of an inch broad, somewhat irregular in form, yellow when ripe, and contain five seeds.

Season intermediate. Early plantings will blossom in seven weeks, yield pods for the table in eight weeks, and ripen in ninety days.

The ripe seeds are pale greenish-yellow, with an olive-green line encircling the eye; roundish-ovoid, three-eighths of an inch long, and nearly the same in thickness. A quart contains nearly eighteen hundred seeds, and will plant a row, or drill, of two hundred feet, or two hundred and twenty-five hills.

As a string-bean, or for shelling in the green state, it is inferior to many other varieties, and is little cultivated for use in these forms; but as a variety for baking, or for cooking in any form when ripe, it is much esteemed, and recommended for cultivation.

Hardy and productive.

Long Yellow Six-Weeks

(Six-Weeks; Yellow Six-Weeks; Yellow Flageolet)

The plants of this familiar variety are of vigorous, branching habit, and from fourteen to sixteen inches high; the flowers are pale-purple; the pods are five inches long, six-tenths of an inch broad, often curved or sickle-shaped, green at first, gradually becoming paler, cream-yellow when ripe, and contain five (rarely six) beans.

It is one of the earliest of the Dwarf varieties. Spring plantings will blossom in six weeks, produce pods for the table in seven weeks, and ripen in eighty-seven days. Summer plantings will afford pods for the table in about six weeks, and ripen in sixty-three days. When planted as late in the season as the last of July or first of August, the variety will afford an abundant supply of tender pods from the middle to the last of September.

The ripe seeds are pale yellowish-drab, with an olive-green line about the eye; the drab rapidly changing by age to dull yellowish-brown. They are kidney-shaped, rather straight, three-fourths of an inch long, and three-tenths of an inch thick. About fourteen hundred beans are contained in a quart, and will plant a row of two hundred feet, or a hundred and fifty hills.

It is quite productive, and an excellent early string-bean, but less valuable as a green shelled-bean, or for cooking when ripe. On account of the tender and delicate character of the pods, the ripe seeds are often injured by damp or continued rainy weather. A popular, early garden-bean, much cultivated both in this country and in Europe.

Mohawk

(Early Mohawk)

Stem about eighteen inches high, sturdy and branching; foliage large, deep-green, wrinkled, and blistered; flowers pale-lilac; the pods are five inches and a half long, five-eighths of an inch wide, and generally contain five seeds,—while young they are green, and nearly straight; as they approach maturity they become paler; and, when ripe, are frequently streaked and spotted with purple.

The ripe seeds are variegated with drab, dull purple, and different shades of brown; the brown and dull purple prevailing: they are kidney-shaped, and measure nearly three-fourths of an inch in length, and three-eighths of an inch in width. A quart contains about fourteen hundred and fifty seeds, and will plant a hundred and seventy-five feet of drill, or a hundred and seventy-five hills.

It is about a week later than the earliest varieties. Spring plantings will blossom in about seven weeks, produce pods for the table in eight weeks, and ripen in a hundred days, from the time of sowing. In ordinary seasons, the variety will ripen perfectly if planted the last week in June; and will yield an abundance of pods for the table, if the planting be made as late in the season as the last of July.

The Early Mohawk is quite productive, and one of the hardiest of the Dwarf varieties. It is well adapted for early planting, and is extensively grown by market-gardeners as an early string-bean. The young pods are comparatively tender, and of good quality; and, if gathered as they become of suitable size, the plants will continue to yield them in great abundance. The shelled-beans, green or dry, are less esteemed, and considered inferior to many other varieties.

Newington Wonder

A healthy, vigorous variety, with deep-green foliage and bright-purple flowers. The plants often produce slender, barren runners, eighteen inches or two feet in length; but they are generally of short duration, and the variety is treated as other Dwarfs.

The pods are small and straight; usually about four inches long, and nearly half an inch broad. They are pale-green at first; and afterwards change to yellowish-white, tinted or washed with bright pink. At maturity they are dusky-drab, sometimes clouded or shaded with purple, and contain six or seven beans.

The ripe seeds are pale brownish-drab, with a yellowish-brown line about the eye; oblong, flattened, shortened at the ends, nearly half an inch long, and a fourth of an inch deep: about thirty-six hundred are contained in a quart. As the seeds are comparatively small, and the plants of spreading habit, this amount of seeds will plant a row four hundred feet in length, or four hundred hills.

The variety is not early, and, when cultivated for its seeds, should have the benefit of the whole season; though, with favorable autumnal weather, the crop will ripen if planted the middle of June. Spring plantings will blossom in eight weeks, produce young pods in nine weeks, and ripen in a hundred and six days.

The Newington Wonder is remarkably prolific; and, in its manner of growth and general character, resembles the Tampico or Turtle-soup. As a string-bean, it is one of the best. The pods, though not large, are crisp, succulent, and tender, and produced in great abundance throughout most of the season. The seeds, in their green state, are small, and of little value for the table: when ripe, they afford an excellent substitute for the Tampico or Turtle-soup; the difference, aside from the colour, being scarcely perceptible.

The Newington Wonder of English and French authors appears to be, in some respects, distinct from the American variety. It is described as very dwarf, about a foot high, early and productive; pods dark-green, moderately long, not broad, thick and fleshy; seeds quite small, light chestnut-coloured.

Pea-Bean

Plant vigorous, much branched, and, like the Blue Pod and White Marrow, inclined to send up running shoots; foliage comparatively small, deep-green; flowers white; the pods are about four inches long, half an inch wide, nearly straight, green when young, paler as they approach the season of ripening, yellowish when fully ripe, and contain five beans.

It is comparatively a late variety. When planted in spring, it will blossom in fifty days, afford green pods in fifty-eight days, and ripen in about fifteen weeks. In favorable autumns, it will ripen if planted as late as the 20th of June; but it is not so early as the Blue Pod or White Marrow, and, when practicable, should have the advantage of the entire season.

The ripe seeds of the pure variety are quite small, roundish-ovoid, five-sixteenths of an inch long, a fourth of an inch in width and thickness, and of a pure yet not glossy white colour: about forty-four hundred seeds are contained in a quart.

As a garden variety, it is of little value, though the young pods are crisp and tender. It is cultivated almost exclusively as a field-bean. If planted in rows or drills two feet apart, three pecks of seeds will be required for an acre; or eighteen quarts will seed this quantity of land, if the rows are two feet and a half apart. When planted in hills, eight seeds are allowed to a hill; and, if the hills are made three feet apart, eight quarts will plant an acre. The yield varies from fourteen to twenty bushels, according to soil, season, and cultivation.

The Pea-bean, the White Marrow, and the Blue Pod are the principal if not the only kinds of much commercial importance; the names of other varieties being rarely, if ever, mentioned in the regular reports of the current prices of the markets. If equally well ripened, and, in their re[Pg 468]spective varieties, equally pure, the Pea-bean and the White Marrow command about the same prices; the former, however, being more abundant in the market than the latter. By many, and perhaps by a majority, the Pea-bean is esteemed the best of all baking varieties.

Pottawottomie

The plants of this variety are remarkable for their strong, vigorous habit, and large, luxuriant foliage. The flowers are flesh-white; the pods are six inches long, green at first, then mottled and streaked with lively rose-red on a cream-white ground (the markings changing to purple at maturity), and contain five (rarely six) seeds.

The variety is comparatively late. If sown early in the season, the plants will flower in seven weeks, afford pods for shelling in eleven weeks, and ripen in a hundred days, from the time of planting.

The ripe seeds are of a light creamy-pink colour, streaked and spotted with a red or reddish-brown: the soft, flesh-like colour, however, soon becomes duller and darker, and at last gives place to a dull, cinnamon-brown. They are kidney-shaped, fully three-fourths of an inch long, and about three-eighths of an inch broad. About a thousand will measure a quart, and will plant a row two hundred feet in length, or a hundred and twenty-five hills. On account of the large size and spreading habit of the plants, five seeds will be sufficient for a hill; and, in the rows, they should be dropped five or six inches from each other.

The young pods are inferior to most varieties in crispness, and tenderness of texture; and are comparatively but little used. The seeds are remarkably large, separate easily from the pods, and, green or ripe, are remarkably farinaceous and well flavoured, nearly or quite equalling the Dwarf and Running Horticultural.

Red Flageolet

(Scarlet Flageolet)

A half-dwarf, French Bean, two to three feet high; flowers pale-purple; the pods are six inches and a half long, somewhat curved, green while young, pale-yellow at maturity, and contain five or six seeds.

It is one of the latest of the Dwarf varieties. If sown early, the plants will blossom in seven weeks, and pods may be gathered for use in about nine weeks; in thirteen weeks the pods will be sufficiently advanced for shelling, and the crop will be ready for harvesting in a hundred and ten days. It requires the whole season for its full perfection; but, for its young pods or for green beans, plantings may be made to the last week in June.

The ripe beans are blood-red when first harvested, but gradually change by age to deep-purple: they are kidney-shaped, nearly straight, slightly flattened, three-fourths of an inch long, three-eighths of an inch broad, and nearly the same in thickness. Fifteen hundred seeds are contained in a quart.

The Red Flageolet yields abundantly; and the young pods are not only of good size, but remarkably crisp and tender. If plucked as they become fit for use, the plants continue to produce fresh pods for many weeks. The green beans are farinaceous, and excellent for table use; but are seldom cooked in their ripened state.

Red-Speckled

Plant branching, and of strong growth,—nearly a foot and a half high; foliage remarkably large; flowers pale-purple; pods five inches and a half long, nearly straight, green while young, paler with occasional marks and spots of purple when more advanced, yellowish-white when ripe, and containing five (rarely six) seeds.

Season intermediate. Plants from seeds sown after settled warm weather will blossom in six weeks, and green pods may be plucked for use in fifty days. For shelling in their green state, pods may be gathered in ten weeks, and the crop will ripen off in ninety days. For its young pods, or for green beans, plantings may be made to the last week in June; but the crop will not mature, unless the weather continues favorable till the 1st of October.

The ripe seeds are variegated with deep-red and pale-drab, the red predominating; kidney-shaped, nearly straight, three-fourths of an inch long, and three-tenths of an inch deep. A quart contains fourteen hundred and fifty seeds, and will plant a row of two hundred and twenty-five feet, or a hundred and fifty hills.

The variety is hardy and productive. It is extensively cultivated as a garden-bean in England and France, and has been common to the gardens of this country for nearly two centuries. The young pods are of medium quality; but the seeds, green or dry, are mealy and well flavoured. On account of the parchment-like character of the pods, the seeds seldom suffer from the effects of wet weather.

Refugee

(Thousand to One)

Plant sixteen to eighteen inches high, and readily distinguished from most varieties by its small, smooth, deep-green, and elongated leaves; flowers purple; pods five inches long, nearly cylindrical, pale-green while young, greenish-white streaked with purple when sufficiently advanced for shelling, yellow when ripe, and usually yielding five beans.

The Refugee is not an early sort. The plants blossom in seven weeks, produce young pods in eight weeks, and ripen in eighty-seven days, from the time of sowing. Plantings for the ripened product may be made till the middle of June; and for the green pods, to the middle of July.

The ripe seeds are light-drab, with numerous spots and broad patches of bright-purple, nearly straight, cylindrical at the middle, tapering to the ends (which are generally rounded), five-eighths of an inch long, and three-tenths of an inch thick. Eighteen hundred and fifty are contained in a quart, and will plant a row two hundred and fifty feet in length, or two hundred hills.

The variety is hardy, yields abundantly, and the young pods are thick, fleshy, and tender in texture. As a string-bean, or for pickling, it is considered one of the best of all varieties, and is recommended for general cultivation. The seeds are comparatively small, and are rarely used either in a green or ripened state.

Rice

Half-dwarf, about two feet high; flowers white; pods very small, scarcely more than three inches in length, and only two-fifths of an inch in width, usually containing six seeds.

The variety requires a full season for its perfection. Plants from seeds sown early in spring will blossom in seven weeks, yield young pods in ten weeks, and ripen in a hundred and twelve days.

The ripe seeds are very small, and of a peculiar yellowish-white, semi-transparent, rice-like colour and appearance. They are quite irregular in form, usually somewhat oblong or ovoid, often abruptly shortened at the ends, three-eighths of an inch long, and a fourth of an inch thick. Nearly five thousand are contained in a quart.

The young pods are tender and excellent; but the green beans are small, and rarely used. The ripe seeds are peculiar, both in consistency and flavour: they are quite brittle and rice-like; and, when cooked, much relished by some, and little esteemed by others.

Garden, Farm & Flower Seeds (1897), Cole's

Rob-Roy

Plant half-dwarf,—early in the season, producing slender, transient, barren runners two or three feet in length; flowers purplish-white; the pods are five inches long, often produced in pairs, yellow as they approach maturity, yellowish-white when ripe, and contain five or six seeds.

It is one of the earliest of the Dwarfs. Spring plantings will blossom in six weeks, produce pods for the table in seven weeks, and ripen in eighty-two days. If planted in June, pods may be plucked for use in six weeks, and the crop will be ready for harvesting in sixty-eight days.

The ripe seeds are clear, bright-yellow; the surface being generally veined, and the eye surrounded with an olive-green line. They are of an oblong form, nearly straight on the side of the eye, rounded at the back, five-eighths of an inch long, and three-tenths of an inch deep. Fifteen hundred seeds are contained in a quart, and will be sufficient to plant a row of two hundred feet, or a hundred and fifty hills.

The Rob-Roy generally matures in great perfection; being seldom stained or otherwise injured by rain or the dampness of ordinary seasons. It is also one of the earliest of the Dwarf varieties, but desirable as a string-bean rather than for its qualities as a green shelled-bean, or for cooking when ripe. If cultivated for its pods only, plantings may be made until the first of August.

Round Yellow Six-Weeks

(Round Yellow; Dwarf Yellow)

Fourteen to sixteen inches high; flowers pale-purple; pods about five inches long, half an inch broad, pale yellowish-green as they approach maturity, and, when fully ripe, remarkably slender, and more curved than in their green state,—they contain five or six beans.

The variety is early; blossoming in six weeks, producing young pods in seven weeks, and ripening in ninety days, from the time of planting. When planted in June, pods may be plucked for use in seven weeks, and the crop will be ready for harvesting in eighty days. For its green pods, plantings may be made to the last of July.

The ripe seeds are orange-yellow, with a narrow, reddish-brown belt, or line, encircling the eye; oblong or ovoid, half an inch long, and three-tenths of an inch thick. A quart contains two thousand seeds, and will plant a row two hundred and twenty-five feet in length, or two hundred and twenty-five hills.

As an early string-bean, the variety is worthy of cultivation, but is little used, and is really of little value, as a shelled-bean, green or ripe. It has been common to the gardens of this country for more than a century; and, during this period, no apparent change has taken place in the character of the plant, or in the size, form, or colour of the seed.

Solitaire

A French variety. The ripe seeds are similar to those of the Refugee; but the plants are quite distinct in foliage and general habit. Its height is about eighteen inches; the flowers are purple; the pods are six inches long, slender, nearly cylindrical, green at first, paler and streaked with purple when more advanced, and contain six seeds.

It is not early. Spring plantings will blossom in sixty days, produce pods for the table in seventy days, and ripen in about fifteen weeks. It may be planted for its green pods until the first of July.

The beans, when ripe, are variegated with light-drab and deep-purple, the purple prevailing. They are often straight, sometimes curved, nearly cylindrical at the eye, usually rounded, but sometimes shortened, at the ends, three-fourths of an inch long, and a fourth of an inch thick: two thousand measure a quart.

On account of the size and branching character of the plants, more space must be allowed in cultivation than is usually given to Common Dwarf varieties. If planted in rows, they should be at least eighteen inches apart, and the plants eight or ten inches from each other in the rows; and, if planted in hills, they should be thinned to four or five plants, and the hills should not be less than three feet apart.

It is not much esteemed as a shelled-bean, either green or ripe. As a string-bean, it is one of the best. Its pods are long, cylindrical, remarkably slender, succulent, and tender. It is also a very prolific variety, and the pods remain for an unusual period without becoming tough or too hard for the table. Recommended for cultivation.

Swiss Crimson

(Scarlet Swiss)

Plant vigorous, often producing running shoots; flowers pale-purple; pods nearly straight, six inches long, pale-green while young, yellow streaked with brilliant rose-red as they approach maturity, and containing five (rarely six) seeds.

It is comparatively a late variety. If planted as early as the weather will permit, the plants will blossom in seven weeks, the young pods will be ready for use in nine weeks, and the crop will be ready for harvesting in a hundred and five days. Planted and grown in summer weather, it will produce young pods in sixty days, and ripen in thirteen weeks. Plantings for the green seeds may be made to the first of July.

The ripe seeds are clear bright-pink, striped and spotted with deep purplish-red: the pink changes gradually to dull, dark-red, and the variegations to dark-brown. They are kidney-shaped, comparatively straight, somewhat flattened, three-fourths of an inch long, and three-eighths of an inch broad. Thirteen hundred seeds are contained in a quart, and will plant a row two hundred feet in length, or a hundred and fifty hills.

It is hardy and productive, and, as a shelled-bean, of excellent quality, either in its green or ripened state. As a variety for stringing, it is not above medium quality.

Turtle-Soup

(Tampico)

Plant vigorous, producing numerous slender, barren runners two feet or more in length; flowers rich deep-purple; pods five inches long, green and sickle-shaped while young, pale greenish-white stained with purple when more advanced, yellow clouded with purple when ripe, and containing five or six seeds.

The variety is quite late, and requires most of the season for its full perfection. Plants from early sowings will blossom in eight weeks, the young pods will be sufficiently grown for use in ten weeks, and the crop will ripen in a hundred and eight days. As the young pods are tender and of excellent quality, and are also produced in great abundance, a planting for these may be made as late as the last week in June, which will supply the table from the last of August till the plants are destroyed by frost.

The ripe seeds are small, glossy-black, somewhat oblong, and much flattened: thirty-six hundred are contained in a quart, and will plant four hundred feet of drill, or three hundred and fifty hills.

It is very productive, and deserving of cultivation for its young and tender pods; but is of little or no value for shelling while green. The ripened seeds are used, as the name implies, in the preparation of a soup, which, as respects colour and flavour, bears some resemblance to that made from the green turtle.

Victoria

This is one of the earliest of the Dwarf varieties. Early plantings will blossom in six weeks, yield pods for the table in seven weeks, produce pods of suitable size for shelling in about ten weeks, and ripen in eighty-four days. When planted after the season has somewhat advanced,—the young plants thus receiving the benefit of summer temperature,—pods may be gathered for the table in about six weeks, and the crop will ripen in sixty-three days.

Stalk fourteen to sixteen inches high, with comparatively few branches; flowers purple; pods four and a half to five inches long, streaked and spotted with purple, tough and parchment-like when ripe, and containing five or six seeds.

The ripe seeds are flesh-coloured, striped and spotted with purple (the ground changing by age to dull reddish-brown, and the spots and markings to chocolate-brown), oblong, somewhat flattened, shortened or rounded at the ends, five-eighths of an inch long, and three-tenths of an inch thick: fourteen hundred are contained in a quart.

The variety is remarkably early; and, on this account, is worthy of cultivation. For table use, the young pods and the seeds, green or dry, are inferior to many other sorts.

White's Early

A remarkably hardy and vigorous variety, eighteen to twenty inches high. Flowers white, tinged with purple; pods five inches and a half long, curved or sickle-shaped, green at first, yellowish-white striped with purple when fully ripe, and containing five seeds.

Early plantings will blossom in about six weeks, young pods may be plucked for use in seven weeks, and the crop will ripen in eighty-two days. If planted as late in the season as the first week in July, the variety will generally ripen perfectly; and, when cultivated for its green pods, plantings may be made at any time during the month.

The ripe seeds are either drab or light-slate,—both colours being common,—marked and spotted with light-drab. In some specimens, drab is the prevailing colour. They are kidney-shaped, irregularly compressed or flattened, nearly three-fourths of an inch long, and three-eighths of an inch deep. A quart contains about sixteen hundred seeds, and is sufficient for planting a row two hundred and fifty feet in length, or two hundred hills.

This variety, as an early string-bean, is decidedly one of the best, and is also one of the hardiest and most prolific. The pods should be plucked when comparatively young; and, if often gathered, the plants will continue a long time in bearing. As a shelled-bean, either in its green or ripened state, it is only of medium quality.

The long peduncles, or stems, that support its spikes of flowers, its stocky habit, and fine, deep-green, luxurious foliage, distinguish the variety from all others.

White Flageolet

From sixteen to eighteen inches high, of strong and branching habit. Flowers white; pods five inches and a half long, sickle-shaped, green while young, yellowish-white at maturity, and containing six (rarely seven) seeds.

It is a half-early variety; blossoming in six weeks, yielding pods for the table in seven weeks, pods for shelling in eleven weeks, and ripening in ninety days, from the time of planting. Later plantings will ripen in a shorter period, or in about eighty days; and, if cultivated as a string-bean, seed sown as late in the season as the last week of July will supply the table from the middle of September with an abundance of well-flavoured and tender pods.

The ripe bean is white, kidney-shaped, flattened, three-fourths of an inch long, and three-tenths of an inch broad: about twenty-two hundred are contained in a quart, and will plant a drill, or row, of two hundred and seventy-five feet, or nearly three hundred hills.

The White Flageolet is very productive, and is recommended for cultivation: the young pods are crisp and tender, and the seeds, green or ripe, are farinaceous, and remarkable for delicacy of flavour.

White Kidney

(Kidney; Large White Kidney; Royal Dwarf)

The plants of this variety are from sixteen to eighteen inches high, and readily distinguishable, from their large and broad leaves, and strong, branching habit of growth; the flowers are white; the pods are somewhat irregular in form, six inches long, green at first, yellow when ripe, and contain five (rarely six) beans.

The White Kidney-bean is not early: it blossoms in seven weeks, produces young pods in nine weeks, pods for shelling in eleven weeks, and ripens in a hundred and ten days, from the time of planting.

The ripe seeds are white, more or less veined, pale-yellow about the hilum, kidney-shaped, nearly straight, slightly flattened, fully three-fourths of an inch long, and about three-eighths of an inch thick: from twelve to thirteen hundred are contained in a quart; and this quantity of seeds will plant a hundred and seventy-five feet of drill, or a hundred and forty hills.

As a string-bean, the variety has little merit; but as a shelled-bean, green or ripe, it is decidedly one of the best of the Dwarfs, and well deserving of cultivation. The seeds are of large size, pure white, separate readily from the pods, and are tender and delicate.

White Marrow

(White Marrowfat; Dwarf White Cranberry; White Egg)

Plants vigorous, much branched, and inclined to produce running shoots; flowers white; pods five inches long, nearly three-fourths of an inch broad, pale-green at first, then changing to clear yellow, afterwards becoming pure waxen-white, cream-yellow when ripe, and containing five seeds.

When planted at the commencement of favorable weather, the variety will blossom in seven weeks, yield pods for the table in eight weeks, and ripen in a hundred and five days. When grown for the ripened product, the planting should not be delayed beyond the 20th of June. Planted at this season, or the last week in June, the crop will blossom the first week in August; and, about the middle of the month, pods may be gathered for the table. By the second week in September, the pods will be of sufficient size for shelling; and, if the season be ordinarily favorable, the crop will ripen the last of the month. It must not, however, be regarded as an early variety; and, when practicable, should be planted before the 10th of June.

The ripe seeds are clear white, ovoid or egg-shaped, nine-sixteenths of an inch long, and three-eighths of an inch thick. In size, form, or colour, they are scarcely distinguishable from those of the White Running Cranberry. If well grown, twelve hundred seeds will measure a quart.

As a string-bean, the White Marrow is of average quality: but, for shelling in the green state, it is surpassed by few, if any, of the garden varieties; and deserves more general cultivation. When ripe, it is remarkably farinaceous, of a delicate fleshy-white when properly cooked, and by many preferred to the Pea-bean.

In almost every section of the United States, as well as in the Canadas, it is largely cultivated for market; and is next in importance to the last named for commercial purposes.

In field-culture, it is planted in drills two feet apart; the seeds being dropped in groups, three or four together, a foot apart in the drills. Some plant in hills two and a half or three feet apart by eighteen inches in the opposite direction, seeding at the rate of forty-four quarts to the acre; and others plant in drills eighteen inches apart, dropping the seeds singly, six or eight inches from each other in the drills.

The yield varies from twenty to thirty bushels to the acre, though crops are recorded of nearly forty bushels.

Yellow-Eyed China

Plant sixteen to eighteen inches high, more branched and of stronger habit than the Black or Red Eyed; flowers white; pods six inches long, nearly straight, pale-green while young, cream-white at maturity, and containing five or six seeds.

It is an early variety. When sown in May, or at the beginning of settled weather, the plants will blossom in six weeks, afford string-beans in seven weeks, pods for shelling in ten or eleven weeks, and ripen in ninety days, from the time of planting. From sowings made later in the season (the plants thereby receiving more directly the influence of summer weather), pods may be plucked for the table in about six weeks, and ripened beans in seventy-five days. Plantings for supplying the table with string-beans may be made until the last week in July.

The ripe beans are white, spotted and marked about the eye with rusty-yellow, oblong, inclining to kidney-shape, more flattened than those of the Red or Black Eyed, five-eighths of an inch long, and three-eighths of an inch in breadth: fifteen hundred and fifty are contained in a quart, and will plant two hundred feet of drill, or a hundred and fifty hills. The plants are large and spreading, and most productive when not grown too closely together.

The Yellow-eyed China is one of the most healthy, vigorous, and prolific of the Dwarf varieties; of good quality as a string-bean; and, in its ripened state, excellent for baking, or in whatever manner it may be cooked. It also ripens its seeds in great perfection; the crop being rarely affected by wet weather, or injured by blight or mildew.

Pole/Running Beans

As a class, these are less hardy than the Dwarfs, and are not usually planted so early in the season. The common practice is to plant in hills three feet or three and a half apart; though the lower-growing sorts are sometimes planted in drills fourteen or fifteen inches apart, and bushed in the manner of the taller descriptions of pease.

If planted in hills, they should be slightly raised, and the stake, or pole, set before the planting of the seeds. The maturity of some of the later sorts will be somewhat facilitated by cutting or nipping off the leading runners when they have attained a height of four or five feet.

Case-Knife

This variety, common to almost every garden, is readily distinguished by its strong and tall habit of growth, and its broad, deep-green, blistered leaves. The flowers are white. The pods are remarkably large; often measuring nine or ten inches in length, and nearly an inch in width. They are of a green colour till near maturity, when they change to yellowish-green, and, when fully ripe, to cream-white. A well-formed pod contains eight or nine seeds.

Early plantings will blossom in seven or eight weeks, yield pods for stringing in about ten weeks, green beans in twelve or thirteen weeks, and ripen in a hundred and five days. Later plantings, with the exclusive advantage of summer weather, will supply string-beans in seven weeks, pods for shelling in eight or nine weeks, and ripen in ninety-six days. Plantings for the green beans may be made till nearly the middle of July; and, for the young pods, to the 25th of the month.

The ripe seeds are clear white, kidney-shaped, irregularly flattened or compressed, often diagonally shortened at one or both of the ends, three-fourths of an inch long, and three-eighths of an inch deep. A quart contains about fifteen hundred seeds, and will plant a hundred and seventy-five hills.

It is one of the most prolific of the running varieties. As a shelled-bean, it is of excellent quality in its green state; and, when ripe, farinaceous, and well flavoured in whatever form prepared. The large pods, if plucked early, are succulent and tender, but coarser in texture than those of many other sorts, and not so well flavoured.

The Case-knife, in its habit and general appearance, much resembles the Sabre, or Cimeter, of the French; and perhaps is but a sub-variety. Plants, however, from imported Sabre-beans, were shorter, not so stocky, a little earlier, and the pods, generally, less perfectly formed.

Corn-Bean

Stem six feet and upwards in height; flowers bright-lilac; the pods are five inches and a half long, green while young, cream-white at maturity, and contain six or seven seeds.

The variety is late, but remarkable for hardiness and productiveness. The shelled-beans, green or ripe, are little used; the young pods are crisp, succulent, and excellent for the table; and the variety deserves more general cultivation. If plucked as fast as they become of suitable size, the plants will continue to produce them in abundance for six or eight weeks.

The ripe seeds are chocolate-brown, somewhat quadrangular, flattened, half an inch long, and three-eighths of an inch broad. In size and form, they somewhat resemble grains of Indian corn: whence the name. Twelve hundred and fifty seeds are contained in a quart, and will plant a hundred and twenty-five hills.

Succotash Beans, very likely the beans mentioned as 'Corn-Beans'

Horticultural

(Marbled Prague; London Horticultural)

Stem six feet or more in height; flowers purple; the pods are from five to six inches long, nearly three-fourths of an inch broad, pale-green while young, greenish-white streaked and blotched with brilliant rose-red when more advanced, much contorted, hard, parchment-like and very tenacious of their contents when ripe, and enclose five or six seeds.

When planted at the commencement of the season, the variety will blossom in about seven weeks, produce pods for stringing in nine weeks, green beans in twelve weeks, and ripen in a hundred days. Plantings made during the last week in June will mature their crop, if the season be favorable. For the green beans, plantings may be made until the last of June; and, for the young pods, until the first of July.

The ripe beans are flesh-white, streaked and spotted with bright-pink, or red, with a russet-yellow line encircling the eye. They are egg-shaped, rather more than half an inch in length, and four-tenths of an inch in width and depth. From the time of ripening, the soft, flesh-like tint gradually loses its freshness, and finally becomes cinnamon-brown; the variegations growing relatively duller and darker. A quart contains about eleven hundred seeds, and will plant a hundred and twenty-five hills.

The Horticultural Bean was introduced into this country from England about the year 1825. It has now become very generally disseminated, and is one of the most popular of the running sorts. As a string-bean, it is of good quality; shelled in its green state, remarkably farinaceous and well flavoured; and, when ripe, one of the best for baking or stewing. It is hardy and productive, but is liable to deteriorate when raised many years in succession from seed saved in the vegetable garden from the scattered pods accidentally left to ripen on the poles. To raise good seed, leave each year a few hills unplucked; allowing the entire product to ripen.

Indian Chief

(Wax-Bean; Butter-Bean; Algerian; D'Alger, of the French)

Stem six or seven feet high, with large, broad foliage and purple flowers; the pods are five inches long, nearly as thick as broad, sickle-shaped, green at first, but soon change to a fine, waxen, semi-transparent cream-white,—the line marking the divisions being orange-yellow. At this stage of growth, the colour indicates approaching maturity; but the pods will be found crisp and succulent, and are in their greatest perfection for the table. When ripe, they are nearly white, much shrivelled, and contain six or seven seeds.

When cultivated for the ripened product, the seed should be planted as early in the season as the weather will permit. The plants will then blossom in eight or nine weeks, afford young pods in about eleven weeks, pods for shelling in thirteen or fourteen weeks, and ripen in a hundred and twenty-four days. Plantings for green pods may be made until the first of July.

At the time of harvesting, the seeds are deep indigo-blue, the hilum being white. They are oblong, often shortened abruptly at the ends, half an inch long, nearly the same in depth, and three-tenths of an inch thick. Fourteen hundred seeds measure a quart, and will plant a hundred and seventy-five hills.

Its fine, tender, succulent, and richly coloured pods are its chief recommendation; and for these it is well worthy of cultivation. They are produced in profuse abundance, and continue fit for use longer than those of most varieties. In moist seasons, the pods remain crisp and tender till the seeds have grown sufficiently to be used in the green state. The ripe seeds are little used.

Mottled Cranberry

A comparatively strong-growing, but not tall variety. The flowers are white; the pods are short and broad, four inches and a half long, three-fourths of an inch wide, yellow at maturity, and contain four or five seeds.

If planted early, the variety will blossom in seven weeks, yield pods for the table in eight or nine weeks, green beans in eleven weeks, and ripen in a hundred days. When planted after settled warm weather, it will ripen in ninety days.

The ripe seeds are white, the eye surrounded with a broad patch of purple, which is also extended over one of the ends: they are of a rounded-oval form, half an inch long, and three-eighths of an inch in width and thickness. A quart contains fourteen hundred and fifty seeds, and will plant a hundred and fifty hills. As the plants are of dwarfish character, the seeds are sometimes sown in drills; a quart being required for two hundred feet.

The Mottled Cranberry is moderately productive, and the young pods are tender and well flavoured: the seeds, while green, are farinaceous, and, though of good quality when ripe, are but little used.

Mottled Prolific

Plant branching, healthy, and vigorous, six feet or more in height; flowers purple; the pods are four inches and a half long, usually produced in pairs, green at first, washed with purple when more advanced, light-brown at maturity, and contain six seeds.

It is a late variety. Plantings made during the first of the season will not produce pods for use until the last of July, or beginning of August; but, if these are plucked as they become of suitable size, the plants will continue in bearing until destroyed by frost.

The ripe beans are drab, thickly and minutely spotted with black, and also distinctly marked with regular lines of the same colour. They are of an oblong form, flattened, often squarely or diagonally shortened at the ends, nearly half an inch in length, and three-tenths of an inch in width. A quart contains thirty-one hundred seeds, and will plant about three hundred hills.

As a shelled-bean, in its green or ripened state, the variety has little merit. Its recommendations are its fine, tender pods, its remarkable productiveness, and its uniformly healthy habit.

Prédhomme

Introduced from France. Plant four or five feet high, with broad, deep-green, blistered foliage and white flowers; the pods are nearly cylindrical, three inches long, green while young, cream-white when ripe, and contain from six to eight seeds, set very closely together.

The ripe beans are dull-white, veined, oblong, often shortened at the ends, a third of an inch long, and nearly a fourth of an inch in width and thickness. A quart contains about thirty-five hundred seeds, and will plant three hundred and fifty hills.

Early plantings will blossom in eight weeks, afford pods for the table in about ten weeks, and ripen in a hundred and eight days. It may be planted for its green pods to the first of July.

It is of little value as a shelled-bean in its green state. When ripe, it is of good quality, and, as a string-bean, one of the best; the pods being very brittle, succulent, and fine flavoured. They remain long upon the plants without becoming tough and hard; and are tender, and good for use, until almost ripe. On account of their thin and delicate character, the seeds, in unfavorable seasons, are often stained and otherwise injured by dampness at the time of ripening.

Princess

A French variety. Plant six feet or more in height, with lively-green foliage and white flowers; the pods are five inches long, pale-green while young, yellow at maturity, and contain six or seven, and sometimes eight, seeds.

The ripe bean is white, egg-shaped, two-fifths of an inch long, and a fourth of an inch thick: nearly three thousand are contained in a quart, and will plant three hundred and fifty hills.

The variety somewhat resembles the Prédhomme; but the seeds are larger and brighter, the pods are longer, the seeds are less close in the pods, and it is some days earlier. It ripens in about three months from the time of planting. A good sort for stringing, and of excellent quality when ripe.

Red Cranberry

This is one of the oldest and most familiar of garden-beans, and has probably been longer and more generally cultivated in this country than any other variety.

The plants are five or six feet high, of medium strength and vigor; flowers pale-lilac. The pods are quite irregular in form; often reversely curved, or sickle-shaped; four inches and a half long; yellowish-green while young; clear-white when suitable for shelling; yellowish-white, shrivelled, and contorted, when ripe; and contain five or six seeds.

Its season is intermediate. If planted early, the variety will blossom in seven weeks, yield young pods in nine weeks, green beans in eleven weeks, and ripen in ninety-five days. In favorable seasons, the crop will ripen if the seeds are planted the last of June; but, for the young pods or for green beans, plantings may be made to near the middle of July.

Seeds clear, deep-purple, the hilum white, round-ovoid, slightly compressed, half an inch long, and about three-eighths of an inch in depth and thickness. Fourteen hundred and fifty seeds are contained in a quart, and will plant a hundred and fifty hills.

It is a hardy and productive variety, principally grown as a string-bean. The pods are succulent and tender; and these qualities are retained to a very advanced stage of growth, or until quite of suitable size for shelling. The dark colour of the bean, which is to some extent imparted to the pods in the process of cooking, is by some considered an objection; and the White Cranberry, though perhaps less prolific, is preferred. As a shelled-bean, it is of good quality in its green state; but, in its ripened state, little used, though dry and farinaceous.

Red Orleans

(Scarlet Orleans)

Five to six feet high; flowers white; the pods are sickle-shaped, five inches long, green when young, often tinged with red when more advanced, yellow at full maturity, and contain five or six seeds, packed closely together.

It is one of the earliest of the running varieties. Spring plantings will blossom in about seven weeks, afford pods for the table in eight weeks, green beans in eleven weeks, and ripen in eighty-five-days. Planted later in the season, pods sufficiently large for stringing may be gathered in six weeks, and the crop will begin to ripen in about seventy days. As a string-bean, the variety may be planted until the first of August.

At the time of harvesting, the ripe seeds are of a bright blood-red colour, but change rapidly by age to brownish-red. They are of an oblong form, often squarely or diagonally shortened at the ends by contact with each other in the pods, half an inch long, and three-tenths of an inch broad. A quart, which contains nearly twenty-four hundred seeds, will plant about two hundred and seventy-five hills.

The Red Orleans is quite prolific, and a desirable sort for soups and stews. The young pods are tender, and well flavoured; but its remarkable precocity must be considered its chief recommendation.

French writers describe the ripe seeds as exceeding the above dimensions; but specimens received from Paris seedsmen correspond in size, form, and colour with the description before given.

Dry Bean Varieties

Rhode-Island Butter

Plant seven feet and upwards in height, with large, broad, deep-green, wrinkled foliage; flowers blush-white; the pods are six inches long, nearly three-fourths of an inch broad, green while young, paler when more advanced, cream-white and much shrivelled when ripe, and contain seven seeds.

If planted early in the season, green pods may be plucked for the table in nine or ten weeks, pods for shelling in twelve weeks, and the crop will ripen in a hundred and twenty-three days. Planted early in June, the pods will generally all ripen; but, if the planting is delayed to the last of the month, the crop will but partially mature, unless the season prove more than usually favorable. The vines will, however, yield a plentiful supply of pods, and also of green beans.

The seeds, at maturity, are cream-yellow, with well-defined spots and stripes of deep yellowish-buff. They are broad-kidney-shaped, flattened, five-eighths of an inch long, and nearly half an inch broad. The cream-yellow gradually changes by age to brown, and the markings become relatively darker. Fourteen hundred seeds are contained in a quart, and will plant a hundred and fifty hills.

The variety yields abundantly; and the large pods are tender, succulent, and excellent for table use. The beans, in their green state, are of good quality, though little used when ripe.

Sabre, or Cimeter

Stem seven or eight feet high; leaves broad, large, deep-green, and much wrinkled or corrugated; flowers white; pods large, broad, and thin, curved at the ends in the form of a sabre, or cimeter, green when young, cream-white when ripe, and contain eight beans.

The variety will blossom in eight weeks, afford young pods for the table in ten weeks, green beans in eleven weeks, and ripen in a hundred days, from the time of planting. If sown in June, the crop will mature in ninety days. Plantings for the green seeds may be made till the last of June, and for the young pods to the middle of July.

The ripe seeds are clear-white, kidney-form, three-fourths of an inch long, and three-eighths of an inch broad. Sixteen hundred are contained in a quart, and will plant a hundred and sixty hills.

The Sabre Bean is remarkably productive; the young pods are crisp and tender, excellent for table use, and good for pickling; the seeds, green or dry, are farinaceous, and of delicate flavour and appearance.

In height and foliage, size and form of the pods, colour and size of the ripe seeds, it resembles the Case-knife. The principal difference between the varieties is in the earlier maturity of the Sabre.

Soissons

Introduced from France. Stem six feet or more high; foliage large, broad, wrinkled; flowers white; the pods are eight inches long, three-fourths of an inch broad, sword-shaped, yellowish-green when near maturity, yellowish-white when ripe, and contain six or seven seeds.

The variety requires the whole season for its full perfection. If planted early, it blossoms in nine weeks, produces young pods in eleven weeks, and ripens off in gradual succession till the plants are destroyed by frost. If cultivated for its young pods, plantings may be made to the last week in June.

The ripe seeds are remarkably large,—often measuring nearly an inch in length and half an inch in breadth,—pure, glossy-white, kidney-shaped, and generally irregularly compressed. Seven hundred are contained in a quart, and will plant about eighty hills.

The young pods, while quite young and small, are crisp and tender, and the ripe seeds are farinaceous and well flavoured. It is also an excellent sort for shelling in the green state; but the plants are not hardy, and thrive well only in warm soil and sheltered situations. Under ordinary culture, many of the pods are imperfect, and frequently contain but two or three seeds.

White Cranberry

Stem five or six feet high; flowers white; the pods are five inches and a half long, pale-green while young, striped and marbled with red when near maturity, yellowish-buff when ripe, and contain five or six beans.

It is not an early variety. From plantings made at the usual season, young pods may be gathered in about nine weeks, pods for shelling green in twelve weeks, and ripened beans in a hundred and five days. For stringing, or for shelling in a green state, the variety may be planted the first of July; but, in ordinary seasons, few of the pods will reach maturity.

The ripe seeds are white, egg-shaped, sometimes nearly spherical, half an inch long, and three-eighths of an inch in breadth and thickness. In size, form, and colour, they strongly resemble the Dwarf White Marrow; and are not easily distinguished from the seeds of that variety. About twelve hundred and fifty are contained in a quart, and will plant a hundred and twenty-five hills.

The White Cranberry is hardy, yields well, and the young pods are tender and well flavoured. For shelling green, it is decidedly one of the best of all varieties; and for baking, or otherwise cooking, is, when ripe, fully equal to the Pea-bean or White Marrow.

Wild-Goose

Plant seven or eight feet high, of healthy, vigorous habit; flowers bright-purple; the pods are sickle-shaped, pale-green at first, cream-yellow streaked and marbled with purple when ripe, and contain six seeds, closely set together.

The variety requires the entire season for its full perfection. When planted early, it will blossom in nine weeks, produce young pods in eleven weeks, green beans in thirteen weeks, and ripen in a hundred and twenty days. If planted and grown under the influence of summer weather, the plants will blossom in seven weeks, yield young pods in nine weeks, green beans in twelve weeks, and ripen in a hundred days. Plantings for the green seeds may be made to the middle of June, and for the young pods to the first of July.

The ripe beans are pale cream-white, spotted with deep purplish-black (the cream-white gradually changing by age to cinnamon-brown), round-ovoid, four-tenths of an inch long, and about three-eighths of an inch in width and thickness. A quart contains nearly seventeen hundred seeds, and will plant two hundred hills.

The variety has been long cultivated both in Europe and this country. It is hardy and productive. The young pods are of fair quality; and the seeds, green or ripe, are excellent for table use, in whatever form prepared.

Yellow Cranberry

Five to six feet high, with yellowish-green foliage and pale-purple flowers: the pods are five inches long, three-fourths of an inch broad, often sickle-shaped; pale-green at first; cream-yellow, shrivelled, and irregular in form, like those of the Red variety, at maturity; and contain five or six seeds.

It is a few days later than the White Cranberry, and nearly two weeks later than the Red. Planted at the commencement of the season, it will blossom in eight weeks, yield pods for the table in about ten weeks, pods for shelling in twelve or thirteen weeks, and ripen in a hundred and ten days. Early summer-plantings will blossom in seven weeks, produce pods for the table in less than nine weeks, and ripen in about a hundred days. When grown for the ripened crop, it should have the advantage of the entire season; but, when cultivated for its young pods, plantings may be made till the first of July.

Seeds yellow, with a narrow, dark line encircling the hilum: round-ovoid, half an inch long, and three-eighths of an inch in breadth and thickness: thirteen hundred and fifty are contained in a quart, and will plant a hundred and twenty-five hills.

The variety is hardy and prolific; of good quality as a string-bean, or for shelling in the green state. When ripe, the seeds are nearly equal to the White Marrow for baking, though the colour is less agreeable.

132

Asparagus Bean (Vigna unguiculata)

(Long-Podded Dolichos; Dolichos sesquipedalis; Metre-Long Bean; Snake Bean)

The Asparagus-bean, in its manner of growth, inflorescence, and in the size and character of its pods, is quite distinct from the class of beans before described. It is a native of Tropical America, and requires a long, warm season for its full perfection.

The stem is from six to seven feet high; the leaves are long, narrow, smooth, and shining; the flowers are large, greenish-yellow, and produced two or three together at the extremity of quite a long peduncle; the pods are nearly cylindrical, pale-green, pendent, and grow with remarkable rapidity,—when fully developed, they are eighteen or twenty inches long, and contain eight or nine seeds.

These should be sown as early in spring as the appearance of settled warm weather; and the plants will then blossom in ten or eleven weeks, afford pods for use in fourteen weeks, and ripen off their crop in gradual succession until destroyed by frost.

The ripe seeds are cinnamon-brown, with a narrow, dark line about the hilum; kidney-shaped, half an inch long, and a fourth of an inch broad: nearly four thousand are contained in a quart, and will plant four hundred and fifty hills.

The seeds are quite small, and are rarely eaten, either in a green or ripe state. The variety is cultivated exclusively for its long, peculiar pods, which are crisp, tender, of good flavor, and much esteemed for pickling. It is, however, much less productive than many of the running kinds of garden-beans, and must be considered more curious than really useful.

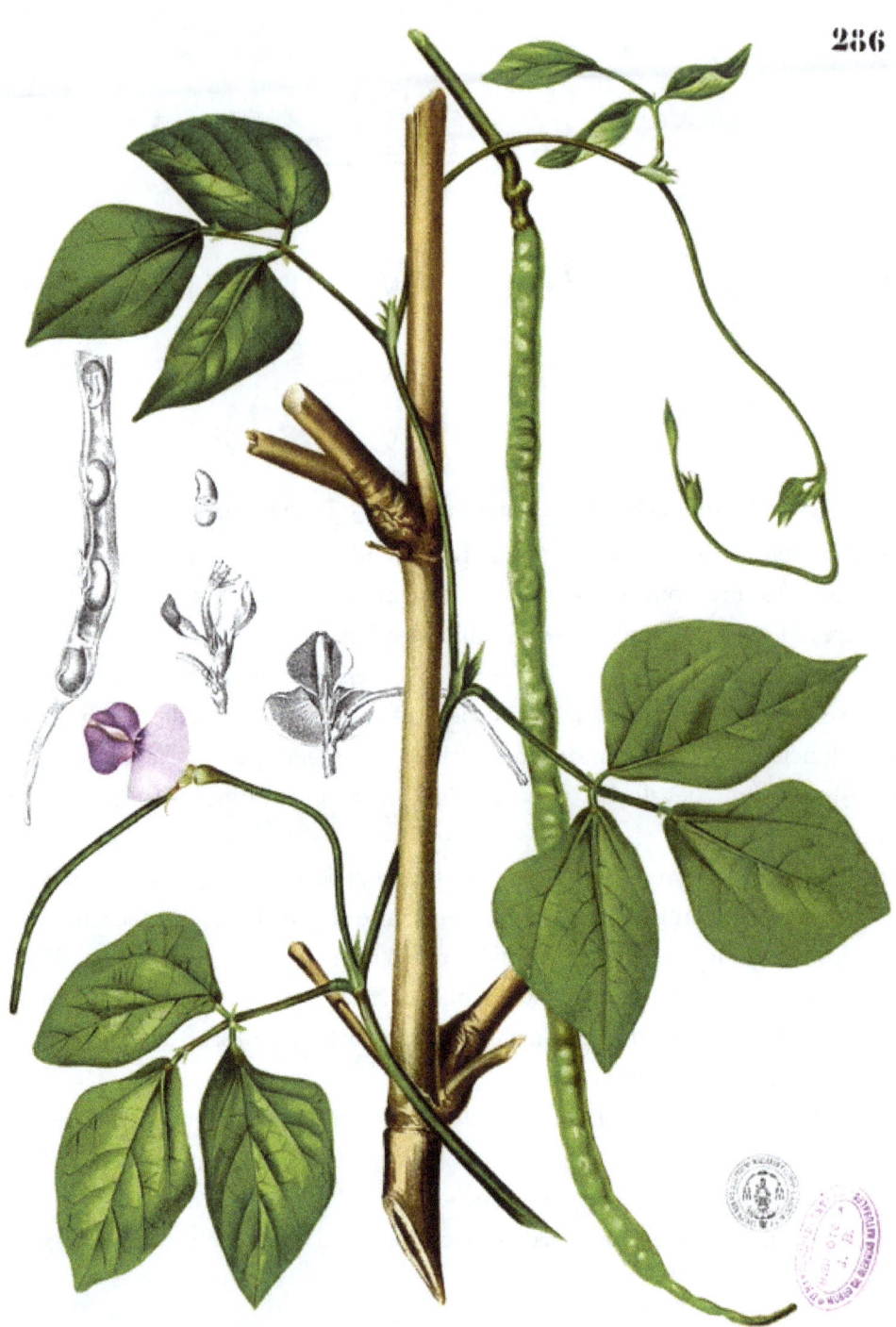

133

Lima Bean (Phaseolus lunatus)

Stem ten feet or more in height; leaves comparatively long and narrow, smooth and shining; flowers small, greenish-yellow, in spikes; the pods are four inches and a half long, an inch and a quarter broad, much flattened, green and wrinkled while young, yellowish when ripe, and contain three or four beans.

The Lima is one of the latest, as well as one of the most tender, of all garden-beans; and seldom, if ever, entirely perfects its crop in the Northern States. Little will be gained by very early planting; as the seeds are not only liable to decay before vegetating, but the plants suffer greatly from cold, damp weather. In the Northern and Eastern States, the seeds should not be planted in the open ground before the beginning of May; nor should the planting be delayed beyond the tenth or middle of the month. In ordinary seasons, the Lima Bean will blossom in eight or nine weeks, and pods may be plucked for use the last of August, or beginning of September. Only a small proportion of the pods attain a sufficient size for use; a large part of the crop being prematurely destroyed by frost.

Seed and Plant Guide (1898), H.W. Buckbee

The ripe seeds are dull-white or greenish-white, with veins radiating from the eye; broad, kidney-shaped, much flattened, seven-eighths of an inch long, and two-thirds of an inch broad. A quart contains about seven hundred seeds, and will plant eighty hills.

The pods are tough and parchment-like in all stages of their growth, and are never eaten. The seeds, green or ripe, are universally esteemed for their peculiar flavor and excellence; and, by most persons, are considered the finest of all the garden varieties. If gathered when suitable for use in their green state, and dried in the pods in a cool and shaded situation, they may be preserved during the winter. When required for use, they are shelled, soaked a short time in clear water, and cooked as green beans: thus treated, they will be nearly as tender and well flavored as when freshly plucked from the plants.

The seeds are sometimes started on a hot-bed, in thumb-pots, or on inverted turf, or sods, cut in convenient pieces; and about the last of May, if the weather is warm and pleasant, transplanted to hills in the open ground.

By the following method, an early and abundant crop may be obtained in comparatively favorable seasons:—

> "As soon in spring as the weather is settled, and the soil warm and in good working condition, set poles about six feet in length, three feet apart each way, and plant five or six beans in each hill; being careful to set each bean with its germ downward, and covering an inch deep. After they have grown a while, and before they begin to run, pull up the weakest, and leave but three of the most vigorous plants to a hill. As these increase in height, they should, if necessary, be tied to the stakes, or poles, using bass-matting, or other soft, fibrous material, for the purpose. When they have ascended to the tops of the poles, the ends should be cut or pinched off; as also the ends of all the branches, whenever they rise above that height. This practice checks their liability to run to vines, and tends to make them blossom earlier, and bear sooner and more abundantly, than they otherwise would do."

In tropical climates, the Lima Bean is perennial.

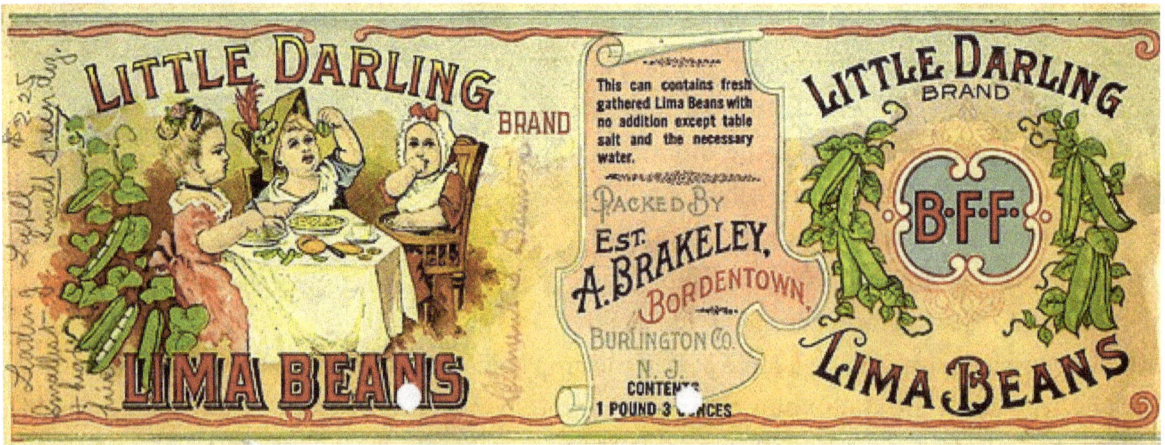

Little Darling Brand Lima Beans Can Label, 1906

Green Lima

A sub-variety of the Common Lima, differing principally in the pea-green color of the seeds.

As generally found in the market, the seeds of the Common and Green Lima are more or less intermixed. By some, the Green is considered more tender, and thought to remain longer on the plants without becoming hard, than the White. The habits of the plants are the same, and there is no difference in the season of maturity. A careful selection of seeds for planting, and skilful culture, would undoubtedly give a degree of permanency to this difference in color; which appears to be the principal, if not the only, point of variation.

Mottled Lima

This, like the Green, is a sub-variety of the Common Lima. The ripe seeds are dull-white or greenish-white, mottled and clouded with purple.

In the habit of the plant, in the foliage, pods, form, or size of the seeds, or season of maturity, there are no marks of distinction when compared with the Common Lima.

Sieva (Small Lima)

(Carolina; Saba; West-Indian; Small Lima; Carolina Sewee)

The Sieva is a variety of the Lima, attaining a height of ten or twelve feet. The leaves and flowers resemble those of the Common Lima. The pods, however, are much smaller, and remarkable for their uniform size; generally measuring three inches in length, and about seven-eighths of an inch in width: they are green and wrinkled while young, pale yellowish-brown when ripe, and contain three, and sometimes four, seeds.

Though several days earlier than the Lima, the Sieva Bean requires the whole season for its complete maturity; and even when planted early, and receiving the advantage of a warm summer and a favorable autumn, it is seldom fully perfected in the Northern States: for, though much of the crop may ripen, a large portion almost invariably is prematurely destroyed by frost.

The variety will blossom in eight weeks from the time of planting, afford pods for shelling in twelve weeks, and ripen from near the middle of September till destroyed by frost.

The seeds are white or dull yellowish-white, broad-kidney-shaped, much flattened, five-eighths of an inch long, and nearly half an inch broad. A quart contains about sixteen hundred, and will plant about a hundred and fifty hills.

The Sieva is one of the most productive of all varieties. The young pods, however, are tough and hard, and are never eaten. The beans, in their green or ripe state, are similar to the Lima, and are nearly as delicate and richly flavored. It is from two to three weeks earlier than the last named, and would yield a certain abundance in seasons when the Lima would uniformly fail. As a shelled-bean, green or dry, it must be classed as one of the best, and is recommended for cultivation.

Mottled Sieva

A sub-variety of the Common Sieva; the principal if not the only mark of distinction being in the variegated character of the seeds, which are dull-white, spotted and streaked with purple.

It is sometimes described as being earlier than the Common variety; but, from various experiments in the cultivation of both varieties, there appears to be little if any difference in their seasons of maturity. The color and form of the flower are the same as the Sieva; the pods are of the same size and shape; and the leaves have the same elongated form, and smooth, glossy appearance.

Lima bean varieties, including many small varieties

134

Scarlet Runner Bean (Phaseolus coccineus)

(Phaseolus multiflorus)

From South America. Though nearly allied to the Common Kidney-bean, it is considered by botanists a distinct species; differing in its inflorescence, in the form of its pods, and particularly in the fact that the cotyledons, or lobes of the planted seed, do not rise to the surface of the ground in the process of germination. It is, besides, a perennial plant. The roots are tuberous, and, though small, not unlike those of the Dahlia.

If taken up before frost in the autumn, they may be preserved in a conservatory, or warm parlour or sitting-room, during winter, and reset in the open ground on the approach of warm weather; when new shoots will soon make their appearance, and the plants will blossom a second time early and abundantly.

The plants are twelve feet or more in height or length, with deep-green foliage and brilliant scarlet flowers; the latter being produced in spikes, on long footstalks. The pods are six inches long, nearly an inch broad, somewhat hairy while young, sickle-shaped and wrinkled when more advanced, light reddish-brown when ripe, and contain four or five seeds.

It requires the whole season for its perfection, and should be planted as early as the weather will admit. The plants will then blossom in seven or eight weeks, produce young pods in nine weeks, green seeds in twelve weeks, and ripen in a hundred and fifteen days.

The ripe seeds are lilac-purple, variegated with black, or deep purplish-brown,—the edge, or border, little, if any, marked; hilum long and white; form broad-kidney-shaped; size large,—if well grown, measuring seven-eighths of an inch long, six-tenths of an inch broad, and three-eighths of an inch thick. About five hundred and fifty are contained in a quart, and will plant eighty hills.

In this country, it is usually cultivated as an ornamental, climbing annual; the spikes of rich, scarlet flowers, and its deep-green foliage, rendering the plant one of the most showy and attractive objects of the garden.

Though inferior to some of the finer sorts of garden-beans, its value as an esculent has not been generally appreciated. The young pods are tender and well flavored; and the seeds, green or ripe, are much esteemed in many localities.

> "In Britain, the green pods only are used; on the Continent, the ripened seeds are as much an object of culture; in Holland, the Runners are grown in every cottage-garden for both purposes; while, in France and Switzerland, they are grown chiefly for the ripened seeds. In England, they occupy a place in most cottage-gardens, and are made both ornamental and useful. They cover arbors, are trained over pales and up the walls of cottages, which they enliven by the brightness of their blossoms; while every day produces a supply of wholesome and nutritious food for the owner. The French, now enthusiastically fond of this legume, at one time held it in utter detestation."

Painted Lady-Runner

A sub-variety of the Scarlet-runner, with variegated flowers; the upper petals being scarlet, the lower white. The ripe seeds are paler, and the spots and markings duller. Cultivation and uses the same.

White-Runner

A variety of the Scarlet-runner. The plants are less vigorous, the pods are longer and less wrinkled, and the flowers and seeds pure white.

The green pods are used in the same manner as those of the Scarlet-runner, and are similar in texture and flavor; but the shelled-beans, either green or ripe, are generally considered superior to those of the Scarlet variety. They are sometimes seen in vegetable markets under the name of the "Lima;" and are probably often cultivated, as well as purchased and consumed, as the Lima. The White-runner beans, however, are easily distinguished by their greater thickness, more rounded form, and especially by their uniform whiteness.

FIELD AND GARDEN VEGETABLES OF THE LATE NINETEENTH CENTURY | 475

Phaseolus coccineus

135

Chick-Pea (Cicer arietinum)

(Egyptian Pea)

The Chick-pea is a hardy, annual plant, originally from the south of Europe, but also indigenous to the north of Africa and some parts of Asia. The stem is two or three feet high, erect and branching; the leaves are pinnate, with from six to nine pairs of oval, grayish, toothed leaflets; the flowers resemble those of the Common Pea, and are produced on long peduncles, generally singly, but sometimes in pairs; the pods are about an inch long, three-fourths of an inch broad, somewhat rhomboidal, hairy, inflated or bladder-like, and contain two or three globular, wrinkled, pea-like seeds.

Sowing and Cultivation—The seed should be sown in April, in the manner of the Garden-pea; making the drills about three feet apart, an inch and a half deep, and dropping the seeds two inches asunder in the drills. All the culture required is simply to keep the ground between the rows free from weeds. The crop should be harvested before the complete maturity of the seeds.

Use—"The Pease, though not very digestible, are largely employed in soups, and form the basis of the *purée aux croutons*, or bread and pea soup, so highly esteemed in Paris." They are also extensively used, roasted and ground, as a substitute for coffee.

There are three varieties, as follow:

Red Chick-Pea

A variety with rose-colored flowers, and red or brownish-red seeds.

White Chick-Pea

Both the flowers and seeds white; plant similar to those of the other varieties.

Yellow Chick-Pea

This variety has white blossoms and yellow seeds. The plant, in height, foliage, or general habit, differs little from the White or the Red Seeded.

136

Chickling Vetch (Lathyrus sativus)

(Grass Pea; Cultivated Lathyrus)

Stem three or four feet high or long, attaching itself to trellises, branches, or whatever may be provided for its support, in the manner of pease; the leaves are small and grass-like; flowers solitary, smaller than those of the Common Pea, and generally bright-blue; the pods are an inch and a half long, three-fourths of an inch broad, flattened, winged along the back, and enclose two compressed but irregularly shaped seeds of a dun or brownish color and pleasant flavor.

Dried Lathyrus sativus seeds

Cultivation and Use—The seeds are sown at the time and in the manner of the taller kinds of garden-pease. The plant is principally cultivated for its seeds, the flour of which is mixed with that of wheat or rye, and made into bread. It is also fed to stock; and, in some localities, the plants are given as green food to horses and cattle.

> "In 1671, its cultivation and use were prohibited on account of its supposed pernicious properties; as it was thought to induce rigidity of the limbs, and to otherwise injuriously affect the system."[1]

White-Flowered Chickling Vetch

A variety with white flowers and seeds. The foliage is also much paler than that of the Common Chickling Vetch.

Other species of the genus also produce farinaceous seeds suitable for food, but in too small quantities to admit of being profitably cultivated in this country.

---

1. The seeds do contain variable amounts of a certain neurotoxins (ODAP) considered the cause of the disease that cause paralysis of the lower body. If eaten irregularly in small amounts, they are not considered harmful, but as a regular part of the diet, can have serious side effects, something witnesses historically after famine periods when this pulse was eaten as a famine food.

137

Broad Bean (Vicia faba)

(Horse-Bean; Garden-Bean, of the English; English Bean)

The English Bean differs essentially from the Common American Garden or Kidney Bean usually cultivated in this country; and is classed by botanists under a different genera, and not as a distinct species, as intimated in the "American Gardener." Aside from the great difference in their general appearance and manner of growth, the soil, climate, and mode of cultivation, required by the two classes, are very dissimilar: the American Garden-bean thriving best in a light, warm soil, and under a high temperature; and the English Bean in stiff, moist soil, and in cool, humid seasons.

English Bean

The English Bean is a native of Egypt, and is said to be the most ancient of all the now cultivated esculents. It is an annual plant, with an upright, smooth, four-sided, hollow stem, dividing into branches near the ground, and growing from two to four feet and upwards in height. The leaves are alternate, pinnate, and composed of from two to four pairs of oval, smooth, entire leaflets; the flowers are large, nearly stemless, purple or white, veined and spotted with purplish-black; the pods are large and downy; the seeds are rounded, or reniform, flattened, and vary to a considerable extent in size and colour in the different varieties,—they will vegetate until more than five years old.

Soil and Planting—As before remarked, the English Bean requires a moist, strong soil, and a cool situation; the principal obstacles in the way of its successful cultivation in this country being the heat and drought of the summer. The seeds should be planted early, in drills two feet asunder for the smaller-growing varieties, and three feet for the larger sorts; dropping them about six inches from each other, and covering two inches deep. A quart of seed will plant about a hundred and fifty feet of row or drill.

Cultivation—

"When the plants have attained a height of five or six inches, they are earthed up slightly for support; and, when more advanced, they are sometimes staked along the rows, and cords extended from stake to stake to keep the plants erect. When the young pods appear, the tops of the plants should be pinched off, to throw that nourishment, which would be expended in uselessly increasing the height of the plant, into its general system, and consequently increase the bulk of crop, as well as hasten its maturity. This often-recommended operation, though disregarded by many, is of very signal importance."—*M'Int.*

Taking the Crop—The pods should be gathered for use when the seeds are comparatively young, or when they are of the size of a marrowfat-pea. As a general rule, all vegetables are most tender and delicate when young; and to few esculents does this truth apply with greater force than to the class of plants to which the English Bean belongs.

Use—The seeds are used in their green state, cooked and served in the same manner as shelled kidney-beans. The young pods are sometimes, though rarely, used as string-beans.

Varieties—

Dutch Long Pod

Plant from four to five feet high, dividing into two or three branches; flowers white; pods horizontal, or slightly pendulous, six or seven inches long, about an inch in width, three-fourths of an inch thick, and containing five or six large white or yellowish-white seeds.

Not early, but prolific, and of good quality.

Dwarf Fan, or Cluster

(Early Dwarf; Bog-bean)

A remarkably dwarfish, early variety, much employed in forcing. Stem about a foot high, separating near the ground into two or three branches; flowers white; the pods, which are produced in clusters near the top of the plant, are almost cylindrical, three inches long, three-fourths of an inch thick, and contain three or four small, oblong, yellow seeds.

It is one of the smallest and earliest of the English Beans, and yields abundantly.

Early Dwarf Crimson-Seeded

(Vilmorin's Dwarf Red-seeded)

Plant sixteen inches high, separating into two or three divisions, or branches; the flowers resemble those of the Common varieties, but are somewhat smaller; the pods are erect, three inches and a half long, three-fifths of an inch[Pg 506] wide, half an inch thick, and contain three or four seeds, closely set together, and nearly as large in diameter as the pod.

The ripe seeds are bright brownish-red or crimson, thick, shortened at the back, and depressed at the sides: six hundred and fifty will measure a quart.

The variety is principally esteemed for its dwarfish habit and early maturity.

Early Mazagan

(Early Malta)

This variety, though originally from Mazagan, on the coast of Africa, is one of the hardiest sorts now in cultivation. Stem from two to three feet high, and rather slender; pods four to five inches long, containing four or five whitish seeds.

The Early Mazagan is much less productive than many other sorts; but its hardiness and earliness have secured it a place in the garden, and it has been cultivated more or less extensively for upwards of a century.

Evergreen Long Pod.

(Green Genoa; Green Long Pod; Green Nonpareil)

This variety grows from three to four feet high. The pods are long, somewhat flattened, and generally contain four rather small, oblong, green seeds. It is an excellent bearer, of good quality, and but a few days later than the Common Long Pod. The variety is much esteemed on account of the fine, green colour of the beans; which, if gathered at the proper time, retain their green colour when dressed.

In planting, make the drills three feet apart, and two inches and a half deep; and allow two plants for each linear foot.

Green China

From two to two feet and a half high; pods long, cylindrical, containing three or four beans, which remain of a green colour when dry. It is recommended for its great productiveness and late maturity.

Green Julienne

Plant about three feet and a half high, usually divided into four branches; the pods are erect, four inches long, three-fourths of an inch thick, and contain two or three small, oblong, green seeds.

Early and of good quality.

Green Windsor

(Toker)

Stem three feet high, separating into two, and sometimes three, branches; flowers white; pods erect, often horizontal, four inches and a half long, an inch and a quarter wide, and containing three large, green, nearly circular, and rather thick seeds.

The latter retain their fresh, green colour till near maturity, and, to a considerable extent, when fully ripe; and, on this account, are found in the market, and used at table, after most other varieties have disappeared.

The variety resembles the Common Broad Windsor; but the seeds are smaller, and retain their green colour after maturity. Eleven or twelve well-developed seeds will weigh an ounce.

Horse-Bean

(Scotch Bean; Faba vulgaris arvensis)

Stem from three to five feet high; flowers variable in colour; the ripe seeds are from a half to five-eighths of an inch in length by three-eighths in breadth, generally slightly compressed on the sides, and frequently a little hollowed or flattened at the end, of a whitish or light-brownish colour, occasionally interspersed with darker blotches, particularly towards the extremities; eye black; average weight per bushel sixty-two pounds.

An agricultural sort, generally cultivated in rows, but sometimes sown broadcast. It is not adapted to the climate of the United States, though extensively and profitably grown in England and Scotland.

Johnson's Wonderful

An improved variety of the Broad Windsor, recently introduced, and apparently of excellent quality. The pods are long, and contain six or eight beans, which are similar in size and form to the Windsor.

Long-Podded

(Lisbon; Hang-Down Long Pod; Early Long Pod; Sandwich; Turkey Long Pod; Sword Long Pod)

Stems from three to five feet high; pods six to seven inches long, an inch and a fourth broad, rather pendulous, and containing four or five whitish, somewhat oblong, flattened seeds, about an inch in length, and five-eighths of an inch in breadth.

The variety has been long in cultivation, is remarkably productive, and one of the most esteemed of the English Beans. It is about a week later than the Early Mazagan.

Marshall's Early Dwarf Prolific

Plant from eighteen inches to two feet high, separating into numerous branches. It resembles the Early Mazagan; but is two weeks earlier, and much more productive. The pods are produced in clusters near the ground, and contain four or five seeds, which are larger than those of the last named.

Red or Scarlet Blossomed

Stem three or four feet high, separating near the ground into four branches; flowers generally bright-red, approaching scarlet, but varying from pale to purplish-red and blackish-purple, and sometimes to nearly jet-black; the pods, which differ from all other varieties in their dark, rusty-brown colour, are erect, four inches long, nearly an inch broad, and contain three and sometimes four seeds.

The variety is remarkably hardy and productive; but less esteemed than many others, on account of its dark colour. It deserves cultivation as an ornamental plant.

Red Windsor

(Scarlet Windsor; Dark-Red)

This variety resembles the Violet or Purple; growing about four feet high. The pods are narrower than those of the Broad Windsor, and contain about the same number of seeds: in the green state, these are darker than those of the Violet, but change to scarlet when fully grown, and to deep-red when ripe.

The Red Windsor is late, but prolific, and of good quality. It is, however, little cultivated, on account of its dark and unattractive appearance. The seed weighs about thirty-one grains.

Royal Dwarf Cluster

A very Dwarf, and comparatively new variety; growing only twelve or fourteen inches high. It produces its pods in clusters, three or four beans in each pod, which are smaller than Marshall's Early Prolific. On account of its branching habit, it should not have less than ten or twelve inches in the line, which is nearly its proper distance between the rows. Much esteemed for the delicacy and smallness of the beans while young, and considered one of the best of the early Dwarf sorts.

Toker

(Large Toker)

Height about five feet; pods rather long, and very broad, containing three or four beans of a whitish colour,—differing from the Common Windsor in being of an elongated, oval form.

This is a medium late sort, and an excellent bearer, but considered somewhat coarse, and therefore not so much esteemed as the Windsor. The ripe seed weighs thirty-six grains.

Violet or Purple

(Violette)

Stem about four feet high, with two or three ramifications; flowers white; pods generally erect, sometimes at right angles, a little curved, four inches or upwards in length, an inch and a fourth in width, four-fifths of an inch thick, containing two and sometimes three seeds. When ripe, the beans are large, not regular in form, rather thin, of a violet-red colour, changing by age to a mahogany-red; the size and shape being intermediate between the Long Pod and Broad Windsor.

The variety is of good quality, and productive; but less desirable than many other sorts, on account of its dark colour.

White-Blossomed Long Pod

The flowers of this sort differ from all others in being pure white; having no spots on the large upper petal, or on the wings or smaller side petals. It is liable to degenerate; but may easily be distinguished, when in flower, by the above characters. Stem about four feet high; pods long, nearly cylindrical, and slightly pendulous, generally containing four and sometimes five seeds, which are black or blackish-brown, three-fourths of an inch long, and half an inch broad.

It is a moderate bearer, and of excellent quality; but not used in an advanced state, on account of its colour. The variety possesses the singular anomaly of having the whitest flowers and the darkest seeds of any of the English Beans. The seed weighs about twelve grains.

Windsor

(*White Broad Windsor; Taylor's Large Windsor; Kentish Windsor; Mumford; Wrench's Improved Windsor*)

Stem about four feet high; flowers white; pods generally horizontal or inclined, five inches long, an inch and a fourth wide, seven-eighths of an inch thick, and containing two or three beans; seeds large, yellowish, of a flat, circular form, an inch broad, but varying in size according to soil, culture, and season. A quart contains from two hundred and fifty to two hundred and seventy-five seeds.

This familiar sort is much esteemed and extensively cultivated. It is considered the earliest of the late Garden varieties; and excellent as a summer bean, on account of its remaining longer fit for use than any other, with the exception of the Green Windsor. It is a sure bearer; and, as the pods are produced in succession, pluckings may be made from day to day for many weeks.

The seeds are the heaviest of all the English Beans; nine well-grown specimens weighing an ounce.

Broad bean plant, flowers, and pods

138

The Lentil (Lens culinaris)

(Ervum lens)

A hardy, annual plant, with an erect, angular, branching stem a foot and a half high. The leaves are winged, with about six pairs of narrow leaflets, and terminate in a divided tendril, or clasper; the flowers are small, numerous, and generally produced in pairs; the pods are somewhat quadrangular, flattened, usually in pairs, and enclose one or two round, lens-like seeds, the size and colour varying in the different varieties,—about four hundred and fifty are contained in an ounce, and their power of germination is retained three years.

Cultivation—"The soil best adapted for the Lentil is that of a dry, light, calcareous, sandy nature."

When cultivated as green food for stock, it should be sown broadcast; but, if grown for ripe seeds, it should be sown in drills,—the last of April or beginning of May being the most suitable season for sowing.

Use—

"The Lentil is a legume of the greatest antiquity, and was much esteemed in the days of the patriarchs. In Egypt and Syria, the seeds are parched, and sold in shops; being considered by the natives as excellent food for those making long journeys. In France, Germany, Holland, and other countries of Europe, it is grown to a considerable extent, both for its seeds and haum. The former are used in various ways, but principally, when ripe, in soups, as split pease. When given as green food to stock, it should be cut when the first pods are nearly full grown."

Varieties—

Common Lentil

(Yellow Lentil)

This variety is considered superior to the Large Lentil, though the seeds are much smaller. In the markets of Paris, it is the most esteemed of all the cultivated sorts. Its season is the same with that of the last named.

Green Lentil

(Lentille verte Du Puy)

The Green Lentil somewhat resembles the Small Lentil, particularly in its habit of growth; though its stem is taller and more slender, and its foliage deeper coloured. The principal distinction is in the colour of the seeds, which are green, spotted and marbled with black.

Large Lentil

Flowers small, white, generally two, but sometimes three, on each peduncle; the pods are three-fourths of an inch long, half an inch broad, flattened, and generally contain a single seed, which is white or cream-coloured, lens-shaped, three-eighths of an inch in diameter, and an eighth of an inch in thickness. The plant is about fifteen inches high.

It is one of the most productive of all the varieties, though inferior in quality to the Common Lentil.

One-Flowered Lentil

(Ervum monanthos)

The stem of this quite distinct species is from twelve to fifteen inches high; the flowers are yellow, stained or spotted with black, and produced one on a foot-stalk; the pods are oval, smooth, and contain three or four globular, wrinkled, grayish-brown seeds, nearly a fourth of an inch in diameter.

About five hundred and fifty seeds are contained in an ounce.

The One-flowered Lentil is inferior to most of the other sorts; but is cultivated to some extent, in France and elsewhere, both for its seeds and herbage.

Red Lentil

Seeds of the size and form of those of the Common Lentil, but of a reddish-brown colour; flowers light-red. Its season of maturity is the same with that of the last named.

Small Lentil

(Lentille petite)

Seeds about an eighth of an inch in diameter; flowers reddish; and pods often containing two seeds.

This is the "Lentille petite" of the French; and is the variety mostly sown for green food in France, although its ripe seeds are also used. It is rather late, and grows taller than any of the other sorts, except the Green Lentil. When sown in drills, they should be from ten to fifteen inches apart, and the plants about four or five inches distant in the rows.

The Lentils are of a close, branching habit of growth; and a single plant will produce a hundred and fifty and often a much greater number of pods.

The lentil plant

139

Lupin (Lupinus)

The Lupines are distinguished among leguminous plants by their strong, erect, branching habit of growth. Of the numerous species and varieties, some are cultivated for ornament, others for forage, and some for ploughing under for the purpose of enriching the soil. The only species grown for their farinaceous seeds, or which are considered of much value to the gardener, are the two following:

White Lupine

(Lupinus albus)

An annual species, with a sturdy, erect stem two feet high; leaves oblong, covered with a silvery down, and produced seven or eight together at the end of a common stem; the flowers are white, in loose, terminal spikes; the pods are straight, hairy, about three inches long, and contain five or six large, white, flattened seeds,—these are slightly bitter when eaten, and are reputed to possess important medical properties.

Lupinus alba seeds prepared in brine, a common snack in Spain, were they are known as 'altramuces'

The White Lupine was extensively cultivated by the Romans for its ripened seeds, which were used for food; and also for its green herbage, which was employed for the support of their domestic animals.

It is of little value as an esculent; and, compared with many other leguminous plants, not worthy of cultivation.

The seeds should be sown where the plants are to remain, as they do not succeed well when transplanted. Sow early in May, in drills sixteen to eighteen inches apart; cover an inch and a half deep, and thin to five or six inches in the rows.

Yellow Lupine

(Lupinus luteus)

The Yellow Lupine is a native of Sicily. It is a hardy annual, and resembles the foregoing species in its general character. The flowers are yellow; the pods are about two inches long, hairy, flattened, and enclose four or five large, roundish, speckled seeds. It blossoms and ripens at the same time with the White, and is planted and cultivated in the same manner.

This species is grown in Italy for the same purposes as the White, but more extensively. It is also grown in some parts of the south of France, on poor, dry grounds, for cutting in a green state, and ploughing under as a fertiliser.

Lupin plants in flower

140

The Pea (Pisum sativum)

The native country of the Pea, like that of many of our garden vegetables, is unknown. It is a hardy, annual plant; and its cultivation and use as an esculent are almost universal.

To give in detail the various methods of preparing the soil, sowing, culture, gathering, and use, would occupy a volume.

The following directions are condensed from an elaborate treatise on the culture of this vegetable, by Charles M'Intosh, in his excellent work entitled "*The Book of the Garden*":[1]

Soil and its Preparation—The Pea comes earliest to maturity in light, rich soil, abounding in humus: hence the practice of adding decomposed leaves or vegetable mould has a very beneficial effect. For general crops, a rich, hazel loam, or deep, rich, alluvial soil, is next best; but, for the most abundant of all, a strong loam, inclining to clay. For early crops, mild manure, such as leaf-mould, should be used. If the soil is very poor, stronger manure should be employed. For general crops, a good dressing may be applied; and for the dwarf kinds, such as Tom Thumb, Bishop's New Long Pod, and the like, the soil can hardly be too rich.

Seed and Sowing—A quart of ripe peas is equal to about two pounds' weight; and contains, of the largest-sized varieties, about thirteen hundred, and of the smaller descriptions about two thousand, seeds. A pint of the small-seeded sorts, such as the Daniel O'Rourke, Early Frame, and Early Charlton, will sow a row about sixty feet in length; and the same quantity of larger-growing sorts will sow a row of nearly a hundred feet, on account of being sown so much thinner. A fair average depth for covering the seed is two and a half or three inches; though some practise planting four or five inches deep, which is said to be a preventive against the premature decay of the vines near the roots.

As to distance between the rows, when peas are sown in the usual manner (that is, row after row throughout the whole field), they should be as far asunder as the length of the stem of the variety cultivated: thus a pea, that attains a height or length of two feet, should have two feet from row to row, and so on to those taller or lower growing.

They are sometimes sown two rows together, about a foot apart, and ten, twenty, or even fifty feet between the double rows; by which every portion of the crop is well exposed to the sun and air, and the produce gathered with great facility. There is no loss of ground by this method; for other crops can be planted within a foot or two of the rows, and this amount of space is necessary for the purpose of gathering.

A common practice in ordinary garden culture is to sow in double rows twelve or fourteen inches apart, slightly raising the soil for the purpose. When so planted, all of the sorts not over two feet in height may be successfully grown without sticking. When varieties of much taller growth are sown, a greater yield will be secured by bushing the plants; which is more economically as well as more strongly done if the planting is made in double rows. The staking, or bushing, should be furnished when the plants are three or four inches high, or immediately after the second hoeing: they should be of equal height, and all straggling side-twigs should be removed for appearance' sake.

Early Crops—The earliest crops produced in the open garden without artificial aid are obtained by judicious selection of the most approved early varieties, choosing a warm, favorable soil and situation, and sowing the seed either in November, just as the ground is closing, or in February or March, at the first opening of the soil; the latter season,[Pg 518]however, being preferable, as the seed then vegetates with much greater certainty, and the crop is nearly or quite as early. Great benefit will be derived from reflected heat, when planted at the foot of a wall, building, or tight fence, running east and west. It is necessary, however, when warm sunshine follows cold, frosty nights, to shade the peas from its influence an hour or two in the morning, or to sprinkle them with cold water if they have been at all frozen.

They are sometimes covered with a narrow glass frame of a triangular form, and glazed on both sides, or on one only, according as they may be used on rows running from north to south, or from east to west. In the latter case, such frames may have glass in the south side only.

Subsequent Cultivation—

"When the crop has attained the height of about five inches, a little earth should be drawn around the stems, but not so closely as to press upon them: it should form a sort of ridge, with a slight channel in the middle. The intention here is not, as in many other cases, to encourage the roots to diverge in a horizontal direction (for they have no disposition to do so), but rather to give a slight support to the plants until they take hold of the stakes that are to support them. Those crops which are not to be staked require this support the most: and they should have the earth drawn up upon one side only, that the vines may be thrown to one side; which will both facilitate the operation of gathering, and keep the ground between them clear at the same time, while it supports the necks of the plants better than if the earth was drawn up on both sides."

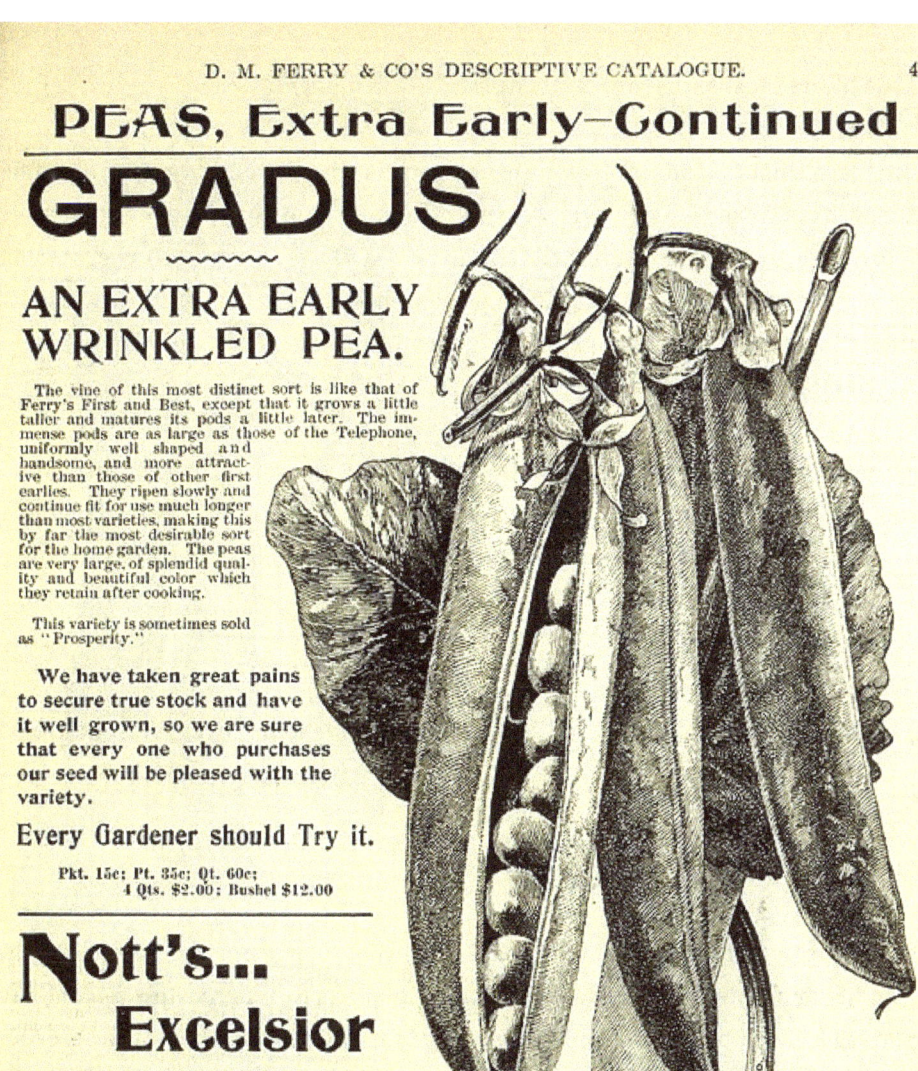

An extra early wrinkled pea
D. M. FERRY & GO'S DESCRIPTIVE CATALOGUE. 49, 1921

Mildew—One of the most successful cultivators (T. A. Knight) says:

"...that the secondary and immediate cause of this disease is a want of a sufficient supply of moisture from the soil, with excess of humidity in the air; particularly if the plants be exposed to a temperature below that to which they have been accustomed. If damp and cloudy weather succeed that which has been warm and bright, without the intervention of sufficient rain to moisten the ground to some depth, the crop is generally much injured by mildew."

"While engaged in the production of those excellent peas which bear his name, he proved this theory by warding off mildew by copious waterings of the roots. The fashionable remedy, at present, is the application of sulphur. This, no doubt, subdues the disease, but does not remove the cause."[2]

Gathering—The crop should be gathered as it becomes fit for use. If even a few of the pods begin to ripen, young pods will not only cease to form, but those partly advanced will cease to enlarge.

Use—

"In a sanitary point of view, peas cannot be eaten too young, nor too soon after they are gathered; and hence people who depend on the public markets for their supply seldom have this very popular vegetable in perfection, and too often only when it is almost unfit for use. This is a formidable objection to the use of peas brought from long distances. It is, of course, for the interest of the producer to keep back his peas till they are fully grown, because they measure better, and, we believe, by many are purchased quicker, as they get greater bulk for their money. This may be so far excusable on the part of such: but it is inexcusable that a gentleman, having a garden of his own, should be served with peas otherwise than in the very highest state of perfection; which they are not, if allowed to become too old, or even too large."[3]

"peas, in a green state, are with difficulty sent to a distance, as, when packed closely together, heat and fermentation speedily take place. This is one of the causes why peas from the South, or those brought by long distances to market, are discoloured, devoid of flavor, and, worst of all, very unwholesome to eat. peas intended for long transportation should be packed in open baskets (not in boxes or tight barrels), and laid in layers not more than two inches thick; and, between such layers, a thick stratum of clean straw or other dry material should be placed."

Varieties—These are very numerous, and, like those of the Broccoli Lettuce, not only greatly confused, but often based on trifling and unimportant distinctions.

From experiments made a few years since in the gardens of the London Horticultural Society, under the direction of Mr. Thompson, who planted no less than two hundred and thirty-five reputed sorts (all of which were then enumerated in seedsmen's catalogues), only twenty-seven of the number were selected as being really useful. About the same time, upwards of a hundred sorts were grown by Mr. M'Intosh,[4] from which twelve were selected as being truly distinct and valuable.

> "New sorts are yearly introduced: and it would be injudicious not to give them a fair trial; for as we progress in pea-culture, as in every other branch of horticulture, we may reasonably expect that really improved and meritorious sorts will arise, and be substituted for others that may be inferior."

Auvergne

(White Sabre; White Cimeter)

The plant is of moderately strong habit of growth, producing a single stem from four to five feet high, according to the soil in which it is grown; and bears from twelve to fifteen pods. These are generally single, but sometimes in pairs; when fully grown, four inches and a half long, and over half an inch broad; tapering to the point, and very much curved. They contain from eight to ten peas, which are closely compressed, and of the size of the Early Frames. Even the small pods contain as many as six or seven peas in each. The ripe seed is white.

Plants from seed sown May 1 were in blossom June 26; and the pods were sufficiently grown for plucking, July 12.

The Auvergne Pea was introduced from France into England some years ago by the London Horticultural Society. Although it very far surpasses most of the varieties of the White Pea, it has never become much disseminated, and is very little known or cultivated. It is, however, a most characteristic variety, and always easily distinguishable by its long, curved pods. It is one of the most productive of all the garden peas.

Batt's Wonder

Plant three feet in height, of robust growth; foliage dark-green; pods narrow, nearly straight, but exceedingly well filled, containing seven or eight peas of medium size, which, when ripe, are small, smooth, and of a bluish-green colour.

Planted May 1, the variety will flower about July 1, and the pods will be fit for use the middle of the month.

The variety withstands drought well, and the pods hang long before the peas become too hard for use. It is an excellent pea for a second crop.

PEAS—Continued

PREMIUM GEM This variety is nearly as early as the American Wonder, and the very productive vine is decidedly larger, growing to a height of from 12 to 15 inches. The pods are large, and crowded with six to nine very large peas of fine quality. The dry peas are green, large, wrinkled, often flattened. Market gardeners use more of this sort than of any other wrinkled pea. Pkt. 10c; Pt. 15c; Qt. 25c; 4 Qts. 75c; Bushel $4.50

Second Early Peas

McLEAN'S ADVANCER A green, wrinkled variety about two and a half feet high, with broad, long pods, which are abundantly produced and well filled to the ends. Considered by some the best of the second early sorts. This pea is used very extensively by the market gardeners in the vicinity of New York, on account of its great productiveness, the fine appearance of its pods, and quality of the peas. It is very largely used among canners, as the skin will stand cooking without breaking. Careful comparison shows that our stock is unequaled. Pkt. 10c; Pt. 15c; Qt. 25c; 4 Qts. 75c; Bushel $5.00

BLISS' EVERBEARING A variety maturing soon after the Gems, and continuing a long time in bearing. Vine stout, about eighteen inches high, bearing at the top six to ten broad pods. If these are removed as they mature, and the season and soil are favorable, the plant will throw out from the axil of each leaf branches bearing pods which will mature in succession, thus prolonging the season. The peas are large and wrinkled, cook very quickly, are tender, of superior flavor, and preferred by many to any other sort. Pkt. 10c; Pt. 15c; Qt. 25c; 4 Qts. 75c; Bushel $4.50

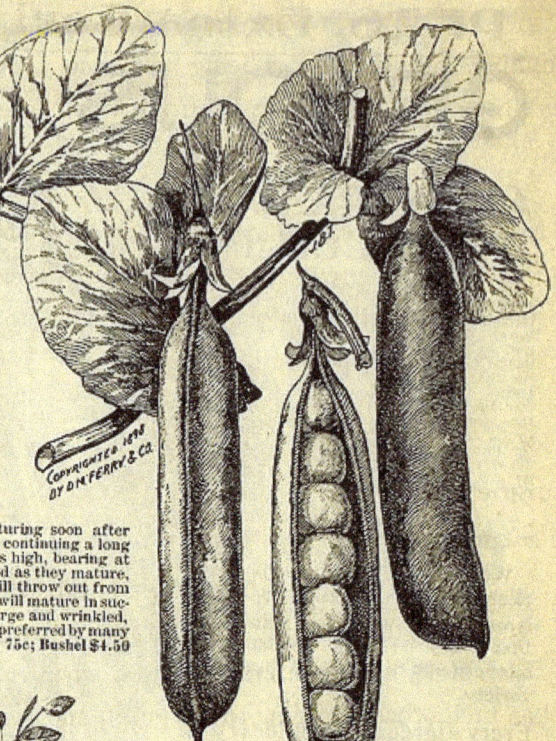

PEAS, PREMIUM GEM.

HORSFORD'S MARKET GARDEN The vine of this variety is of medium height, giving the greatest number of pods of any on our list. Pods contain five to seven medium sized, dark green peas, which retain their color and sweetness well after canning. Dry peas wrinkled and sweet. A very desirable variety for canners' use. Pkt. 10c; Pt. 15c; Qt. 25c; 4 Qts. 75c; Bushel $5.00

THE ADMIRAL Vines vigorous, about four feet high, comparatively slender, little branched. Pods usually borne in pairs and in great abundance; they are about two and one half inches long, thick, curved, bright green, carrying six to nine closely crowded peas of the very best quality and color. We know of no pea which remains palatable longer after it becomes large enough to use. Dry pea much wrinkled, medium sized, cream color. This variety ripens with Telephone or a little later. Owing to its great vigor, productiveness, fine color, quality and suitable size of the green peas it is admirably adapted for canners' use. Pkt. 10c; Pt. 15c; Qt. 25c; 4 Qts. 75c; Bushel $5.00

..TELEPHONE..

See description and illustration page 52.

TELEGRAPH Stronger growing and hardier vine, with darker foliage than the Telephone; green peas very large, sweet and when dry are almost smooth. Pkt. 10c; Pt. 15c; Qt. 25c; 4 Qts. 75c; Bushel $4.50

If peas are wanted by mail or express, prepaid, add 10 cents per pint, 15 cents per quart, for charges.

PEAS, THE ADMIRAL.

Second early peas
D.M. Ferry & Co.; Henry G. Gilbert Nursery and Seed Trade Catalog Collection

Beck's Prize-Taker
(Prize-Taker; Rising Sun)

Plant four and a half to five feet in height; pods roundish, curved or hooked near the end, well filled, containing seven to eight middle-sized peas of a fine green colour when young, and mixed olive and white when ripe.

Sown May 1, the variety will blossom June 25, and the pods will be suitable for plucking about the 12th of July.

It is one of the best varieties for the main crop. Similar to, if not identical with, Bellamy's Early Green Marrow.

Bedman's Imperial

The plant generally produces a single stem, which is from three to four feet high; the pods are usually in pairs, but sometimes single, three inches and a quarter long, five-eighths of an inch broad, somewhat curved, and terminate abruptly at the points. Each pod contains six to seven peas, which are of an ovate form, and about a third of an inch in their greatest diameter. The ripe seed is pale-blue.

Planted May 1, the variety blossomed the last of June, and furnished peas for use about the 18th of July. For many years, this variety stood foremost among the Imperials; but is now giving place to other and greatly superior sorts.

Bellamy's Early Green Marrow

Plant of strong and robust habit of growth, sometimes with a single and often with a branching stem, four and a half or five feet high, and producing from twelve to eighteen pods: these are in pairs, rarely single, three inches and a half long, seven-tenths of an inch broad, slightly curved, thick-backed, and terminate abruptly at the point. The surface is smooth, and of a very dark-green colour. They contain, on an average, from six to seven large bluish-green peas. The ripe seed has a mixed appearance; some being dull yellowish-white, and others light olive-green, in about equal proportions.

Plants from seed sown the first week in May were in blossom the last week in June, and pods were plucked for use about the middle of July.

The variety is highly recommended, both as a good bearer and a pea of excellent quality, whether for private use or for marketing: for the latter purpose it is peculiarly adapted, as the pod is of a fine deep-colour, handsomely and regularly shaped, and always plumply filled.

Bishop's Early Dwarf

Pods single or in pairs, about two inches long, bent back at both ends, and increasing in size towards the middle; pea about a fourth of an inch in diameter, and irregularly shaped, cream-coloured, with blotches of white, particularly about the eye. The plant grows little more than a foot high, and is fairly productive. Early sowings will give a supply for the table in about ten weeks.

This once-popular, Early Dwarf sort is now rapidly giving place to Bishop's New Long-podded,—a more prolific and much superior variety.

Bishop's New Long-Podded

Stem about two feet high; pods nearly straight, almost cylindrical, containing six or seven white peas. It is an early variety, an abundant bearer, of excellent quality, and in all respects much superior to the Common Bishop's Early Dwarf. Planted the 1st of May, it will blossom June 14, and yield peas for the table the 10th of July.

M'Intosh[5] describes it as:

> "a most abundant bearer, producing a succession of pods during most of the pea-season. Like all peas of its class, it requires a rich soil, and from four to six inches between the seed in the line. It is one of the most valuable sorts for small gardens and for domestic use. It originated in England with Mr. David Bishop; and is a hybrid between Bishop's Early Dwarf and one of the Marrowfats, carrying with it the characters of both its parents."

Black-Eyed Marrow

Plant about five feet high, strong and vigorous; pods generally single, sometimes in pairs, three inches and a quarter in length, three-fourths of an inch in breadth, becoming rough or wrinkled on the surface as they approach maturity, and containing about six large, round, cream-white or brownish-white black-eyed seeds, about three-eighths of an inch in diameter.

Its season is nearly the same with the Dwarf and Missouri Marrow. If sown the 1st of May, the plants will blossom the 28th or 30th of June, and yield peas for the table about July 15: the crop will ripen the last of the same month.

This is a very prolific as well as excellent variety. It is little cultivated in gardens at the North, though sometimes grown as a field-pea in the Canadas. In the Middle States, and at the South, it is a popular market-sort, and its cultivation is much more extensive.

The dark colour of the eye of the ripened seed distinguishes the variety from all others.

Blue Cimeter

(Sabre; Dwarf Sabre; Blue Sabre; New Sabre; Beck's Eclipse)

Plant about three feet high; pods generally in pairs, well filled, long, roundish, gradually curved from the stem to the point, or cimeter-shaped; seeds of good quality, larger than those of the Prussian Blue, from which the variety doubtless originated, and to which, when grown in poor soil, it has a tendency to return.

If planted the 1st of May, it will blossom about the 28th of June, and the pods will be suitable for plucking about the middle of July.

It bears abundantly, but not in succession; and, for this reason, is much prized by market-gardeners. The most of the pods being fit to pluck at the same time, the crop is harvested at once, and the land immediately occupied with other vegetables.

Blue Imperial

(Dwarf Blue Imperial)

Plant strong and vigorous, four feet in height, with large, healthy foliage; pods single and in pairs, three inches and a quarter in length, three-fourths of an inch in breadth, containing six or seven large peas.

The ripe seed is somewhat indented and irregularly compressed, three-eighths of an inch in diameter, and of a greenish-blue colour.

With respect to season, the variety is intermediate. If planted the 1st of May, it will blossom the 26th of June, and the pods will attain a size fit for plucking about the 12th of July. It is very hardy; yields abundantly; thrives well in almost any description of soil or situation; and, though not so sweet and tender as some of the more recent sorts, is of good quality. It vegetates with much greater certainty, and its crops are more reliable, than the higher-flavored varieties; and these qualities will still secure its cultivation by those who prefer a certain and plentiful supply of fair quality, to a precarious and limited yield of extraordinary sweetness and excellence. It has long been grown in this country, and is considered a standard variety.

Blue Prussian

(Dwarf Blue Prussian; Prussian Blue; Green Prussian)

Plant of a vigorous but not robust habit of growth, with a single stem about three feet high, which is sometimes branching. The pods are generally produced in pairs, but are also sometimes single, and vary from twelve to sixteen on each plant. They are from two and three-fourths of an inch to three inches long, three-fourths of an inch wide, somewhat curved, and rather broader towards the point, where they terminate abruptly. They contain about seven peas, which are four-tenths of an inch long, seven-twentieths of an inch wide, about the same in thickness, and compressed on the sides, from being so close together. The ripe seed is blue.

Sown the 1st of May, the plants blossomed June 28, and yielded peas for use the middle of July.

It produces abundantly, and is a valuable sort for late summer use. "It is unquestionably the parent of the Blue Imperial and all like varieties."

Blue Spanish Dwarf

(Groom's Superb; Blue Fan)

Plant from a foot and a half to three feet high. The pods are single and in pairs, in about equal proportion, two inches and a half long, containing from six to seven peas each. The ripe seed is pale-blue.

Plants from sowings made the first of May will blossom the last of June, and yield peas for use the middle of July.

It is a useful variety for small gardens, as it is a low grower and a fair bearer; but it is now much surpassed by Bishop's Long-podded and Burbridge's Eclipse, both of which are considered more prolific and better flavored.

British Queen

(Hair's Defiance; Tall White Mammoth; Erin's Queen)

The plant is of a showy and robust habit of growth, from six to seven feet high, sometimes with a single stem, but generally branching within nine inches or a foot of the ground, and frequently furnished with two and even three laterals, which are of the same height as the whole plant. The pods begin to be produced at the first joint above the first lateral shoot, and are in number from thirteen to eighteen on each plant. They are generally single, but frequently in pairs, from three inches and a quarter to three inches and three-quarters long, rather flattened and broad when first fit to gather, but becoming round and plump when more advanced. They are quite smooth, of a bright-green colour, slightly curved, wavy on the upper edge, and contain from five to seven exceedingly large peas, which are not so close together as to compress each other. The ripe seed is white, large, and wrinkled.

Sown the 1st of May, the plants will blossom about the 30th of June, and pods may be plucked for use about the 15th of July. They will ripen off about the 1st of August.

This is one of the best late peas in cultivation. It belongs to the class known as Wrinkled, or Knight's Marrow; but is much superior in every respect to all the old varieties usually called Knight's Marrows, being much more prolific and richly flavored. As an intermediate variety, it deserves a place in every garden.

Burbridge's Eclipse

(Stubbs's Dwarf)

Plant a robust grower, always with a simple stem, attaining the height of a foot and a half to about two feet; pods in pairs, rarely single, and from three inches to three inches and a quarter long, seven-tenths of an inch broad, perfectly straight, and of equal width throughout, with a slight waving on the upper edge,—they contain from five to seven peas, which are ovate, nearly half an inch long, a third of an inch broad, and the same in thickness.

Seed was planted May 1, the plants blossomed June 26, and pods were plucked for use July 14.

This may be classed among the valuable contributions which have been made to the list of peas during the last few years. Unlike most of the dwarf varieties, it is a most productive sort; and thus its dwarf character is not its chief recommendation. For private gardens, or for cultivation for market, few peas surpass this and Bishop's Long-podded.

Carter's Victoria

(Carter's Eclipse)

Plant six to seven feet high; pods large, slightly curved, containing seven or eight large peas, which are sweet and of excellent quality. The ripe seeds are white, and much shrivelled or wrinkled.

Plants from seeds sown May 1 blossomed July 1, and the pods were fit for plucking the 18th of the month.

The variety continues long in bearing, and the peas exceed in size those of Knight's Tall White Marrow. It is one of the best late tall peas.

Charlton

(Early Charlton)

The original character of this variety may be described as follows:

Plant about five feet high, and of vigorous growth; leaves large, with short petioles; tendrils small; pods broad, containing six or seven peas of excellent quality. They are rather larger than those of the Early Frame, with which this is often confounded. The Early Charlton may, however, be distinguished by its stronger habit of growth, flat pods, larger seeds, and by being fit for use about a fortnight later than the Early Frame; so that, when sown at the same time, it forms a succession.

According to the Messrs. Lawson, this is the oldest, and for a long period was the best known and most extensively cultivated, of all the varieties of white garden-peas. Its history can be traced as far back as 1670; and from that time till about 1770, or nearly a century, it continued to stand first in catalogues as the earliest pea, until it was supplanted by the Early Frame about 1770. It is further said by some to be the source from which the most esteemed early garden varieties have arisen; and that they are nothing else than the Early Charlton Pea, considerably modified in character from the effects of cultivation and selection. Although this idea may seem far-fetched, it is not improbable, especially when we take into consideration the susceptibility of change, from cultivation and other causes, which the Pea is ascertained to possess. Thus if the Early Charlton, or any other variety, be sown for several years, and only the very earliest and very latest flowering-plants selected for seed each season, the difference in the time of ripening between the two will ultimately become so great as to give them the appearance of two distinct varieties; and by sowing the earlier portion on light, early soils, and the later on strong, black, coarse, or low soils, the difference will become materially increased. It is therefore probable, that the Early Frame, with its numerous sub-varieties (including the Dan O'Rourke, Prince Albert, Early Kent, and a multitude of others), may have originated in the Charlton, though some of them differ essentially in their habit of growth.

The various names by which it has been known are Reading Hotspur, Master's or Flander's Hotspur, Golden Hotspur, Brompton Hotspur, Essex Hotspur, Early Nicol's Hotspur, Charlton Hotspur, and finally Early Charlton; the last name becoming general about 1750.

An English writer remarks:

"...that the variety now exists only in name. That which is sold for the Early Charlton is often a degenerated stock of Early Frames, or any stock of Frames which cannot be warranted or depended upon, but which are, nevertheless, of such a character as to admit of their being grown as garden varieties. The Early Charlton, if grown at all by seed-growers as a distinct variety, is certainly cultivated to a very limited extent."

Of the popular American improved early sorts, the Hill's Early, Hovey's Extra Early, Landreth's Extra Early, are hardy, as well as very prolific; and are not only well adapted for private gardens, but may be recommended as the most profitable kinds for cultivating for early marketing. In an experimental trial of these kinds with the Early Daniel O'Rourke, and some of the most approved of the earliest foreign varieties, they proved to be nearly or quite as early, fully as prolific, continued longer in bearing, and were much more stocky and vigorous in habit.

The Marchioness Peas

PEAS One pound of seed plants 100 feet of drill; 100 pounds of the dwarf varieties, or 50 pounds of the tall varieties, plant an acre

The Marchioness/World's Record Pea
112 superb varieties for market gardeners season of 1926

Champion of Paris

(Excelsior; Paradise Marrow; Stuart's Paradise)

Plant of vigorous growth, with a simple stem five to six feet high, rarely branched, producing from eight to ten pods. These last are generally single, but sometimes in pairs, from three inches and a quarter to three inches and three-quarters long, and five-eighths of an inch wide. They are curved almost as much as those of the Cimeter; and, when near maturity, become quite fleshy, wrinkled, and thick-backed. They contain from six to seven large peas, which are close together without being compressed. The ripe seed is white, medium-sized, somewhat flattened and pitted. If sown May 1, the plants will blossom June 28, and the pods will be ready for plucking July 16.

This is a very excellent pea, an abundant cropper, and considerably earlier than the Auvergne and Shillings Grotto; to both of which it is also greatly superior.

Climax

(Napoléon)

Plant three feet and a half high, of robust habit; pods single or in pairs, three inches long, containing five or six peas; when ripe, these are of medium size, pale-blue or olive, sometimes yellowish, shaded with blue, and, like the Eugénie, much wrinkled and indented.

If sown the beginning of May, the variety will blossom about the 15th of June, pods may be plucked for use the 10th of July, and the crop will ripen the 25th of the same month.

English catalogues represent the Napoléon as being "the earliest blue pea in cultivation, podding from the bottom of the haum to the top, with fine large pods." In a trial growth, it proved early and productive; not only forming a great number of pods, but well filling the pods after being formed. In quality it is tender, very sweet and well flavored, resembling the Champion of England. Its season is nearly the same with that of the Eugénie, and the variety is well deserving of cultivation.

Mr. Harrison, the originator of the Eugénie and Napoléon, states that both of the peas were originally taken from one pod.

Dantzic

Plant six to seven feet high, branching; pods in pairs, two and a half inches long, half an inch broad, compact, and slightly bent. When ripe, the seed is the smallest of all the light peas, quite round or spherical, of a bright-yellow colour, beautifully transparent, with whitish eyes.

If sown the 1st of May, the plants will blossom the 8th of July, afford peas for the table about the 25th of the same month, and ripen from the 10th to the middle of August.

It is not a productive variety, and is seldom cultivated in England or in this country; but is grown extensively on the shores of the Baltic, and exported for splitting, or boiling whole.

Dickson's Favourite

(Dickson's Early Favourite)

Plant five feet high, stocky, vigorous, and very prolific; pods ten to twelve on a stalk, long, round when fully grown, curved, hooked at the extremity, but not so much so as in the Auvergne, —to which, in many respects, it bears a strong resemblance. The pods are remarkably well filled, containing from eight to ten peas of medium size, round, and very white.

Planted the 1st of May, the variety blossomed June 25, and pods were gathered for use the 12th of July.

This pea is highly deserving of cultivation as a second early variety.

Dillistone's Early

The plant is of slender habit of growth, produces a single stem two feet high, and bears, on an average, from seven to nine pods: these are smaller than those of the Dan O'Rourke, generally single, but occasionally in pairs, almost straight, and contain seven peas each. The seed, when ripe, is white.

Sown at the time of the Dan O'Rourke, the plants were a mass of bloom three days before the last named had commenced blossoming, and the crop was ready for gathering seven days before the Dan O'Rourke.

This is undoubtedly the earliest pea known, and is quite seven or eight days earlier than the Dan O'Rourke, which has hitherto been regarded as the earliest variety. A striking feature of Dillistone's Early is, that its changes take place at once. It blooms in a mass, its pods all appear together, and the whole crop is ready to be gathered at the same time.

In the Chiswick Garden, England, where a hundred and sixteen varieties were experimentally cultivated, during the season of 1860, under the supervision of Robert Hogg, LL.D., this variety was beginning to die off, when the Dan O'Rourke was yet green and growing.

Dwarf Marrow

(Dwarf White Marrow; Dwarf Marrowfat; Early Dwarf Marrowfat)

Plant from three to four feet in height, generally with a single stem, but sometimes branching; pods somewhat flattened, generally single, but sometimes produced in pairs, three inches to three inches and a half long, three-fourths of an inch broad at the middle, tapering with a slight but regular curve to both ends, and containing about six closely-set peas: these are cream-coloured and white; the white prevailing about the eye, and at the union of the two sections of the pea; not perfectly round, but more or less compressed, slightly wrinkled, and measuring nearly three-eighths of an inch in diameter.

Planted the 1st of May, the variety blossomed the last of June, and afforded peas for the table the 15th of July.

The Dwarf Marrow is hardy and productive. Though not so sweet or well flavored as some of the more recent sorts, its yield is abundant and long continued; and, for these qualities, it is extensively cultivated. The variety, however, is rarely found in an unmixed state; much of the seed sown under this name producing plants of stronger habit of growth than those of the true Dwarf Marrow, and more resembling the Tall White variety.

Early Dan O'Rourke

(Dunnett's First Early; Waite's Dan O'Rourke; Carter's Earliest; Sangster's Number One)

Plant from three and a half to four feet high,—in general habit not unlike the Early Frame, of which it is probably an improved variety; pods usually single, two inches and three-fourths long, containing five or six peas.

When fully ripe, the pea is round, cream-coloured, white at the eye and at the junction of the cotyledons, and nearly a fourth of an inch in diameter.

Plants from seeds sown May 1 were in bloom June 7, and pods were gathered for use from the 25th of the month.

The Dan O'Rourke is remarkable for its precocity; and, with the exception of Dillistone's Early and one or two American varieties, is the earliest of all the sorts now in cultivation. It is hardy, prolific, seldom fails to produce a good crop, appears to be well adapted to our soil and climate, is excellent for small private gardens, and one of the best for extensive culture for market.

Its character as an early pea can be sustained only by careful culture, and judicious selection of seeds for propagation. If grown in cold soil, from late-ripened seeds, the variety will rapidly degenerate; and, if from the past any thing can be judged of the future, the Dan O'Rourke, under the ordinary forms of propagation and culture, will shortly follow its numerous and once equally popular predecessors to quiet retirement as a synonyme of the Early Frame or Charlton.

Early Frame

(Early Dwarf Frame; Early Double-Blossomed Frame; Essex Champion; Single-Blossomed Frame)

Plant three to four feet in height; pods in pairs, slightly bent backwards, well filled, terminating rather abruptly at both ends, and about two and a half inches long by from three-eighths to half an inch in breadth. The peas, when fully ripe, are round and plump, cream-coloured, white towards the eye and at the union of the cotyledons, and measure nearly a fourth of an inch in diameter.

Sown the 1st of May, the variety blossomed June 20, and the pods were ready for plucking the 6th of July.

This well-known pea, for a long period, was the most popular of all the early varieties. At present, it is less extensively cultivated; having been superseded by much earlier and equally hardy and prolific sorts.

> "The flowers sometimes come single, and sometimes double; the stalk from the same axil dividing into two branches, each terminating in a flower: hence the names of 'Single-blossomed' and 'Double-blossomed' have both been occasionally applied to this variety."

Early Hotspur

(Early Golden Hotspur; Golden Hotspur; Superfine Early; Reading Hotspur)

Similar to the Early Frame. Mr. Thompson represents it as identical. The Messrs. Lawson describe it as follows: "Pods generally in pairs, three inches long, half an inch broad, nearly straight, and well filled; pea similar to the Double-blossomed Early Frame, but rather larger."

Early Warwick

(Race-Horse)

Once at the head of early peas: now considered by the most experienced cultivators to be identical with the Early Frame.

Early Washington

(Cedo nulli)

A sub-variety of the Early Frame; differing slightly, if at all, either in the size or form of the pod, colour and size of the seed, or in productiveness.

Once popular, and almost universally cultivated: now rarely found on seedsmen's catalogues.

Eugénie

Plant about three feet in height, with pale-green foliage; pods single or in pairs, three inches long, containing five or six peas. When ripe, the peas are of medium size, cream-coloured, and much shrivelled and indented.

Plants from sowings made May 1 were in blossom June 14, green peas were plucked July 10, and the pods ripened from the 18th to the 25th of the same month. English catalogues describe the variety as being: "the earliest white, wrinkled marrow-pea in cultivation; podding from the bottom of the stalk to the top, with fine large pods." In a trial-growth, it proved hardy and very prolific; and the peas, while young, were nearly as sweet as those of the Champion of England. The pods were not remarkable for diameter; but, on the contrary, were apparently slender. The peas, however, were large; and, the pods being thin in texture, the peas, when shelled, seemed to be equal in diameter to the pods themselves. As a new variety, it certainly promises well, and appears to be worthy of general cultivation. It will come to the table immediately after the earliest sorts, and yield a supply till the Marrows are ready for plucking.

Fairbeard's Champion of England

(Champion of England)

Plant of strong and luxuriant habit of growth, with a stem from five to six feet in height, which is often undivided, but also frequently branching. The laterals are produced within about eighteen inches of the ground, and sometimes assume a vigorous growth, and attain as great a height as the main stem. They produce pods at the first joint above the lateral, and are continued at every succeeding joint to the greatest extremity of the plant. The pods are generally single, but frequently in pairs, about three inches and a half long, slightly curved, and terminate abruptly at the point; the surface is quite smooth, and the colour light-green till maturity, when they become paler and shrivelled. They contain six or seven quite large peas, which are closely packed together and compressed. The ripe seed is wrinkled, and of a pale olive-green.

Sown the 1st of May, the plants were in flower June 25, and pods were gathered for use the 12th of July.

This variety was originated in England, by Mr. William Fairbeard, in 1843; and, with the Early Surprise, came out of the same pod,—the produce of a plant found in a crop of the Dwarf White Knight's Marrows, to which class it properly belongs. It is, without doubt, one of the most valuable acquisitions which have been obtained for many years; being remarkably tender and sugary, and, in all respects, of first-rate excellence. The rapid progress of its popularity, and its universal cultivation, are, however, the best indications of its superiority.

The variety was introduced into this country soon after it was originated, and was first sold at five dollars per quart.

Fairbeard's Nonpareil

Stem branching, three and a half to four feet high, with a habit of growth and vigor similar to the Early Frames. The pods are full and plump, but do not become thick-backed and fleshy as they ripen, like those of the Frames. They contain from six to eight peas, which are

Champion of England and other peas
Seed annual 1903

close together, much compressed, and of that sweet flavor which is peculiar to the Knight's Marrows. The ripe seed is small and wrinkled, and of the same colour as the other white, wrinkled peas.

The variety was originated by Mr. William Fairbeard, who also raised the Champion of England. It is earlier than the last-named sort, nearly as early as the Frames, and a most valuable acquisition.

Fairbeard's Surprise

(Early Surprise; Surprise)

The plant of this variety is of a free but not robust habit of growth, and always with a simple stem, which is about four feet high. The pods are produced at every joint, beginning at about two feet and a half from the ground. They are generally single, but sometimes in pairs, three inches long, slightly curved, but not quite so much as those of the Champion of England. They contain from six to seven peas, which are of good size, but not so sweet as those of the last-named sort. The ripe seed is somewhat oval, and of a pale, olive-green colour.

The variety is a day or two earlier than the Champion of England. It originated from the Dwarf White Knight's Marrow, and was taken from the pod in which was found the Champion of England.

Flack's Imperial

(Flack's Victory; Flack's Victoria; Flack's New Large Victoria)

The plant is of a robust habit of growth, with a stem which is always branching, and generally about three feet in height; the pods are numerous, varying from twelve to eighteen on a plant, generally produced in pairs, but often singly, three inches and a half long, three-fourths of an inch broad, and considerably curved,—terminating abruptly at the point, where they are somewhat broader than at any other part. Each pod contains from six to eight very large peas, which are of an ovate shape, half an inch long, seven-twentieths of an inch broad, and the same in thickness. The ripe seed is blue.

Plants from seed sown May 1 will blossom June 28, and supply the table July 15.

It is one of the most prolific peas in cultivation; grows to a convenient height; and, whether considered for private gardens or for market supplies, is one of the most valuable varieties which has been introduced for years.

General Wyndham

The plant is of a robust habit, six to seven feet high, and frequently branched; the foliage is dark-green and blotched; the pods are either single or in pairs, and number from ten to fourteen on each plant,—they contain eight very large peas, which are of the deep, dull-green colour of the Early Green Marrow. The ripe seed is white and olive mixed.

This is a valuable acquisition, and was evidently procured from the Ne Plus Ultra; but it is a more robust grower, and produces much larger pods.

The plant continues growing, blooming, and podding till very late in the season; and, when this is in the full vigor of growth, the Ne Plus Ultra is ripening off. The peas, when cooked, are of a fine, bright-green colour, and unlike those of any other variety.

Hair's Dwarf Mammoth

Plant strong and vigorous, from three to three feet and a half high, branching, with short joints; pods single or in pairs, broad, comparatively flat, containing about six very large peas, which are sugary, tender, and excellent. The ripe seeds are shrivelled, and vary in colour; some being cream-white, and others bluish-green.

Sown May 1, the plants will blossom July 1, and the pods will be ready for use the 15th of the same month.

Very prolific, and deserving of cultivation.

Harrison's Glory

Plant three feet high, of a bushy, robust habit of growth; pods rather short, nearly straight, and flattish, containing five or six medium-sized peas, of good quality: when ripe, the seeds are light-olive, mixed with white, and also slightly indented

If planted May 1, the variety will flower June 23, and the pods will be fit for gathering about the 10th of July.

A good variety; but, like Harrison's Glory, the pods are frequently not well filled.

Harrison's Perfection

Plant three feet in height, of vigorous habit; pods small, straight, containing five peas of good size and quality.

Sown the 1st of May, the variety will flower June 23, and the pods will be fit for plucking about the 12th of July.

The only defect in this variety is, that the pods are often not well filled. When growing, it is scarcely distinguishable from Harrison's Glory; but, in the mature state, the seeds of the former are smooth and white, while those of the latter are indented, and of an olive-colour.

King of the Marrows.

Plant six feet in height, stocky, and of remarkably vigorous habit; pods single or in pairs, containing five or six large seeds, which, when ripe, are yellowish-green, and much shrivelled and indented, like those of the Champion of England.

If planted May 1, the variety will blossom the last of June, and pods for the table may be plucked about the 15th of July.

Though comparatively late, it is one of the best of the more recently introduced sorts, and well deserving of general cultivation. When the pods are gathered as fast as they become fit for use, the plants will continue to put forth new blossoms, and form new pods for an extraordinary length of time; in favorable seasons, often supplying the table for five or six weeks.

It is very tender and sugary, and little, if at all, inferior to the Champion of England.[Pg 541]

In common with most of the coloured peas, the ripe seeds, when grown in this country, are much paler than those of foreign production; and, when long cultivated in the climate of the United States, the blue or green is frequently changed to pale-blue or yellowish-green, and often ultimately becomes nearly cream-white.

Knight's Dwarf Blue Marrow

A dwarfish sub-variety of Knight's Marrows, with wrinkled, blue seeds.

Knight's Dwarf Green Marrow

(Knight's Dwarf Green Wrinkled)

Plant about three feet high; pods in pairs, three inches long, three-fourths of an inch wide, flattish, and slightly bent. The ripe peas are of a light bluish-green colour. It differs from the foregoing principally in the height of the plant, but also to some extent in the form of the pods.

Knight's Dwarf White Marrow

(Knight's Dwarf White Wrinkled Marrow)

Plant three feet high; pods in pairs, three inches long, three-fourths of an inch wide, straight, or nearly so, well filled, and terminating abruptly at both ends; pea, on an average, about three-eighths of an inch in diameter, flattened, and very much wrinkled; colour white, and sometimes of a greenish tinge. It is a few days earlier than the Dwarf Green.

Knight's Tall Blue Marrow

A sub-variety of Knight's Tall Marrows, with blue, wrinkled, and indented seeds. It resembles the Tall White and Tall Green Marrows.

Knight's Tall Green Marrow

Plant from six to seven feet in height, of strong growth; pods large, broad, and well filled; the seed, when ripe, is green, and much wrinkled or indented.

If planted the first of May, the variety will blossom towards the last of June, and supply the table the middle of July.

The peas are exceedingly tender and sugary; the skin also is very thin. "From their remarkably wrinkled appearance, together with the peculiar sweetness which they all possess, Knight's Marrows may be said to form a distinct class of garden-peas; possessing qualities which, together with their general productiveness, render them a valuable acquisition, both to cultivators and consumers."

If planted not less than six feet apart, these peas will bear most abundantly from the ground to the top: they also yield their pods in succession, and are the best for late crops.

Knight's Tall White Marrow

(Knight's Tall White Wrinkled Marrow)

Height and general character of the plant similar to Knight's Tall Green Marrow. Pods in pairs. The ripe seed is white. Very productive and excellent.

Matchless Marrow

This is a good marrow-pea, but now surpassed by the improved varieties of the Early Green Marrow. It possesses no qualities superior to that variety, and is not so early. The plant grows from five to six feet in height; and the pods contain about seven large peas, which are closely compressed together.

Milford Marrow

The plant is of a strong and robust habit of growth, always with a single stem, attaining the height of four and a half or five feet, and producing from twelve to sixteen pods, which are almost always in pairs, three inches and three-quarters long, and three-quarters of an inch wide. They do not become broad-backed, thick, or fleshy, but rather shrivelled, and contain from six to seven very large peas, which are roundish and somewhat compressed, half an inch long, nearly the same broad, and nine-twentieths thick.

Its season is near that of Bellamy's Early Green Marrow; if planted May 1, blossoming June 28, and being fit for plucking about the middle of July.

Missouri Marrow

(Missouri Marrowfat)

Plant three feet and a half or four feet high, strong and vigorous, generally simple, but sometimes divided into branches; pods single and in pairs, three inches long, wrinkled on the surface as they ripen, nearly straight, and containing about six peas, rather closely set together. When ripe, the pea is similar to the Dwarf Marrow in form, but is larger, paler, more wrinkled, and much more regular in size.

Plants from seed sown May 1 were in blossom the 30th of June, and pods were gathered for use the 14th of July. It is a few days later than Fairbeard's Champion of England, and nearly of the season of the Dwarf Marrowfat, of which it is probably but an improved or sub-variety.

It is of American origin, very productive, of good quality, and well deserving of cultivation.

Ne Plus Ultra

(Jay's Conqueror)

This is comparatively a recent variety. It belongs to the wrinkled class of peas; is as early as Bellamy's Green Marrow; and possesses, both in pod and pea, the same fine, deep, olive-green colour.

The plant is of strong and robust habit of growth, six to seven feet high, with a branching stem. It begins to produce pods at two or two and a half feet from the ground; and the number, in all, is from twelve to eighteen. The pods are generally in pairs, three inches and a half long, three-fourths of an inch wide, very plump and full, almost round, slightly curved, and terminate abruptly at the end. Their colour is deep, bright-green, and the surface smooth. They contain seven very large peas, each of which is half an inch long, nearly the same broad; and, although they are not so closely packed as to compress each other, they fill the pods well.

When sown the first of May, the variety will blossom the last of June, and afford peas for use the 15th of July.

It is one of the best tall Marrows in cultivation. The ripe seed is mixed white and olive.

Noble's Early Green Marrow

A sub-variety of Bellamy's Early Green Marrow. It is a much more abundant bearer; producing from eighteen to twenty pods on a plant, which are singularly regular in their size and form.

Prince Albert

(Early Prince Albert; Early May; Early Kent)

Plant from two and a half to three feet in height, usually without branches; pods generally in pairs, two inches and a half in length, half an inch broad, tapering abruptly at both[Pg 545]ends, slightly bent backwards, and well filled; pea, when fully ripe, round, cream-coloured, approaching to white about the eye and at the line of the division of the lobes, and measuring about a fourth of an inch in diameter.

Sown May 1, the plants blossomed June 15, and pods were plucked for use July 6.

The Prince Albert was, at one period, the most popular of all the early varieties, and was cultivated in almost every part of the United States. As now found in the garden, the variety is not distinguishable from some forms of the Early Frame; and it is everywhere giving place to the Early Dan O'Rourke, Dillistone's Early, and other more recent and superior sorts.

Queen of the Dwarfs

A very dwarfish variety, from six to nine inches high. Stem thick and succulent; foliage dark bluish-green. Each plant produces from four to six pods, which are of a curious, elliptic form, and contain three or four large peas. Ripe seed white, of medium size, egg-shaped, unevenly compressed.

The plants are tender; the pods do not fill freely; and the variety cannot be recommended for cultivation.

Ringwood Marrow

(Flanagan's Early; Early Ringwood; Beck's Gem)

Plant three and a half to four feet high, usually simple, but sometimes sending out shoots near the ground. The pods are single and in pairs; and, as they ripen, become thick and fleshy, with a rough, pitted, and shrivelled surface: they contain from six to seven large peas, which are nearly round, and about seven-tenths of an inch in diameter in the green state. The ripe seed is white.

The variety is comparatively early. If planted May 1, it will blossom about the 25th of June, and the pods will be ready to pluck about the 10th of July.

A very valuable sort, producing a large, well-filled pod, and is a most abundant bearer. It has, however, a peculiarity, which by many is considered an objection,—the pod is white, instead of green, and presents, when only full grown, the appearance of over-maturity. This objection is chiefly made by those who grow it for markets, and who find it difficult to convince their customers, that, notwithstanding the pod is white, it is still in its highest perfection. So far from being soon out of season, it retains its tender and marrowy character longer than many other varieties.

A new sort, called the "Lincoln Green," is said to possess all the excellences of the Ringwood Marrow, without the objectionable white pod.

Royal Dwarf or White Prussian
(Dwarf Prolific; Poor Man's Profit)

Plant of medium growth, with an erect stem, which is three feet high, generally simple, but occasionally branching. The pods are usually single, but sometimes in pairs, nearly three inches long, half an inch broad, almost straight, and somewhat tapering towards the point. The surface is quite smooth, and the colour bright-green. They are generally well filled, and contain from five to six peas, which are ovate, not compressed, four-tenths of an inch long, a third of an inch broad, and the same in thickness. The ripe seed is white.

Plants from seed sown the 1st of May will blossom June 25, and supply the table about the middle of July. The crop will ripen the 25th of the same month.

This is an old and prolific variety, well adapted for field culture, and long a favorite in gardens, but now, to a great extent, superseded.

Sebastopol

Plant of rather slender habit, three feet and a half in height; pods usually single, two inches and three quarters in length, containing from five to seven peas, which, when ripe, are nearly round and smooth, cream-coloured, and scarcely distinguishable, in their size, form, or colour, from the Early Frame and kindred kinds.

If planted May 1, the variety will blossom June 16, afford pods of sufficient size for shelling about July 7, and ripen the 20th of the same month.

It is early, very productive, of superior quality, and an excellent sort for growing for market, or in small gardens for family use. In an experimental cultivation of the variety, it proved one of the most prolific of all the early sorts.

Shillings Grotto. *Cot. Gard.*

Plant with a simple stem, four feet and a half to five feet high; the pods are generally single, but frequently in pairs, three inches and a half long, about half an inch wide, slightly curved, and, when fully matured, assuming a thick-backed and somewhat quadrangular form. Each pod contains, on an average, seven large peas. The ripe seed is white.

A great objection to this variety is the tardiness with which it fills; the pods being fully grown, and apparently filled, when the peas are quite small and only half grown. Though considered a standard sort, it is not superior to the Champion of England; and will probably soon give place to it, or some other of the more recent varieties.

Spanish Dwarf

(Early Spanish Dwarf; Dwarf Fan; Strawberry)

Plant about a foot high, branching on each side in the manner of a fan; and hence often called the "Dwarf Fan." The pods are sometimes single, but generally in pairs, two inches and a half long, half an inch broad, terminate rather abruptly at the point, and contain from five to six rather large peas. The ripe seed is cream-white.

Sown May 1, the plants were in blossom June 26, and pods were plucked for use July 14.

The Spanish Dwarf is an old variety, and still maintains its position as an Early Dwarf for small gardens, though it can hardly be considered equal to Burbridge's Eclipse or Bishop's Long-podded.

There is a variety of this which is called the Improved Spanish Dwarf, and grows fully nine inches taller than the old variety; but it possesses no particular merit to recommend it.

Tall White Marrow

(Large Carolina; Tall Marrowfat)

Plant six to seven feet in height, seldom branched; pods three to three inches and a half long, three-fourths of an inch broad, more bluntly pointed than those of the Dwarf variety, and containing six or seven peas. When ripe, the pea is nearly of the colour of the Dwarf Marrow, but is more perfectly spherical, less wrinkled, and, when compared in bulk, has a smoother, harder, and more glossy appearance.

Planted May 1, the variety will blossom near the 1st of July, and will come to the table from the 15th to the 20th of the same month. It is a few days later than the Dwarf.

In this country, it has been longer cultivated than any other sort; and, in some of the forms of its very numerous sub-varieties, is now to be found in almost every garden. It is hardy, abundant, long-continued in its yield, and of excellent quality. In England, the variety is cultivated in single rows three feet apart. In this country, where the growth of the pea is much less luxuriant, it may be grown in double rows three feet and a half apart, and twelve inches between the single rows.

Taylor's Early

Similar in habit, production, and early maturity, to the Early Dan O'Rourke.

Thurston's Reliance

Plant strong and robust, six to seven feet high; pods generally single, but occasionally in pairs, and from three inches and a half to four inches and a quarter long. They are broad and flat, shaped like the pods of the Blue Cimeter, and contain seven or eight very large peas. Ripe seed white, large, and unevenly compressed.

This is a quite distinct and useful pea; an abundant bearer; and the pods are of a fine deep-green colour, which is a recommendation for it when grown for market. It comes in at the same time as the Auvergne and Shillings Grotto, but is of a more tender constitution.

Tom Thumb

(Beck's Gem; Bush Pea; Pois Nain Hatif Extra, of the French)

Plant of remarkably low growth, seldom much exceeding nine inches in height, stout and branching; pods single, rarely in pairs, two inches and a half in length, half an inch broad, containing five or six peas, which are cream-yellow, and measure about a fourth of an inch in diameter.

Planted the 1st of May, the variety blossomed the 12th of June, and the pods were of suitable size for plucking July 4.

In the colour of its foliage, its height and general habit, the variety is very distinct, and readily distinguishable from all other kinds. It is early, of good quality, and, the height of the plant considered, yields abundantly. It may be cultivated in rows ten inches apart.

Mr. Landreth, of Philadelphia, remarks as follows:

> "For sowing at this season (November, in the Middle States), we recommend trial of a new variety, which we have designated 'Tom Thumb,' in allusion to its extreme dwarfness. It seldom rises over twelve inches, is an abundant bearer, and is, withal, quite early. It seems to be admirably adapted to autumn sowings in the South, where, on apprehended frost, protection may be given: it is also equally well suited to early spring planting for the same reason. It is curious, as well as useful; and, if planted on ground well enriched, will yield as much to a given quantity of land as any pea known to us."

It is a desirable variety in the kitchen garden; as, from its exceeding dwarfish habit, it may be so sown as to form a neat edging for the walk or border.

Veitch's Perfection

Plant three feet and a half to four feet high, of strong, robust growth, somewhat branched; pods ten or twelve on a stalk or branch, large, flat, straight, containing six or eight large peas, which are very sugary and excellent. The ripe seeds are large, of a light olive-green colour; some being nearly white.

Planted the 1st of May, the variety will be in flower June 28, and the pods will be fit for use about the middle or 20th of July. It is one of the best peas for main or late crops.

Victoria Marrow

Plant from six to seven feet high; pods remarkably large, nearly four inches in length, generally in pairs, straight, roundish, well filled, containing from six to eight peas of extraordinary size and of good quality. The ripe peas are olive-green.

The Victoria Marrow is not early. Planted May 1, it will blossom the last of June, and be fit for the table from the middle of July.

This variety bears some resemblance to Knight's Tall Marrow; but, like nearly all others, it is less sugary. Those who have a fancy for large peas will find this perhaps the largest.

Warner's Early Emperor

(Warner's Early Conqueror; Early Railway; Early Wonder; Deck's Morning-Star; Early Emperor)

This variety grows somewhat taller, and is a few days earlier, than the Prince Albert: the pods and peas are also somewhat larger. It is an abundant bearer; and, on the whole, must be considered a good sub-variety of the Early Frame.

Woodford's Marrow

(Nonpareil)

Plant of strong and robust habit of growth, like a vigorous-growing Marrow; rising with a stem three feet and a half high, which is sometimes simple, but generally branching at about half its height from the ground. The pods begin to be produced at little more than half the height of the plant; and, from that point to the top, every joint produces single or double pods, amounting, in all, to ten or twelve on each. They are single or in pairs, in nearly equal proportions, about three inches and a half long, seven-tenths of an inch broad, quite smooth, and of a dark-green colour. When ready to gather, they are rather flattened, but become round as they ripen. They contain, on an average, seven peas, which are of a dark olive-green colour, rather thick in the skin, and closely packed; so much so as to be quite flattened on the sides adjoining.

Sown May 1, the variety blossomed June 28, and peas were gathered for the table July 17.

This is a very characteristic pea, and may at once be detected from all others, either by the ripe seed or growing plants, from the peculiar dark-green colour, which, when true, it always exhibits. It is well adapted for a market-pea; its dark-green colour favoring the popular prejudices.

1,2,3,4,5. The Book of the Garden. By **Charles M'Intosh**. 2 vols. Edinburgh and London, 1855.

141

Sugar/Snap Pea (Pisum macrocarpum)

(String-Peas; Skinless Pea)

In this class are included such of the varieties as want the tough, inner film, or parchment lining, common to the other sorts. The pods are generally of large size, tender and succulent, and are used in the green state like string-beans; though the seeds may be used as other peas, either in the green state or when ripe. "When not ripe, the pods of some of the sorts have the appearance of being swollen or distended with air; but, on ripening, they become much shrivelled, and collapse closely on the seeds." The varieties are not numerous, when compared with the extensive catalogue of the kinds of the Common Pea offered for sale by seedsmen, and described by horticultural writers. The principal are the following:

Common Dwarf Sugar

(Dwarf Crooked-Podded Sugar)

Stalk about two feet high, dividing into branches when cultivated in good soil; flower white; pods single or in pairs, six-seeded, three inches long by five-eighths of an inch broad, crooked or jointed-like with the seeds, as in all of the Sugar peas, very prominent, especially on becoming ripe and dry; pea fully a fourth of an inch in diameter, white, and slightly wrinkled.

The variety is quite late. Sown the beginning of May, the plants blossomed the last week in June, and pods were gathered for use July 17.

It is prolific, of good quality as a shelled-pea, and the young pods are tender and well flavored.

Early Dwarf Dutch Sugar

(Early Dwarf de Grace)

Plant about twenty inches high, branching; leaves of medium size, yellowish-green; flowers white; pods two inches and three-quarters in length, half an inch wide, somewhat sickle-shaped, swollen on the sides, flattened at the lower end, and containing five or six peas, which, when ripe, are roundish, often irregularly flattened or indented, wrinkled, and of a yellowish-white colour.

The variety is the lowest-growing and earliest of all the Eatable-podded kinds. If sown at the time of the Common Dwarf Sugar, it will be fit for use twelve or fourteen days in advance of that variety. It requires a good soil; and the pods are succulent and tender, but are not considered superior to those of the Common Dwarf Sugar.

Giant Eatable-Podded

(Giant Sugar)

Stalk four to five feet high; leaves large, yellowish-green, stained with red at their union with the stalk of the plant; flower reddish; pods transparent yellowish-green, very thick and fleshy, distended on the surface by the seeds, which are widely distributed, curved, and much contorted, six inches long, and sometimes nearly an inch and a half in diameter,—exceeding in size that of any other variety. They contain but five or six seeds, which, when ripe, are irregular in form, and of a greenish-yellow colour, spotted or speckled with brown.

It is about a week later than the Large Crooked Sugar.

Large Crooked Sugar

(Broadsword; Six-Inch-Pod Sugar)

Plant nearly six feet in height, and branching when grown in good soil; the leaves are large, yellowish-green; flowers white; pods very large,—measuring from four to five inches in length and an inch in width,—broad, flat, and crooked. When young, they are tender, and easily snap or break in pieces, like the young pods of kidney-beans; and are then fit for use. The sides of the pods exhibit prominent marks where pushed out by the seeds, even at an early stage of growth. The ripe peas are somewhat indented or irregularly compressed, and of a yellowish-white colour.

It is one of the best of the Eatable-podded sorts, and is hardy and productive. It is, however, quite late; blossoming, if sown May 1, about the last of June, and producing pods for use in the green state about the 20th of July.

Purple-Podded or Australian

(Blue-Podded; Botany-Bay Pea)

Plant five feet high, generally without branches; pods usually in pairs, flattened, with thick, fleshy skins, and commonly of a dark-purple colour; but this characteristic is not permanent, as they are sometimes found with green pods; in which case, they are, however, easily distinguished from those of other peas by their thick and fleshy nature. When ripe, the peas are of medium size, often much indented and irregularly compressed, and of a light, dunnish, or brown colour. Season intermediate.

It is very productive, and seems possessed of properties which entitle it to cultivation.

Red-Flowered Sugar

(Chocolate)

Stem four or five feet in height, generally simple, but branching when grown in rich soil; leaves long, yellowish-green, tinged with red where they connect with the stalk of the plant; flowers pale-red; pods three inches long, seven-tenths of an inch broad, more or less contorted, containing six to eight peas; seed comparatively large, pale-brown, marbled with reddish-brown.

Season nearly the same as that of the Common Dwarf Sugar. It is productive, remarkably hardy, and may be sown very early in spring, as it is little affected by cool and wet weather; but the green peas are not much esteemed, as they possess a strong and rather unpleasant flavor. The green pods are tender and good; and, for these, the variety may be worthy of cultivation.

Tamarind Sugar

(Late Dwarf Sugar; Tamarind Pea)

Plant similar to the Common Dwarf Sugar, but of more luxuriant habit, and with larger foliage; flowers white; pods single or in pairs, six to eight seeded, very long and broad,—often measuring four inches in length and an inch in breadth,—succulent, and generally contorted and irregular in form. A few days later than the Common Dwarf Sugar.

Hardy, prolific, and deserves more general cultivation.

White-Podded Sugar

Stem four to five feet high; leaves yellowish-green, and, like those of the Giant Eatable-podded, stained with red at their insertion with the stalk; flowers purple; pods nearly three inches long, five-eighths of an inch wide, sickle-shaped and contorted, of a yellowish-white colour, containing five or six peas. The ripe seeds are irregularly flattened and indented, of a greenish-yellow colour, marbled or spotted with brown or black.

The variety is quite late. Sown May 1, the pods were not fit for use till July 24.

The pods are crisp and succulent, though inferior in flavor to most of the Eatable-podded varieties.

Yellow-Podded Sugar

Stem three to four feet high; leaves large, yellowish-green; flowers white, tinted with yellow; pods four inches long, tapering slightly at the ends, greenish-yellow, thick and fleshy, containing six or seven peas, widely separated. The ripe seeds are oblong, rather regular in form, and of a creamy-white colour.

It is one of the earliest of the Eatable-podded sorts; coming to the table, if planted May 1, about the middle of July. It is of good quality, but not hardy or productive; and seems to have little to recommend it, aside from the singular colour of its pods.

142

Peanut (Arachis hypogaea)

(Ground Bean; Earth Nut; Pindar Nut; Ground Nut)

A native of Africa, and also of Central and Tropical America. It is an annual plant; and the stem, when full grown, is about fifteen inches in height. The leaves are pinnate, with four leaflets, and a leafy, emarginate appendage at the base of the petioles; the flowers are yellow, and are produced singly, in the axils of the leaves; the fruit, or pod, is of an oblong form, from an inch to an inch and a half in length, rather more than three-eighths of an inch in diameter, often contracted at the middle, but sometimes bottle-formed, reticulated, and of a yellowish colour; the kernels, of which the pods contain from one to three, are oblong, quite white, and enclosed in a thin, brown skin, or pellicle.

A remarkable peculiarity of this plant is, that the lower blossoms (which alone produce fruit), after the decay of the petals, insinuate their ovaries into the earth; beneath which, at the depth of several inches, the fruit is afterwards perfected.

The seed, or kernel, retains its germinative property but a single season; and, when designed for planting, should be preserved unbroken in the pod, or shell.

The peanut plant
Franz Eugen Köhler, 1887

Soil and Cultivation—The Pea-nut succeeds best in a warm, light, loamy soil. This should be deeply ploughed and well pulverized, and afterwards laid out in slightly raised ridges two feet apart. As the plants require the whole season for their perfection, the seed should be planted as early in spring as the weather becomes suitable. Drop nine inches apart in the drills, and cover an inch and a half or two inches deep. Weeding must be performed early in the season; as, after the blossoming of the plants, they are greatly injured if disturbed by the hoe, or if weeds are removed about the roots.

It is rather tropical in its character, and cannot be cultivated with success either in the Northern or Middle States.

> "The seeds are sometimes dibbled in rows, so as to leave the plants a foot apart each way. As soon as the flowers appear, the vines are earthed up from time to time, so as to keep them chiefly within the ground. When cultivated alone, and there is sufficient moisture, the yield of nuts is from sixty to seventy-five bushels to the acre. If allowed to grow without earthing up, the vines will yield half a ton of hay to the acre. They are killed by the first frost; when the nuts will be mature, and ready for use."

Varieties—

African Peanut

A comparatively small, smooth, and regularly formed sort. Shell thin, usually enclosing two kernels.

Wilmington Peanut

(Carolina)

Similar to the African. The pods, however, are longer, and the shell is thicker and paler. They rarely contain less than two, and often enclose three, kernels. Extensively cultivated in the Carolinas and Gulf States.

Tennessee Peanut

Pods large, thick, and irregular in form; the reticulations very coarse and deep. The pods usually contain two kernels. Less esteemed than either of the preceding varieties.

143

Vetch /Tare (Vicia sativa)

The Vetch, or Tare, in its properties and habits, somewhat resembles the Common Pea. There are numerous species as well as varieties, and the seeds of all may be used for food; but they are generally too small, or produced too sparingly, to repay the cost of cultivation.

The only variety of much importance to the garden is the following:

White Tare, or Vetch

(Lentil, of Canada; Napoléon Pea)

Annual; stem slender and climbing, about three feet high, the leaves terminating in a branching tendril, or clasper; flowers purplish; pods brown, slender, containing from eight to twelve seeds, or grains, which are globular, sometimes slightly flattened, smooth, and of a yellowish-white colour; they retain their germinative quality three years; an ounce contains about six hundred seeds.

Vicia sativa

In France and Canada, the seeds are used as a substitute for peas, both green and ripe, in soups and other dishes. They are also ground, and made into bread; but in this case their flour is generally mixed with that of wheat, or other of the edible grains.

The seeds may be sown in drills, in April or May, in the manner of garden-pease, or broadcast with oats for agricultural purposes.

Varieties—

Summer Tare, or Vetch

An agricultural variety, grown at the north of England and in Scotland. It is sown broadcast, and cultivated as wheat or barley. Both the haum and seed are used.

Winter Tare, or Vetch

Extensively grown in England and Scotland; usually sown in autumn, mixed with rye, for early spring food for stock. The seeds are smaller than those of the summer variety.

Not sufficiently hardy to survive the winters of the Northern States.

144

Winged Pea (Lotus tetragonolobus)

(Red Birdsfoot Trefoil; Asparagus Pea)

A hardy, creeping, or climbing, annual plant, fifteen or eighteen inches in height, or length; leaves trifoliate; flowers large, solitary, bright-scarlet; pods three inches and a half long, with four longitudinal, leafy membranes, or wings; seeds globular, slightly compressed, yellowish-white.

Lotus tetragonolobus flower

Use—The ripened seeds are sometimes used as a substitute for coffee; and the pods, while young and tender, form an agreeable dish, not unlike string-beans. It is often cultivated as an ornamental plant; and, for this purpose, is generally sown in patches, four or five seeds together on the border, where the plants are intended to remain.

When grown as an esculent, sow in double drills an inch and a half deep, and two feet apart; the single rows being made twelve inches from each other.

Part Ten: Medicinal Plants

Bene-plant; Camomile; Coltsfoot; Elecampane; Hoarhound; Hyssop; Liquorice; Pennyroyal; Poppy; Palmate-leaved or Turkey Rhubarb; Rue; Saffron; Southernwood; Wormwood.

Hyssop (Hyssopus officinalis)

145

Sesame (Sesamum)

(Bene-Plant)[1]

This plant is said to have been introduced into this country from Africa by the negroes. It is cultivated in the south of Europe, and in Egypt is grown to a considerable extent for forage and culinary purposes.

It is a hardy annual, with an erect, four-sided stem from two to four feet high, and opposite, lobed, or entire leaves; the flowers terminate the stalk in loose spikes, and are of a dingy-white color; the seeds are oval, flattened, and produced in an oblong, pointed capsule.

Propagation and Cultivation—It is propagated from seeds, which should be sown in spring, as soon as the ground has become well settled. They may be sown where the plants are to remain; or in a nursery-bed, to be afterwards transplanted. The plants should be grown in rows eighteen inches or two feet apart, and about a foot apart in the rows. The after-culture consists simply in keeping the ground loose, and free from weeds. The plant is said to yield a much greater amount of herbage if the top is broken or cut off when it is about half grown.

Use—

"The seeds were at one time used for food; being first parched, then mixed with water, and afterwards stewed with other ingredients. A sort of pudding is made of the seeds, in the same manner as rice; and is by some persons much esteemed. From the seeds of the first-named sort an oil is extracted, which will keep many years without having any rancid smell or taste. In two years, the warm taste which the new oil possesses wears off, and it becomes quite mild and pleasant, and may be used as a salad-oil, or for all the purposes of olive-oil. Two quarts of oil have been extracted from nine pounds of the seeds."

The properties of the plant are cooling and healing, with some degree of astringency. A few of the leaves, immersed a short time in a tumbler of water, give it a jelly-like consistence, without imparting color or flavor; and in this form it is generally used.

There are three varieties:

Biformed-Leaved

Plant larger than that of the Oval-leaved; the lower leaves are three-parted, while those of the upper part of the stalk are oval or entire.

Oval-Leaved

Stem about two feet high, with a few short branches; the leaves are oblong, and entire on the borders.

Trifid-Leaved

Taller and more vigorous than either of the preceding. The upper as well as the lower leaves are trifid, or three-parted.

1. 'Bene-Plant' was the original identification given by the author. It is clear that what is he referring to is the plant generally known as 'sesame'.

The sesame plant
Franz Eugen Köhler, 1897

146

Chamomile (Anthemis nobilis)

This is a half-hardy, herbaceous, perennial plant, growing wild in various parts of England, by roadsides and in gravelly pastures. Its stems rest upon the surface of the ground, and send out roots, by which the plants spread and are rapidly increased.

Soil and Culture—Camomile flourishes best in light, poor soil; and is generally propagated by dividing the roots, and setting them in rows a foot apart, and eight or ten inches from each other in the rows. They will soon entirely occupy the ground.

Gathering—The flowers should be gathered in a dry day, and when they are fully expanded. They are generally spread in an airy, shady situation for a few days, and afterwards removed to a heated apartment to perfect the drying.

Common Camomile

The flowers of this variety are single. Though considered more efficacious for medicinal purposes, it is not so generally cultivated as the Double-flowering. Its leaves are finely cut, or divided; and, when bruised, emit a peculiar, pungent odor. It may be grown from seeds, or slips, and from divisions of the plants, or roots.

Double-Flowering Camomile

A variety of the foregoing, with large, white, double flowers. The leaves are of the same form, but milder in their odor and taste. It is equally hardy with the Single-flowering, and much more ornamental. Though generally considered less efficacious than the last named, it is generally cultivated for use and the market on account of the greater bulk and weight of its flowers.

It is propagated by slips, with a few of the small roots attached. Both of the sorts are classed as hardy perennials; but, in the Northern and Eastern States, the plants are frequently destroyed in severe winters.

Use—

"The flowers, which are the parts principally used, have long been in high repute, both in the popular and scientific Materia Medica, and give out their properties by infusion in either water or alcohol. The flowers are also sometimes used in the manufacture of bitter beer, and, along with Wormwood, made, to a certain extent, a substitute for hops. In many parts of England, the peasants have what they call a 'Camomile seat' at the end of their gardens, which is constructed by cutting out a bench in a bank of earth, and planting it thickly with the Double-flowering variety; on which they delight to sit, and fancy it conducive to health."[1]

It is considered a safe bitter, and tonic; though strong infusions, when taken warm, sometimes act as an emetic.

1. The Book of the Garden. By Charles M'Intosh. 2 vols. Edinburgh and London, 1855.

147

Common Coltsfoot (Tussilago farfara)

A hardy, herbaceous, perennial plant. The leaves are all radical, roundish-heart-shaped, and from five to seven inches in diameter; the flower-stem (scape) is six or seven inches high, imbricated, and produces a solitary yellow flower, which is about an inch in diameter. The plants blossom in February and March, before the appearance of the leaves, and often while the ground is still frozen and even covered with snow.

Propagation and Culture—Coltsfoot thrives best in rich, moist soil. It may be propagated from seeds, but is generally increased by dividing its long, creeping roots. The plants require little attention, and will soon occupy all the space allotted.

Tussilago Farfara
Lydia Penrose

Gathering and Use—The leaves are the parts of the plant used, and are generally cut in July and September. They should not be exposed to the sun for drying, but spread singly in an airy, shaded situation. They are esteemed beneficial in colds and pulmonary disorders.

148

Elecampane (Inula helenium)

A hardy, herbaceous, perennial plant, introduced from Europe, but growing spontaneously in moist places, by roadsides, and in the vicinity of gardens where it has been cultivated. Stem from three to five feet high, thick and strong, branching towards the top; the leaves are from nine inches to a foot in length, ovate, toothed on the margin, downy beneath; the flowers are yellow, spreading, and resemble a small sunflower; the seeds are narrow, four-sided, and crowned with down.

The plants blossom in July and August, and there is but one variety cultivated.

Propagation and Culture—It is generally propagated by dividing the roots; but may be grown from seeds, which are sown just after ripening. The plants should be set in rows two feet asunder, and a foot from each other in the rows.

Inula helenium
Franz Eugen Köhler, 1897

Use—Elecampane is cultivated for its roots, which are carminative, sudorific, tonic, and alleviating in pulmonary diseases. They are in their greatest perfection when of two years' growth.

149

Hoarhound (Marrubium vulgare)

Hoarhound is a hardy, herbaceous, perennial plant, introduced from Europe, and naturalized to a considerable extent in localities where it has been once cultivated. Stem hoary, about two feet high; leaves round-ovate; flowers white; seeds small, of an angular-ovoid form and grayish-brown color.

Propagation and Cultivation—The plant prefers a rich, warm soil; and is generally propagated by dividing its long, creeping roots, but may also be raised from seeds. When once established, it will grow almost spontaneously, and yield abundantly.

Gathering and Use—The plants are cut for use as they come into flower; and, if required, the foliage may be cut twice in the season.

The leaves possess a strong and somewhat unpleasant odor, and their taste is "bitter, penetrating, and durable." The plant has long been esteemed for its efficacy in colds and pulmonary consumption.

Marrubium vulgare

150

Hyssop (Hyssopus officinalis)

Hyssop is a hardy, evergreen, dwarfish, aromatic shrub, from the south of Europe. Three kinds are cultivated, as follow:

Common or Blue-Flowering

More generally found in gardens than either of the following varieties. The stems are square and tender at first, but afterwards become round and woody; the leaves are opposite, small, narrow, with six or eight bract-like leaves at the same joint; the flowers are blue, in terminal spikes; seeds small, black, oblong.

Blue hyssop flower

Red-Flowering Hyssop

Quite distinct from the Common or Blue-flowering. The stem is shorter, the plants are more branching in their habit, and the spikes more dense or compact; flowers fine red. It is not so hardy as the White or the Blue Flowering, and is often injured by severe winters.

White-Flowering Hyssop

This is a sub-variety of the Common Blue-flowering; the principal if not the only mark of distinction being its white flowers. Its properties, and modes of culture, are the same.

Soil and Cultivation—The plants require a light, warm, mellow soil; and are propagated from seeds, cuttings, or by dividing the roots. The seeds are sown in April; and, when the seedlings are two or three inches high, they are transplanted to rows eighteen inches apart, and a foot from each other in the rows. The roots may be divided or the slips set in spring or autumn.

Use—The plant is highly aromatic. The leaves and young shoots are the parts used, and are cut, dried, and preserved as other pot-herbs.

"Hyssop has the general virtues ascribed to aromatic plants; and is recommended in asthmas, coughs, and other pulmonary disorders."[1]

1. The Vegetable Cultivator. By John Rogers. London, 1851.

151

Liquorice (Glycyrrhiza glabra)

Liquorice is a hardy, perennial plant. The roots are fleshy, creeping, and, when undisturbed, attain a great length, and penetrate far into the earth; the stem is herbaceous, dull-green, and about four feet high; leaves pinnate, composed of four or five pairs of oval leaflets; flowers pale-blue, in terminal spikes. The fruit consists of short, flattened pods, each containing two or three kidney-shaped seeds.

Soil, Propagation, and Culture—"Liquorice succeeds best in deep, rich, rather sandy, or in alluvial soil. The ground should be well enriched the year previous to planting: and it should either be trenched three feet deep in autumn, laid in ridges, and allowed to remain in that state till spring; or it may be trenched immediately before planting. The former method is the preferable one.

"Liquorice is propagated by portions of the creeping stem (commonly termed 'the creeping root'), from four to six inches in length, each having two or three buds. These are planted in March or April, or as soon as the ground can be well worked, in rows three feet apart, and eighteen inches from each other in the rows; covering with earth to the depth of two or three inches. Every year, late in autumn, when the sap has gone down and the leaves have turned yellow, the old stems should be cut down with a pruning-knife to a level with the ground. At this time, also, the creeping stems are forked up, cut off close to the main stems, and preserved in sand, or in heaps covered with straw and earth, for future plantations. The roots will be ready for taking up three years after planting. This should be done towards winter, after the descent of the sap. A trench three feet must then be thrown out, and the roots extracted; after which, they may be stored in sand for use."[1]

Use—The roots are the parts of the plant used, and these are extensively employed by porter-brewers. "The sweet, mucilaginous juice extracted from the roots by boiling is much esteemed as an emollient in colds."

1. The Gardener's Assistant. By Robert Thompson.

152

Pennyroyal (Hedeoma pulegioides)

The American Pennyroyal is a small, branching, annual plant, common to gravelly localities, and abounding towards autumn among stubble in dry fields from whence crops of wheat or rye have been recently harvested. The stem is erect, branching, and from six to twelve inches high; the leaves are opposite, oval, slightly toothed; flowers bluish, in axillary clusters; seeds very small, deep blackish-brown.

Hedeoma pulegioides

Sowing and Cultivation—In its natural state, the seeds ripen towards autumn, lie dormant in the earth during winter, and vegetate the following spring or summer. When cultivated, the seeds should be sown soon after ripening, as they vegetate best when exposed to the action of frost during winter. They are sown broadcast, or in drills ten or twelve inches asunder. When the plants are in full flower, they are cut off, or taken up by the roots, and dried in an airy, shaded situation.

Use—Pennyroyal possesses a warm, pungent, somewhat aromatic taste, and is employed exclusively for medical purposes. An infusion of the leaves is stimulating, sudorific, tonic, and beneficial in colds and chills.

This plant must not be confounded with the Pennyroyal (*Mentha pulegium*) of English writers, which is a species of Mint, and quite distinct from the plant generally known as Pennyroyal in this country.

153

Poppy (Papaver somniferum)

A hardy annual, growing naturally in different parts of Europe, and cultivated to a considerable extent in Germany for its seeds, which, under the name of "Maw-seed," are an article of some commercial importance. Stem five or six feet high, branching; leaves smooth, glaucous, clasping, and much cut or gashed on the borders; flowers large, terminal, purple and white; the bud pendent, or drooping, until the time of flowering, when it becomes erect. The petals soon fall to the ground, remaining on the plant but a few hours after their expansion; and are succeeded by large, roundish heads, or capsules, two inches and upwards in diameter, filled with the small, darkish-blue seeds for which the plant is principally cultivated.

Soil, Sowing, and Culture—

"The soils best suited to the growth of the Poppy are such as are of medium texture and in the highest state of fertilization. As the seeds are small, and consequently easily buried, the land should be well pulverized by harrowing and rolling. The seeds are sown in April, in drills about half an inch in depth, and twenty inches or two feet distant from each other. The young plants are afterwards thinned out to from six to ten inches' distance in the rows, and the whole crop kept free from weeds by frequent hoeing.

"The period of reaping is about the month of August, when the earliest and generally the largest capsules begin to open. The plants are then cut or pulled, and tied in small bundles, taking care not to allow the heads to recline until they are carried to the place allotted for the reception of the seed; which is then shaken out, and the sheaves again set upon their ends for the ripening of the remaining capsules.

"In Germany and Flanders, a mode of obtaining the first crop is to spread sheets by the side of the row, into which the seeds are shaken by bending over the tops of the plants: these are then pulled, tied in bundles, and removed; when the sheets are drawn forward to the next row, and so on, until the harvesting is completed."[1]

Use—Maw-seed is imported to some extent from different parts of Europe, and is principally used in this country for feeding birds.

Oil-Poppy

(Gray Poppy; Papaver somniferum olifer)

Stem three feet high, smooth and branching; flowers dull-red, or grayish; capsules very large, oblong; seeds of a brownish color, and produced in great abundance.

It is chiefly cultivated in Italy, the south of France, Germany, and Flanders.

Use—The oil of the seeds of the Poppy is of an agreeable flavour; and, in Europe, is chiefly applied to domestic purposes, for which it is esteemed nearly equal to that of the Olive. Its consumption in this country is comparatively trifling; being principally used for the finer kinds of oil-painting and by druggists.

Opium, or White Poppy

(Somniferum, album vel candidum)

Plant strong and vigorous,—the stem, in favorable situations, reaching a height of five or six feet; flowers large, white, and of short duration; seed-pods globular, of large size, often measuring upwards of two inches in diameter; seeds small, white, ripening in August and September.

Sowing and Cultivation—

"Being an annual plant, the Poppy, when sown in spring, matures its seed the last of summer or early in autumn. It is of easy culture, and can be successfully grown in any section of the Northern or Middle States. It may be sown at any time during the month of April, or the first week in May. The best method of cultivating the plant is in rows two feet and a half apart; and, on the poppies attaining a few inches in height, they are hoed out to a distance from one another of six or eight inches.

"Opium is obtained from the capsules or heads of seed, and is extracted after they are fully formed, but while yet green. The process is simple, and may be taught to children in an hour.

"Two or more vertical incisions are made in the capsule with a sharp knife or other instrument, about an inch in length, and not so deep as to penetrate through the capsule. As soon as the incisions are made, a milky juice will flow out, which, being glutinous, will adhere to the capsule. This may be collected by a small hair-brush such as is used by painters, and squeezed into a small vessel carried by the person who collects the juice. The incisions are repeated at intervals of a few days all round the capsule, and the same process of collecting the exuded juice is also repeated.

"The juice thus collected is Opium. In a day or two, it is of the consistence to be worked up into a mass. The narcotic matter of the plant may also be collected by boiling; but it is only the exuded juice that forms pure Opium.

"In the opium countries of the East, the incisions are made at sunset by several-pointed knives or lancets. On the following day the juice is collected, scraped off with a small iron scoop, and deposited in earthen pots; when it is worked by the hand until it becomes consistent. It is then formed in globular cakes, and laid in small earthen basins to be further dried. After the opium is extracted from the capsule, the plant is allowed to stand, and ripen its seeds.

"The seeds of the Poppy have nothing of the narcotic principle, and are eaten by the people of the East as a nourishing and grateful food; and they yield, by expression, an oil which is regarded as inferior only to that of the olive."[2]

The expense of labor forms the principal objection to the cultivation of the Poppy in the United States for its opium. As, however, the plants succeed well, and can be easily and extensively grown in any section of the country; and as the process of extraction, though minute, is yet simple,—the employment of females or children might render its production remunerative.

1,2.The Agriculturist's Manual. By **Peter Lawson** and **Son**. Edinburgh, 1836.

Decorative poppy varieties
Spring 1897 (1897), John A. Salzer Seed Co.

154

Palmate-Leaved Rhubarb (Rheum palmatum)

(Turkish Rhubarb)

This species is readily distinguished by its deeply divided or palmate leaves, and is generally considered as that from which the dried roots chiefly used in medicine are obtained. Like the Pie Rhubarb, it requires a deep, rich soil, which should be thoroughly stirred, and put in as fine a state of cultivation as possible, before setting the plants. These should be placed about three feet apart in each direction, and kept free from weeds during the summer. They will not be ready for taking up until five or six years old.

The roots are thick and succulent, with a brownish skin and bright-yellow flesh, streaked or variegated with red. After being dug, they are washed clean, cut in rather large pieces, and dried either by the sun, or in kilns formed for the purpose; when they are ready for use.

Rhubarb from Turkey and the neighboring countries is generally preferred; but it is said its superiority, to a great degree, is attributable to the manner in which it is dried and prepared for market. It is propagated by seed, or by a division of the roots.

Turkish rhubarb (Rheum palmatum): flowering and fruiting stem with leaf.
M. A. Burnett, c. 1842.

155

Rue (Ruta graveolens)

Rue is a hardy, shrubby, nearly evergreen plant, and thrives best in poor but dry and warm soil. It is propagated by seeds, or slips, and by dividing the roots. The seeds are sown in April, and the roots may be separated in spring or autumn. The plants should be set about eighteen inches apart in each direction. When extensively cultivated, they are set in rows eighteen inches apart, and a foot asunder in the rows.

Common Rue
Franz Eugen Köhler, 1897

Use—Rue has a strong, unpleasant odour, and a bitter, pungent, penetrating taste. The leaves are so acrid as to irritate and inflame the skin, if much handled. Its efficacy as a vermifuge is unquestioned; but it should be used with caution. It was formerly employed in soups; and the leaves, after being boiled, were eaten pickled in vinegar. The plant is rarely used in this country, either as an esculent or for medical purposes.

The kinds cultivated are the following:

Broad-Leaved Rue

Stem shrubby, four or five feet high; leaves compound, of a grayish-green color and strong odor; flowers yellow, in terminal, spreading clusters; the fruit is a roundish capsule, and contains four rough, black seeds.

At one period, this was the sort principally cultivated, and is that referred to in most treatises on medicine. More recently, however, it has given place to the Narrow-leaved, which is much hardier, and equally efficacious.

Narrow-Leaved Rue

Stem three or four feet high; foliage narrower than that of the preceding, but of the same grayish color, and strong, peculiar odor; the flowers are produced in longer and looser clusters than those of the Broad-leaved, and the seed-vessels are smaller. Now generally cultivated because of its greater hardiness.

156

Safflower (Carthamus tinctorius)

(Saffron)[1]

A hardy, annual plant, with a smooth, woody stem, two and a half or three feet high; leaves ovate, spiny; flowers large, compound, bright-orange, or vermilion; seeds ovate, whitish, or very light-brown, a fifth of an inch long, and a tenth of an inch thick.

Carthamus tinctorius

Soil and Cultivation—It grows best on soils rather light, and not wet; and the seed should be sown the last of April, or early in May, in drills about two feet apart and an inch deep. When the plants are two inches high, they should be thinned to six inches apart in the rows, and afterwards occasionally hoed during the summer, to keep the earth loose, and free the plants of weeds.

Use—

"It is cultivated exclusively for its flowers, from which the coloring-matter of Saffron, or Safflower, is obtained. These are collected when fully expanded, and dried on a kiln, under pressure, to form them into cakes; in which state they are sold in the market. It is extensively cultivated in the Levant and several countries of Europe, particularly France, Spain, and Germany; in the latter of which, the first gathering of flowers is obtained in the beginning of September; and others, for six or eight weeks following, as the flowers expand. It flowers somewhat earlier in this country, and seems well adapted to our climate.

"Though the color of the petals is of a deep-orange, they are used for dying various shades of red; the yellow matter being easily separated from the other. The flowers of Saffron are employed in Spain and other countries for coloring dishes and confectioneries; and from the seed a fixed oil is obtained, somewhat similar to that of the Sunflower: for which purpose alone, it does not, however, seem deserving of cultivation."

It was formerly much used in medicine in cases of humours and diseased blood.

1. This plant was identified by the author as 'saffron', however, it is a completely different plant from the *Crocus sativus* from which real saffron is produced.

157

Southernwood (Artemesia abrotanum)

A hardy, shrubby plant, about three feet high. The leaves are pale-green, and cut, or divided, into narrow, thread-like segments; the flowers are numerous, small, yellow, drooping; the seeds resemble those of the Common Wormwood, and retain their germinative properties two years.

The plant is generally propagated by dividing the roots in the manner of other hardy shrubs.

Use—The leaves have a strong, resinous, somewhat aromatic and rather pleasant odor, and are quite bitter to the taste. The root is seldom used; but the leaves and young branches are employed in the same manner and for the same purposes as those of the Common Wormwood.

Artemisia abrotanum
Vietz, Ferd Bern, 1800

158

Wormwood (Artemesia)

The cultivated species are as follow:

Common Wormwood

(Artemisia absinthium)

This species, everywhere common to gardens in this country, is a native of Great Britain. It is a hardy, perennial, shrubby plant, two or three feet in height. The leaves are deeply cut, or divided, pale-green above, and hoary beneath; the flowers are small, numerous, pale-yellow; the seeds are quite small, and retain their powers of germination two years.

The leaves, when bruised, have a strong, somewhat pungent, yet aromatic odor, and are proverbial for their intense bitterness.

Common Wormwood (Artemisia absinthium)

Roman Wormwood

(Artemesia pontica)

This species somewhat resembles the foregoing: but the roots are smaller, less woody, and more fibrous, and the stalks are shorter, and more slender; the leaves are smaller, more finely cut, or divided, pale-green above, and hoary on the under surface, like those of the Common Wormwood; the flowers, which are produced on the upper branches, are small, and of a pale-yellow color; seeds similar to those of the above species, retaining their vitality two years.

It is generally preferred to the Common Wormwood for medicinal purposes, as the taste is more agreeable, and its odor less pungent.

(Sea Wormwood)

(Artemesia maritima)

Indigenous to Great Britain, and common to the seacoast of Holland and the low countries of Europe. Roots creeping, tough, and fibrous; stalks two or three feet high, and, like the roots, tough and woody; leaves numerous, long, narrow, and hoary; flowers yellow, produced on the small branches towards the top of the plant; seeds similar to those of the Common Wormwood.

The leaves are somewhat bitter to the taste, and, when bruised, emit a strong, pleasant, aromatic odor.

Soil and Cultivation — All the species are hardy, aromatic perennials; and, though they will thrive in almost any soil, their properties are best developed in that which is warm, dry, and light. They are generally propagated, as other hardy shrubs, by dividing the plants; but may be raised from seeds, or slips. The seeds are sown in April, in shallow drills; and the seedlings afterwards transplanted to rows two feet apart, and a foot from each other in the rows.

Use — "An infusion of the leaves and tops of the Common Wormwood is used as a vermifuge, tonic, and stomachic; and the leaves are found to be beneficial to poultry."[1]

Most of the other species possess the same properties in a greater or less degree, and are used for the same purposes.

---

1. The Gardener's Assistant. By Robert Thompson.

Part Eleven: Mushrooms/ Esculent Fungi

Agaricus; Boletus; Clavaria; Morchella, or Morel; Tuber, or Truffle

Basket of Edible Mushrooms, Ukraine

159

Varied Mushroom Types

(Agaricus; Boletus; Clavaria; Morchella, or Morel; Tuber, or Truffle)

Although many experiments have been made in the culture of different species of edible Fungi,

"...only one has yet been generally introduced into the garden, though there can be no doubt the whole would finally submit to and probably be improved by cultivation. Many of them are natives of this country, abounding in our woods and pastures; and may be gathered wild, and freely enjoyed by those who have not the means of raising them artificially. In Poland and Russia, there are about thirty sorts of edible Fungi in common use among the peasantry. They are gathered in all the different stages of their growth, and used in various ways,—raw, boiled, stewed, roasted; and being hung up, and dried in stoves or chimneys, form a part of their winter's stock of provisions.

"Mushrooms are not, however, everywhere equally abundant, owing as well to climate as to the more general cultivation of the soil: the character of many of the sorts is, therefore, not perfectly known, and most of them are passed over as deleterious. Indeed, the greatest caution is requisite in selecting any species of this tribe for food; and we can advise none but an experienced botanist to search after any but the common and familiar sort (*Agaricus campestris*) for food."[1]

COMMON MUSHROOM

(Champignon; Agaricus campestris)

This Mushroom, when it first appears, is of a rounded or button-like form, of a white color, and apparently rests on the surface of the ground. When fully developed:

"...the stem is solid, two or three inches high, and about half an inch in diameter; its cap measures from an inch to three and sometimes even upwards of four inches in diameter, is of a white colour, changing to brown when old, and becoming scurfy, fleshy, and regularly convex, but, with age, flat, and liquefying in decay; the gills are loose, of a pinkish-red, changing to liver-colour, in contact with but not united to the stem, very thick-set, some forked next the stem, some next the edge of the cap, some at both ends, and generally, in that case, excluding the intermediate smaller gills."

Loudon says that it is most readily distinguished, when of middle size, by its fine pink or flesh-colored gills and pleas

ant smell. In a more advanced stage, the gills become of a chocolate color; and it is then more liable to be confounded with other kinds of dubious quality: but the species which most nearly resembles it is slimy to the touch, and destitute of the fine odor, having rather a disagreeable smell. Further, the noxious kind grows in woods, or on the margin of woods; while the true Mushroom springs up chiefly in open pastures, and should be gathered only in such places.

The common mushroom

Cultivation—

"This is the only species that has as yet been subjected to successful cultivation; though there can be little doubt that all or most of the terrestrial-growing sorts would submit to the same process, if their natural habitats were sufficiently studied, and their spawn collected and propagated. In this way, the Common Mushroom was first brought under the control of man.

"The seeds of the Common Mushroom, in falling from the gills when ripe, are no doubt wafted by the wind, and become attached to the stems and leaves of grasses and other herbage; and notwithstanding they are eaten by such animals as the horse, deer, and sheep, pass through their intestines without undergoing any material change in their vegetative existence: and hence, in the dung of these animals, when placed together, and kept moderately dry, and brought to a slight state of fermentation, we discover the first stage of the existence of the future brood of mushrooms. This is practically called 'spawn,' and consists of a white, fibrous substance, running like broken threads through the mass of dung, which appears to be its only and proper *nidus*."[2]

It is prepared for use as follows:

"In June and July, take any quantity of fresh horse-droppings,—the more dry and high-fed the better,—mixed with short litter, one-third of cow's dung, and a good portion of mould of a loamy nature; cement them well together, and mash the whole into a thin compost, and spread it on the floor of an open shed, to remain until it becomes firm enough to be formed into flat, square bricks; which done, set them on an edge, and frequently turn them till half dry; then, with a dibble, make two or three holes in each brick, and insert in each hole a piece of good old spawn about the size of a common walnut. The bricks should then be left till they are dry. This being completed, level the surface of a piece of ground, under cover, three feet wide, and of sufficient length to receive the bricks; on which lay a bottom of dry horse-dung six inches thick; then form a pile by placing the bricks in rows one upon another, with the spawn-side uppermost, till the pile is three feet high; next cover it with a small portion of warm horse-dung, sufficient in quantity to diffuse a gentle glow of heat through the whole. When the spawn has spread itself through every part of the bricks, the process is ended, and the bricks may then be laid up in a dry place for use. Mushroom-spawn thus made will preserve its vegetative power many years, if well dried before it is laid up; but, if moist, it will grow, and exhaust itself."[3]

The next step to be taken is the formation of the bed; in the preparation of which, no dung answers so well as that of the horse, when taken fresh from the stable: the more droppings in it, the better. The process recommended by Rogers[4] is as follows:

"About July or August is the general season for making mushroom-beds, though this may be done all the year round. A quantity of the dung mentioned should be collected and thrown together in a heap, to ferment and acquire heat; and, as this heat generally proves too violent at first, it should, previously to making the bed, be reduced to a proper temperature by frequently turning it in the course of the fortnight or three weeks; which time it will most likely require for all the parts to get into an even state of fermentation. During the above time, should it be showery weather, the bed will require some sort of temporary protection, by covering it with litter or such like, as too much wet would soon deaden its fermenting quality. The like caution should be attended to in making the bed, and after finishing it. As soon as it is observed that the fiery heat and rank steam of the dung have passed off, a dry and sheltered spot of ground should be chosen on which to make the bed. This should be marked out five feet broad; and the length, running north and south, should be according to the quantity of mushrooms likely to be required. If for a moderate family, a bed twelve or fourteen feet long will be found, if it takes well, to produce a good supply of mushrooms for some months, provided proper attention be paid to the covering.

"On the space marked for making the bed, a trench should be thrown out about six inches deep. The mould may be laid regularly at the side; and, if good, it will do for earthing the bed hereafter: otherwise, if brought from a distance, that of a more loamy than a sandy nature will be best.

"Whether in the trench, or upon the surface, there should be laid about four inches of good litter, not too short, for forming the bottom of the bed; then lay on the prepared dung a few inches thick, regularly over the surface, beating it as regularly down with the fork; continue thus, gradually drawing in the sides to the height of five feet, until it is narrow at the top like the ridge of a house. In that state it may remain for ten days or a fortnight, during which time the heat should be examined towards the middle of the bed by thrusting some small sharp sticks down in three or four places; and, when found of a gentle heat (not hot), the bed may be spawned: for which purpose, the spawn-bricks should be broken regularly into pieces about an inch and a half or two inches square, beginning within six inches of the bottom of the bed, and in lines about eight inches apart. The same distance will also do for the pieces of spawn, which are best put in by one hand, raising the manure up a few inches, whilst with the other the spawn can be laid in and covered at the same time.

"After spawning the bed, if it is found to be in that regular state of heat before mentioned, it may be earthed. After the surface is levelled with the back of the spade, there should be laid on two inches of mould,—that out of the trench, if dry and good, will do; otherwise make choice of a rich loam, as before directed. After having been laid on, it is to be beaten closely together; and, when the whole is finished, the bed must be covered about a foot thick with good oat or wheat straw; over which should be laid mats, for the double purpose of keeping the bed dry, and of securing the covering from being blown off. In the course of two or three days, the bed should be examined; and, if it is considered that the heat is likely to increase, the covering must be diminished for a few days, which is better than taking it entirely off.

"In about a month or five weeks,—but frequently within the former time, if the bed is in a high state of cultivation,—mushrooms will most likely make their appearance; and, in the course of eight and forty hours afterwards, they will have grown to a sufficient size for use. In gathering, instead of cutting them off close to the ground, they should be drawn out with a gentle twist, filling up the cavity with a little fine mould, gently pressed in level with the bed. This method of gathering is much better than cutting, as the part left generally rots, and breeds insects, which are very destructive, both in frames and on mushroom-beds.

"Where a mushroom-bed is to remain permanently, a covered shed will be found convenient.

"Sometimes it happens that a bed suddenly ceases to produce any mushrooms. This arises from various causes, but principally from the cold state of the bed in winter, or from a too dry state in summer. In the former case, a slight covering of mulchy hay laid over the bed, and on that six or eight inches of well-worked, hot dung, and the whole covered lightly with the straw that was taken off, will most likely bring it about again. In the latter instance, moisture, if required, should be given moderately, two or three mornings; when, after lying about an hour, the whole may be covered up, and be found of much service. In summer, most mushroom-beds in a bearing state require more or less slight waterings. Soft water should be used for the purpose: spring water is of too hard and too cold a nature; and, when at any time applied, checks vegetation. In summer time, a gentle shower of rain, on open beds that are in bearing and seem dry, will add considerably to their productiveness.

"A mushroom-bed seldom furnishes any abundance after two or three months: it has often done its best in six or seven weeks. Heavy rains are most destructive to mushrooms: therefore care should be taken to remove the wet straw, or litter, and directly replace it with dry. Hence the utility of a covered shed, or mushroom-house."

In addition to the foregoing, the following native species may be eaten with perfect safety, if gathered young and used while fresh:

Agaricus Comatus

An excellent species, much employed for making catchup; but should be used in a young state. It is found growing abundantly on stumps of trees, appearing both in spring and autumn.

Agaricus Deliciosus

(Sweet Mushroom)

Found in September and October, growing under fir and pine trees. It is of medium size, yellowish, zoned, with deep orange on the top, somewhat resembling *A. torminosus* (a deleterious species), but readily distinguished from it, as its juice is, when fresh cut, quite red, afterwards turning green, while that of the latter is white and unchangeable.

Sir James Edward Smith says it well deserves its name, and is really the most delicious mushroom known; and Mr. Sowerby is equally high in its praise, pronouncing it very luscious eating, full of rich gravy, with a little of the flavor of mussels.

Agaricus Exquisitus

(St. George's Mushroom; Agaricus Georgii)

This species often attains a weight of five or six pounds. It is generally considered less delicate than the common cultivated mushroom (*A. campestris*); but in Hungary it is regarded as a special gift from the saint whose name it bears. Persoon describes it as superior to *A. campestris* in smell, taste, and digestibility; on which account, he says, it is generally preferred in France.

It is found abundantly in many places, generally growing in rings, and re-appearing for many successive years on the same spot; and, though sometimes met with in old pastures, is generally found in thickets, under trees.

Agaricus Personatus

(Blewits; Blue Hats)

This is one of the species occasionally sold in Covent-Garden Market, London. When mature, it has a soft, convex, moist, smooth pileus, with a solid, somewhat bulbous stem, tinted with lilac. The gills are dirty-white, and rounded towards the stem.

The *Agaricus personatus* constitutes one of the very few mushrooms which have a market value in England. It is quite essential that it should be collected in dry weather, as it absorbs moisture readily, and is thereby injured in flavor, and rendered more liable to decay.

Agaricus Prunulus

This is found only in spring, growing in rings on the borders of wood-lands; at which time abundance of its spawn may be procured, and may be continued in the same way that the spawn of the common cultivated Mushroom is; namely, by transplanting it into bricks of loam and horse-dung, in which it will keep for months.

This mushroom is used both in its green and dried state. In the latter it constitutes what is called "Funghi di Genoa," and is preserved by being simply cut into four pieces, and dried in the air for a few days; when it is strung up, and kept for use.

Agaricus Oreades

(Fairy-Ring Agaricus)

There is little difficulty in distinguishing this mushroom, which is found growing in rings. The pileus is of a brownish-ochre color at first; becomes paler as it grows older, until it fades into a rich cream-yellow.

Dr. Badham says:

"Independent of the excellent flavour of this little mushroom, two circumstances make it valuable in a domestic point of view,—the facility with which it is dried, and its extensive dissemination."

It may be kept for years without losing any of its aroma or goodness.

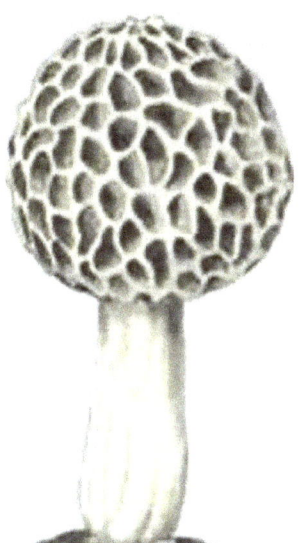

The morel

BOLETUS

Of this, two species are considered eatable,—the *B. edulis* and the *B. scaber*; the former resembling the Common Mushroom in taste, and the latter of good quality while in a young, fresh state, but of little value when dried, as it loses much of its odor, and becomes insipid, and unfit for use.

CLAVARIA

All the species are edible, and many of them indigenous to our woods; being usually found in damp, shady places.

THE MOREL

(*Morchella esculenta*)

In its natural state, the Morel is found growing in orchards, damp woods, and in moist pastures. Its height is about four inches. It is distinguished by its white, cylindrical, hollow, or solid, smooth stem; its cap is of a pale-brown or gray color, nearly spherical, hollow, adheres to the stem by its base, and is deeply pitted over its entire surface. It is in perfection early in the season; but should not be gathered soon after rain, or while wet with dew. If gathered when dry, it may be preserved for several months.

Use—The Morels are used, like the Truffle, as an ingredient to heighten the flavor of ragouts, gravies, and other rich dishes. They are used either fresh or in a dried state.

Cultivation—Its cultivation, if ever attempted, has been carried on to a very limited extent. Of its capability of submitting to culture, there can be little doubt. If the spawn were collected from its natural habitats in June, and planted in beds differently formed, but approximating as nearly as possible to its natural conditions, a proper mode of cultivation would assuredly be in time arrived at. Persoon remarks that:

> "...it prefers a chalky or argillaceous soil to one of a sandy nature; and that it not infrequently springs up where charcoal has been burned, or where cinders have been thrown."
>
> "The great value of the Morel—which is one of the most expensive luxuries furnished by the Italian warehouses, and which is by no means met with in the same abundance as some others of the Fungi—deserves to be better known than it is at present."

The genus comprises a very few species, and they are all edible.

COMMON TRUFFLE

(Tuber cibarium)

On the authority of our most distinguished mycologists, the Common Truffle has not yet been discovered within the limits of the United States. It is said to be found abundantly in some parts of Great Britain, particularly in Wiltshire, Kent, and Hampshire. It is collected in large quantities in some portions of France, and is indigenous to other countries of Europe.

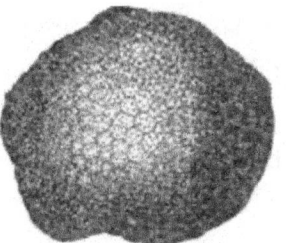

The truffle

The following description by Mascall, in connection with the engraving, will give an accurate idea of its size, form, color, and general character: "The size rarely much exceeds that of a large walnut. Its form is rounded, sometimes kidney-shaped, and rough with protuberances. The surface, when the truffle is young, is whitish; but, in those that are full grown, it is either blackish or a deep-black. The color of the inside is whitish, with dark-blue and white, gray, reddish, light-brown or dark-brown veins, of the thickness of a horse-hair, which are usually variously entangled, and which form a kind of network, or mat. Between the veins are numerous cavities, filled with mucilage, and small, solid grains. These scarcely visible glands were formerly said to be the seeds, or germs, of the young truffles. The less the inside of the Truffle is colored with dark veins, the more tender and delicious is its flesh.

"The blackish, external rind is hard, and very rough, by means of fine fissures, grains, and protuberances; and forms, with its small facets (which are almost hexagonal), an appearance by which it somewhat resembles the fir-apples of the larch. Whilst the truffle is young, its smell resembles that of putrid plants, or of moist, vegetable earth. When it has nearly attained its full growth, it diffuses an agreeable smell, which is peculiar to it, resembling that of musk, which lasts only a few days: it then becomes stronger; and the nearer the fungus is to its dissolution, which speedily ensues, so much the more unpleasant is its odour, till at last it is quite disagreeable and putrid. Whilst young, the flesh is watery, and the taste insipid: when fully formed, its firm flesh, which is like the kernel of the almond, has an extremely aromatic and delicious taste; but as soon as the fungus begins to decay, and worms and putrescence to attack it, its taste is bitter and disagreeable."

Many attempts have been made in Great Britain, as well as in other parts of Europe, to propagate the Truffle by artificial means; but all experiments thus far, if they have not totally failed, have been attended by very unsatisfactory results.

Use—Like the Common Mushroom, it is used principally in stuffings, gravies, and sauces, and in other very highly seasoned culinary preparations. It has long been held in high esteem by epicures and the opulent; but, from its extreme rarity, has always commanded a price which has effectually prohibited its general use. It has been truthfully remarked, "that few know how to raise it, and fewer still possess the proper knowledge to prepare it for the table."

Piedmontese Truffle

(Tuber magnatu)

This species is the most celebrated of all the truffles, and always commands an enormous price. It occurs abundantly in the mountains of Piedmont, and probably nowhere else.

Tuber Melanosporum

This is the Truffle of the Paris markets. It is richly scented, and also greatly superior in flavour to the common sorts.

Other genera and species of Fungi are considered harmless, and are occasionally used for food. Some of the edible kinds, however, in size, form, color, and organization, so closely approach certain poisonous or deleterious species, as to confuse even the most experienced student. None of the family (not excepting even the common cultivated Mushroom) should therefore be gathered for use, except by those who may possess a thorough knowledge of the various species and their properties.

1. Encyclopædia of Gardening. By J. C. Loudon. London, 1850.

2. The Book of the Garden. By Charles M'Intosh. 2 vols. Edinburgh and London, 1855.

3. The Transactions of the London Horticultural Society. Commenced 1815, and continued at intervals to the present time.

4. The Vegetable Cultivator. By John Rogers. London, 1851.

Part Twelve: Miscellaneous Vegetables

Alkekengi, or Ground Cherry; Corn; Egg-plant; Martynia; Oil Radish; Okra, or Gumbo; Pepper; Rhubarb, or Pie-plant; Sunflower; Tobacco; Tomato

Maul's Seed Catalogue for 1887, Wm. Henry Maule

160

Cape Gooseberry/Peruvian Groundcherry (Physalis edulis)

(Strawberry Tomato; Winter Cherry; Ground Cherry; Barbados Gooseberry; Alkekengi)

A hardy annual plant from Central or Tropical America. Stem angular, very much branched, but not erect,—in good soils, attaining a length or height of more than three feet; leaves large, triangular; flowers solitary, yellow, spotted or marked with purple, and about half an inch in diameter; fruit rounded or obtuse-heart-shaped, half an inch in diameter, yellow, and semi-transparent at maturity, enclosed in a peculiar thin, membranous, inflated, angular calyx, or covering, which is of a pale-green color while the fruit is forming, but at maturity changes to a dusky-white or reddish-drab. The pedicel, or fruit-stem, is weak and slender; and most of the berries fall spontaneously to the ground at the time of ripening.

Physalis edulis fruits

The seeds are small, yellow, lens-shaped, and retain their germinative properties three years.

The plants are exceedingly prolific, and will thrive in almost any description of soil. Sow at the same time, and thin or transplant to the same distance, as practised in the cultivation of the Tomato. On land where it has been grown, it springs up spontaneously in great abundance, and often becomes troublesome in the garden.

Use—The fruit has a juicy pulp, and, when first tasted, a pleasant, strawberry-like flavor, with a certain degree of sweetness and acidity intermixed. The after-taste is, however, much less agreeable, and is similar to that of the Common Tomato.

By many the fruit is much esteemed, and is served in its natural state at the table as a dessert. With the addition of lemon-juice, it is sometimes preserved in the manner of the plum, as well as stewed and served like cranberries.

If kept from the action of frost, the fruit retains its natural freshness till March or April.

Purple Alkekengi

(Purple Ground Cherry; Purple Strawberry Tomato; Purple Winter Cherry; Physalis sp.)

This species grows naturally and abundantly in some of the Western States. The fruit is roundish, somewhat depressed, about an inch in diameter, of a deep purple color, and enclosed in the membranous covering peculiar to the genus.

Compared with the preceding species, the fruit is more acid, less perfumed, and not so palatable in its crude state, but by many considered superior for preserving. The plant is less pubescent, but has much the same habit, and is cultivated in the same manner.

Tall Alkekengi

(Tall Ground Cherry; Tall Strawberry Tomato; Physalis pubescens)

Stem about four feet high, erect and branching; leaves oval, somewhat triangular, soft and velvety; flowers yellow, spotted with deep purple; fruit yellow, of the size of the Common Yellow Alkekengi, enclosed in an angular, inflated calyx, and scarcely distinguishable from the last named.

It is grown from seeds, which are sown like those of the Tomato. It is later, and much less prolific, than the species first described.

161

Corn (Zea mays)

Garden and Table Varieties

Adams's Early White

A distinct and well-marked table variety. Ears seven to eight inches in length, two inches in diameter, twelve or fourteen rowed, and rather abruptly contracted at the tips; kernel white, rounded, somewhat deeper than broad, and indented at the exterior end, which is whiter and less transparent than the interior or opposite extremity. The depth and solidity of the kernel give great comparative weight to the ear; and, as the cob is of small size, the proportion of product is unusually large.

In its general appearance, the ear is not unlike some descriptions of Southern or Western field-corn; from which, aside from its smaller dimensions, it would hardly be distinguishable. In quality, it cannot be considered equal to some of the shrivelled-kernelled, sweet descriptions, but will prove acceptable to those to whom the peculiar, sugary character of these may be objectionable. Though later than the Jefferson or Darlings, it is comparatively early, and may be classed as a good garden variety.

Much grown for early use and the market in the Middle States, but less generally known or cultivated in New England.

Black Sweet

(Slate Sweet)

Plant, in height and general habit, similar to Darling's Early; ears six to eight inches in length, uniformly eight-rowed; kernels roundish, flattened, deep slate-colour, much shrivelled at maturity. Early.

The variety is sweet, tender, and well flavored; remains a long period in condition for use; and, aside from its peculiar color (which by some is considered objectionable), is well worthy of cultivation.

Burr's Improved

(Burr's Sweet)

An improved variety of the Twelve-rowed Sweet. The ears are from twelve to sixteen rowed, rarely eighteen, and, in good soils and seasons, often measure eight or ten inches in length, nearly three inches in diameter, and weigh, when in condition for the table, from eighteen to twenty-two ounces; cob white; kernel rounded, flattened, pure white at first, or while suitable for use,—becoming wrinkled, and changing to dull, yellowish, semi-transparent white, when ripe.

The variety is hardy and productive; and, though not early, usually perfects its crop. For use in its green state, plantings may be made to the 20th of June.

The kernel is tender, remarkably sugary, hardens slowly, is thin-skinned, and generally considered much superior to the Common Twelve-rowed.

It is always dried or ripened for seed with much difficulty; often moulding or decaying before the glazing or hardening of the kernel takes place. If the crop is sufficiently advanced as not to be injured by freezing, it will ripen and dry off best upon the stalks in the open ground; but if in the milk, or still soft and tender at the approach of freezing weather, it should be gathered and suspended, after being husked, in a dry and airy room or building, taking care to keep the ears entirely separate from each other.

Darling's Early

(Darling's Early Sweet)

Stalk about five feet in height, and comparatively slender; the ears are from six to eight inches in length, an inch and a half in diameter, and, when the variety is unmixed, uniformly eight-rowed; the kernels are roundish, flattened, pure white when suitable for boiling,—much shrivelled or wrinkled, and of a dull, semi-transparent yellow, when ripe; the cob is white.

The variety is early, very tender and sugary, yields well, produces little fodder, ears near the ground, and is one of the best sorts for planting for early use, as it seldom, if ever, fails to perfect its crop. In the Middle States, and in the milder sections of New England, it may be planted for boiling until near the beginning of July.

The hills are made three feet apart in one direction by two feet and a half in the opposite; or the seeds may be planted in drills three feet apart, dropping them in groups of three together every eighteen inches.

Early Jefferson

Stalk five to six feet high, producing one or two ears, which are of small size, eight-rowed, and measure six or eight inches in length, and about an inch and a half in diameter at the largest part; cob white; kernel white, roundish, flattened,—the surface of a portion of the ear, especially near its tip, often tinged with a delicate shade of rose-red. The kernel retains its color, and never shrivels or wrinkles, in ripening.

Men, Women and Children Wanted to Plant Good Seeds, Ratekin's Seed House

The variety is hardy and productive, but is principally cultivated on account of its early maturity; though, in this respect, it is little, if at all, in advance of Darling's. The quality is tender and good, but much less sugary than the common shrivelled varieties; on which account, however, it is preferred by some palates. It remains but a short time tender and in good condition for boiling; soon becoming hard, glazed, and unfit for use.

Golden Sweet

(Golden Sugar)

Stalk and general habit similar to Darling's Early; ears six to eight inches long, an inch and a half or an inch and three-fourths in diameter, regularly eight-rowed; the kernel, when ripe, is semi-transparent yellow.

The variety is apparently a hybrid between the Common Yellow or Canada Corn and Darling's Early. In flavor, as well as appearance, both of these varieties are recognized. It does not run excessively to stalk and foliage, yields well, is hardy, and seldom fails to ripen perfectly in all sections of New England. For boiling in its green state, plantings may be made until the last week of June or first of July.

In respect to quality, it is quite tender, sweet, and well flavored, but less sugary than most of the other sugar or sweet varieties.

Old Colony

This variety was originated by the late Rev. A. R. Pope, of Somerville, Mass. At the time of its production, he was a resident of Kingston, Plymouth County, Mass.; and, in consequence of the locality of its origin, it received the name above given. In a communication at the close of the sixteenth volume of the "Magazine of Horticulture," Mr. Pope describes it as follows:

> "It is a hybrid, as any one can readily perceive by inspection, between the Southern White and the Common Sweet Corn of New England; and exhibits certain characteristics of the two varieties, combining the size of the ear and kernel and productiveness of the Southern with the sweetness and tenderness of the Northern parent.
>
> "The stalks are from ten to twelve feet in height, and of corresponding circumference. They are also furnished with brace-roots (seldom found upon the common varieties of Sweet Corn); and the pistils are invariably green, and not pink, as in the Southern White."

The ears are from five to seven inches in length, and the number of rows varies from twelve to twenty; the kernels are very long or deep; and the cob, which is always white, is quite small compared with the size of the ear. When ripe, the kernels are of a dull, semi-transparent, yellowish white, and much shrivelled. The ears are produced on the stalk, four or five feet from the ground. It is very productive, but late; and though it will rarely fail in the coldest seasons to yield abundant supplies in the green state for the table, yet it requires a long and warm season for its complete maturity.

For cultivation in the Southern States and tropical climates, it has been found to be peculiarly adapted; as it not only possesses there the sweetness and excellence that distinguish the Sweet Corn of the temperate and cooler sections, but does not deteriorate by long cultivation, as other sweet varieties almost invariably are found to do.

Parching Corn (White Kernel)

(Pop-Corn)

Stalk six feet high, usually producing two ears, which are from six to eight inches long, quite slender, and uniformly eight-rowed; cob white; kernel roundish, flattened, glossy, flinty, or rice-like, and of a dull, semi-transparent, white color. When parched, it is of pure snowy whiteness, very brittle, tender, and well flavoured, and generally considered the best of all the sorts used for this purpose.

In some parts of Massachusetts, as also in New Hampshire, the variety is somewhat extensively cultivated for commercial purposes. Its peculiar properties seem to be most perfectly developed in dry, gravelly, or silicious soils, and under the influence of short and warm seasons. In field culture, it is either planted in hills three feet apart, or in drills three feet apart, and eighteen inches apart in the drills. The product per acre is usually about the same number of bushels of ears that the same land would yield of shelled-corn of the ordinary field varieties.

Maiz: Seikei Zusetsu vol. 19

Increase of size is a sure indication of deterioration. The cultivator should aim to keep the variety as pure as possible by selecting slender and small-sized but well-filled ears for seed, and in no case to plant such as may have yellow or any foreign sort intermixed. The value of a crop will be diminished nearly in a relative proportion to the increase of the size of the ears.

Parching Corn (Yellow)

A yellow variety of the preceding. It retains its color to some extent after being parched; and this is considered an objection. It is tender, but not so mild flavored as the white, and is little cultivated. The size and form of the ears are the same, and it is equally productive.

Red-Cob Sweet

Ears about eight inches in length by a diameter of two inches,—usually twelve but sometimes fourteen rowed; kernels roundish, flattened, white when suitable for boiling, shrivelled, and of a dull, semi-transparent white when ripe; the cob is red, which may be called its distinguishing characteristic. Quality good; the kernel being tender and sweet. It remains long in good condition for the table, and is recommended for general cultivation. Season intermediate.

A sub-variety occurs with eight rows; the form and size of the ear and kernel resembling Darling's Early.

Rice (Red Kernel)

This is a variety of the White Rice, with deep purplish-red or blood-red kernels. The ears are of the same size and form. Its quality, though inferior to the white, is much superior to the yellow. Productiveness, and season of maturity, the same.

Rice (White Kernel)

Stalk six feet or more in height; ears five or six inches long, an inch and a half in diameter, somewhat conical, broadest at the base, and tapering to the top, which is often more or less sharply pointed; the cob is white; the kernels are long and slender, angular, sharply pointed at the outward extremity, as well as to some extent at the opposite, and extremely hard and flinty. They are not formed at right angles on the cob, as in most varieties of corn, but point upward, and rest in an imbricated manner, one over the other.

The variety is hardy and prolific; and, though not late, should have the benefit of the whole season. For parching, it is inferior to the Common Parching Corn before described, though it yields as much bulk in proportion to the size of the kernel, and is equally as white: but the sharp points often remain sound; and it is, consequently, less crisp and tender.

Rice (Yellow Kernel)

Another sub-variety of the White Rice; the ear and kernel being of the same form and size. It is equally productive, and matures as early; but, when parched, is inferior to the White both in crispness and flavour.

Stowell's Evergreen

(Stowell's Evergreen Sweet)

Stalk from six to seven feet in height, and of average diameter; ears of a conical form, six or seven inches long, and two inches and a quarter in diameter at the base; kernels long or deep, pure white when suitable for boiling, of a dull, yellowish-white, and much shrivelled when ripe; cob white, and, in consequence of the depth of the kernels, small in comparison to the diameter of the ear.

The variety is intermediate in its season; and, if planted at the same time with Darling's or equally early kinds, will keep the table supplied till October. It is hardy and productive, very tender and sugary, and, as implied by the name, remains a long period in a fresh condition, and suitable for boiling.

Tuscarora

(Turkey Wheat)

Plant five to six feet in height, moderately strong and vigorous; ears eight-rowed, and of remarkable size,—exceeding, in this respect, almost every sort used for the table in the green state. In good soil, they are often a foot and upwards in length, and from two inches and three-fourths to three inches in diameter at the base. The kernel, which is much larger than that of any other table variety, is pure white, rounded, flattened, and, when divided in the direction of its width, apparently filled with fine flour of snowy whiteness; the cob is red, and of medium size.

In point of maturity, the Tuscarora is an intermediate variety. In its green state, it is of fair quality, and considered a valuable sort by those to whom the sweetness of the sugar varieties is objectionable. In their ripened state, the kernels, to a great extent, retain their fresh and full appearance, not shrivelling in the manner of the sugar sort, though almost invariably indented at the ends like some of the Southern Horse-toothed field varieties.

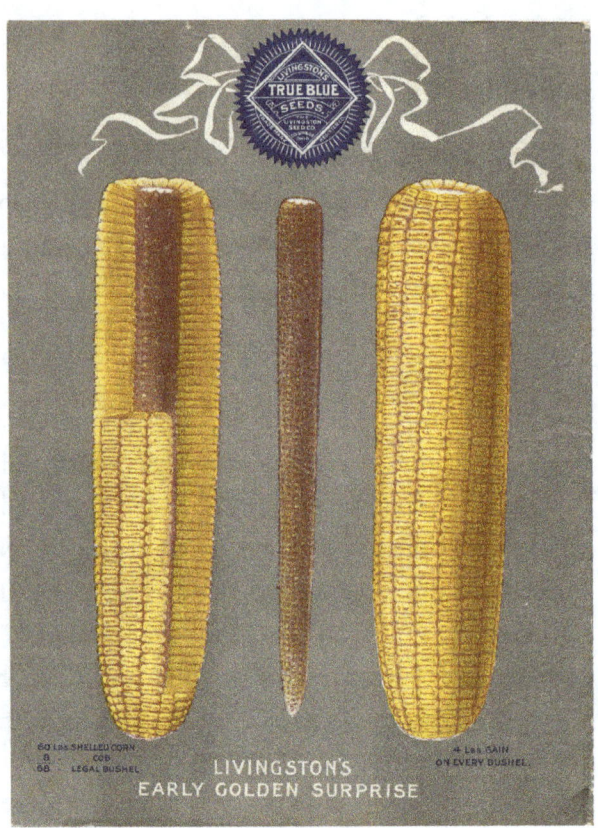

Livingston's Seeds 1899 Annual, The Livingston Seed Company

When ground in the ripe state, it is much less farinaceous and valuable for cooking or feeding stock than the fine, white, floury appearance of the kernel, when cut or broken, would seem to indicate.

Twelve-Rowed Sweet

A large, comparatively late variety. Stalk seven feet high; the ears are from ten to fourteen rowed, seven to nine inches long, often two inches and a half in diameter in the green state, and taper slightly towards the top, which is bluntly rounded; cob white; the kernels are large, round or circular, sometimes tooth-shaped, pure white when suitable for the table, dull white and shrivelled when ripe.

The variety is hardy, yields a certain crop, and is sweet, tender, and of good quality. It is the parent of one or two varieties of superior size and excellence, to which it is now gradually giving place.

Field Varieties

Canada Yellow

(Early Canada)

Ear small, about seven inches in length, symmetrical, broadest at the base, and tapering to the tip, uniformly eight-rowed, in four double rows; kernel roundish, smooth, and of a rich, glossy, orange-yellow color; cob small, white; stalk four to five feet high, slender; the leaves are not abundant, and the ears, of which the plant very rarely produces more than two, near the ground.

On account of the small size of the ear, the yield per acre is much less than that of almost any other field variety; twenty-five or thirty bushels being an average crop. The dwarfish character of the plants, however, admits of close culture,—three feet in one direction by two or two and a half in the opposite,—affording ample space for their full development; four plants being allowed to a hill.

Its chief merit is its early maturity. In ordinary seasons, the crop will be fully ripened in August. If cultivated for a series of years in the Eastern or Middle States, or in a latitude much warmer than that of the Canadas, the plant increases in size, the ears and kernels grow larger, and it is slower in coming to maturity.

Dutton

(Early Dutton)

Ears nine or ten inches long, broadest at the base, tapering slightly towards the tip, ten or twelve rowed, and rarely found with the broad clefts or longitudinal spaces which often mark the divisions into double rows in the eight-rowed varieties,—the outline being almost invariably smooth and regular; kernel as broad as deep, smooth, and of a rich, clear, glossy, yellow color; cob comparatively large, white; stalk of medium height and strength, producing one or two ears.

One of the handsomest of the field varieties, nearly as early as the King Philip, and remarkable for the uniformly perfect manner in which, in good seasons, the ears are tipped, or filled out. In point of productiveness, it compares favorably with the common New-England Eight-rowed; the yield per acre varying from fifty to seventy bushels, according to soil, culture, and season.

Much prized for mealing, both on account of its quality, and its peculiar, bright, rich color. In cultivation, the hills are made three feet and a half apart in each direction, and five or six plants allowed to a hill.

Hill

(Whitman; Whitman's Improved; Webster; Smutty White; Old-Colony Premium)

Stalk six feet or more in height, moderately strong at the ground, but comparatively slender above the ear; foliage not abundant; the ears are produced low on the stalk, often in pairs, are uniformly eight-rowed, well filled at the tips, and, when fully grown, ten or eleven inches in length; cob white, and comparatively small; kernel dusky, transparent-white, large and broad, but not deep.

The Hill Corn is nearly of the season of the Common New-England Eight-rowed, and is unquestionably the most productive of all field varieties. In Plymouth County, Mass., numerous crops have been raised of a hundred and fifteen bushels and upwards to the acre; and, in two instances, the product exceeded a hundred and forty.

This extraordinary yield is in a degree attributable to the small size of the plant, and the relative large size of the ear. The largest crops were obtained by planting three kernels together, in rows three feet asunder, and from fifteen to eighteen inches apart in the rows.

No variety is better adapted for cultivation for farm consumption; but for market, whether in the kernel or in the form of meal, its dull, white color is unattractive, and it commands a less price than the yellow descriptions.

From the most reliable authority, the variety was originated by Mr. Leonard Hill, of East Bridgewater, Plymouth County, Mass.; and was introduced to public notice in 1825-6. Though at present almost universally known as the "Whitman," it appears to have been originally recognised as the "Hill;" and, of the numerous names by which it has since been called, this is unquestionably the only true and legitimate one.

Illinois Yellow

(Western Yellow)

Stalk ten feet or more high; foliage abundant; ears high on the stalk, single or in pairs, twelve to sixteen rowed, eleven to thirteen inches long, broadest at the base, and tapering gradually towards the tip, which is bluntly rounded; kernel bright-yellow, long and narrow, or tooth-formed, paler at the outer end, but not indented; cob white.

The variety ripens perfectly in the Middle States, but is not suited to the climate of New England.

Illinois White

(Western White)

Similar in its general character to the Illinois Yellow. Kernel rice-white; cob generally white, but sometimes red.

King Philip, or Brown

(Improved King Philip)

Ears ten to twelve inches in length, uniformly eight-rowed when the variety is pure or unmixed; kernel copper-red, rather large, somewhat broader than deep, smooth and glossy; cob comparatively small, pinkish-white; stalk six feet in height, producing one or two ears, about two feet and a half from the ground.

In warm seasons, it is sometimes fully ripened in ninety days from the time of planting; and may be considered as a week or ten days earlier than the Common New-England Eight-rowed, of which it is apparently an improved variety.

Very productive, and recommended as one of the best field sorts now in cultivation. In good soil and favourable seasons, the yield per acre is from seventy-five to ninety bushels; although crops are recorded of a hundred and ten, and even of a hundred and twenty bushels.

Northern Crown, Tested Seeds (1892), Northrup, Braslan & Goodwin

As grown in different localities, and even in the product of the same field, there is often a marked variation in the depth of color, arising either from the selection of paler seed, or from the natural tendency of the variety toward the clear yellow of the New-England Eight-rowed. A change of color from yellowish-red to paler red or yellow should be regarded as indicative of degeneracy.

Said to have originated on one of the islands in Lake Winnipiseogee, N.H.

New-England Eight-Rowed

Stalk six or seven feet high, producing one or two ears, which are from ten to eleven inches long, and uniformly eight-rowed; kernel broader than deep, bright-yellow, smooth and glossy; cob comparatively small, white.

The variety is generally grown in hills three feet and a half apart in each direction, and five or six plants allowed to a hill; the yield varying from fifty to seventy bushels to the acre, according to season, soil, and cultivation. It is a few days later than the King Philip, but ripens perfectly in the Middle States and throughout New England; except, perhaps, at the extreme northern boundary, where the Canada Yellow would probably succeed better.

It often occurs with a profuse intermixture of red, sometimes streaked and spotted, sometimes copper-red, like the King Philip, and occasionally of a rich, bright, clear blood-red. As the presence of this color impairs its value for marketing, and particularly for mealing, more care should be exercised in the selection of ears for seed; and this, continued for a few seasons, will restore it to the clear yellow of the Dutton or Early Canada.

Many local sub-varieties occur, the result of selection and cultivation, differing in the size and form of the ear; size, form, and color of the kernel; and also in the season of maturity. The Dutton, Early Canada, King Philip, and numerous other less important sorts, are but improved forms of the New-England Eight-rowed.

Parker

A variety remarkable for the extraordinary size of the ears, which, if well grown, often measure thirteen or fourteen inches in length: they are comparatively slender, and uniformly eight-rowed. Cob white and slim; kernels bright-yellow, rounded, broader than deep.

Productive, but some days later than the Common New-England Eight-rowed.

White Horse-Tooth

(Southern White)

Stalk twelve feet or more in height, with large, luxuriant foliage; ears single, often in pairs, short and very thick, sixteen to twenty-two rowed; kernel remarkably large, milk-white, wedge-formed, indented at the outer end; cob red.

Yellow Horse-Tooth

(Southern Yellow)

Plant similar to that of the White Horse-tooth; kernel very large, bright-yellow, indented; cob red.

Extensively cultivated throughout the Southern States, but not adapted to the climate of the Middle or Northern.

162

Eggplant (Solanum melongena)

The Egg-plant is a native of Africa, and is also indigenous to Tropical America. It is a tender annual, with an erect, branching stem, and oblong, bluish-green, powdered leaves. The flowers are one-petaled, purple, and produced on short stems in the axils of the branches; the fruit is often somewhat oblong, but exceedingly variable in form, size, and colour; the seeds are small, yellowish, reniform, flattened, and retain their germinative properties seven years.

Soil—The Egg-plant will thrive well in any good garden soil, but should have the benefit of a sheltered situation.

The eggplant

Sowing and Culture—The seed should be sown in a hot-bed in March, at the time and in the manner of sowing tomato seed. The young plants are, however, more tender; and should not be allowed to get chilled, as they recover from its effects very slowly. The plant being decidedly tropical in character, the seedlings should not be transplanted into the open ground until the commencement of summer weather; when they may be set out in rows two feet apart, and two feet asunder in the rows. Keep the ground free from weeds, earth up the plants a little in the process of cultivation, and by the last of August, or beginning of September, abundance of fruit will be produced for the table.

If no hot-bed is at hand, sufficient seedling plants for a small garden may be easily raised by sowing a few seeds in March in common flower-pots, and placing them in the sunny window of the sitting-room or kitchen.

In favorable seasons, a crop may be obtained by sowing the seeds in May in the open ground, and transplanting the seedlings, when two or three inches high, in a warm and sheltered situation.

Use—

"It is used both boiled and stewed in sauces like the Tomato. A favourite method among the French is to scoop out the seeds, fill up the cavity with sweet herbs, and fry the fruit whole."[1]

A common method of cooking and serving is as follows: Cut the fruit in slices half an inch thick; press out as much of the juice as possible, and parboil; after which, fry the slices in batter, or in fresh butter in which grated bread has been mixed; season with pepper, salt, and sweet herbs, to suit; or, if preferred, the slices may be broiled as steaks or chops.

Varieties—

American Large Purple

Fruit remarkably large,—often measuring eight inches in depth, seven inches in diameter, and weighing four or five pounds; skin deep-purple, with occasional stripes of green about the stem; plant hardy and stocky.

The American Large Purple is more generally cultivated in this country than any other variety. The plants produce two (and rarely three) fruits; but the first formed are invariably the best developed.

It is similar to, if not identical with, the Round Purple of English and French authors.

Chinese Long White

Quite distinct from the Common White or the Purple. Plant of low growth, with comparatively pale foliage; fruit white, eight or nine inches long, two inches and a half in diameter, and often more or less curved, particularly when the end is in contact with the ground.

It is later than the White or Purple varieties, and nearly of the season of the Scarlet-fruited. To obtain the fruit in full perfection, the plants must be started in a hot-bed.

Guadaloupe Striped

Fruit nearly ovoid, smaller than the Round or Long Purple; skin white, streaked and variegated with red.

Long Purple

The plants of this variety are of the height of the Round Purple, but are subject to some variation in the color of the branches and in the production of spines; flowers large, purple, with a spiny calyx; the fruit is oblong, somewhat club-shaped, six or eight inches in length, sometimes straight, but often slightly bent; at maturity, the skin is generally deep-purple, but the color varies much more than the Large Round; it is sometimes pale-purple, slightly striped, sometimes variegated with longitudinal, yellowish stripes, and always more deeply colored on the exposed side.

It is early, of easy culture, hardy and productive, excellent for the table, thrives well in almost any section of the Northern States, and, if started in a hot-bed, would perfect its fruit in the Canadas.

Eggplant varieties
Seikei Zusetsu vol. 26, 1804

New-York Improved

A sub-variety of the Large Round, producing the same number of fruits, which are generally of a deeper color, and average of larger size. The leaves are often spiny; and, if the variety is genuine, the plants will be readily distinguished from those of the last named by their more dense or compact habit of growth.

It is, however, comparatively late, and better suited to the climate of the Middle States than to that of New England; though it is successfully cultivated in the vicinity of Boston, Mass., by starting the plants in a hot-bed, and setting them in a warm and sheltered situation.

Round Purple

(Large Round Purple)

Plant from two to three feet high, branching, generally tinged with purple, producing two and sometimes three fruits; the leaves are large, downy, oblong, lobed on the borders, with scattered spines on the midribs; flowers large, pale-purple,—the flower-stem and calyx invested with purple spines; the fruit is obovate, four or five inches in diameter, six or seven inches deep, slightly indented at the apex, and of a fine deep-purple when well matured,—specimens sometimes occur slightly striped or rayed with yellowish-green.

The American Large Purple, if not the same, is but an improved form of this variety.

Scarlet-Fruited Egg-Plant

A highly ornamental variety, introduced from Portugal. The plant attains the height of three feet, with leaves about six inches long. In general appearance, it resembles the Common Egg-plant; but the fruit, which is about the size of a hen's egg, is of a beautiful scarlet.

It is rarely if ever used for food, but is principally cultivated for its peculiar, richly colored, and ornamental fruit, which makes a fine garnish.

The variety is late, and comparatively tender. The seeds should be started early in a hot-bed, and the plants grown in a warm and sheltered situation.

White Egg-Plant

Fruit milk-white, egg-shaped, varying from three to five inches in length, and from two inches and a half to three inches and a half in diameter.

It is the earliest, hardiest, and most productive of all varieties. The plants frequently produce five or six fruits each; but the first formed are generally the largest.

If sown in the open ground early in May, the plants will often perfect a portion of their fruit; but they are most productive when started in a hot-bed.

The fruit is sometimes eaten cooked in the manner of the Purple varieties, but is less esteemed.

1. The Book of the Garden. By Charles M'Intosh. 2 vols. Edinburgh and London, 1855.

163

Martynie (Martynia proboscidea)

(Unicorn Plant)

The Martynia

A hardy, annual plant, with a strong, branching stem two feet and a half or three feet high. The leaves are large, heart-shaped, entire or undulated, downy, viscous, and of a peculiar, musk-like odor when bruised or roughly handled; the flowers are large, bell-shaped, somewhat two-lipped, dull-white, tinged or spotted with yellow and purple, and produced in long, leafless racemes, or clusters; the seed-pods are green, very downy or hairy, fleshy, oval, an inch and a half in their greatest diameter, and taper to a long, comparatively slender, incurved horn, or beak. The fleshy, succulent character of the pods is of short duration: they soon become fibrous, the elongated beak splits at the point, the two parts diverge, the outer green covering falls off, and the pod becomes black, shrivelled, hard, and woody. The seeds are large, black, wrinkled, irregular in form, and retain their germinative properties three years.

Sowing and Cultivation—The Martynia is of easy cultivation. As the plants are large and spreading, they should be two feet and a half or three feet apart in each direction. The seeds may be sown in April or May, in the open ground where the plants are to remain; or a few seeds may be sown in a hot-bed, and the seedlings afterwards transplanted.

Gathering and Use—The young pods are the parts of the plant used. These are produced in great abundance, and should be gathered when about half grown, or while tender and succulent: after the hardening of the flesh, they are worthless. They are used for pickling, and by many are considered superior to the Cucumber, or any other vegetable employed for the purpose.

164

Oil Radish (Raphanus sativus olifer)

A variety of the Common Radish, particularly adapted for the production of oil, and distinguished by the name *R. sativus olifer,* or Oil Radish. Its stems are dwarf, from a foot and a half to two feet in height, much branched, spreading, and produce more seed-pods than the Common Radish. It is grown rather extensively in China for its oil; from whence it has been introduced into and cultivated in some parts of Europe: but it does not appear with any particular success, though much has been said and written in its favor.

It seems best suited for southern latitudes, where it may be sown in September, and harvested the following May or June: but, in the northern portions of the United States, it will be found too tender to withstand the winter; and the seed will therefore require to be sown in spring.

The oil is obtained from the seed, and is considered superior to rape-seed oil, but is extracted with greater difficulty.

165

Okra/Gumbo (Hibiscus esculentus)

Dwarf okra

Okra is a half-hardy annual, from Central America. Stem simple, sometimes branched at the top, and from two to six feet in height, according to the variety; the leaves are large, palmate, deep-green; the flowers are large, five-petaled, yellowish on the border, purple at the centre; the seed-pods are angular, or grooved, more or less sharply pointed, an inch or an inch and a half in diameter at the base, and from four to eight inches in length; the seeds are large, round-kidney-shaped, of a greenish-drab color, black or dark-brown at the eye, and retain their power of germination five years.

Soil, Sowing, and Cultivation—Okra may be raised in any common garden soil, and is propagated by seeds sown in April or May. The Dwarf varieties may be grown in rows two feet apart, and a foot from each other in the rows; but the taller sorts require a space of at least three feet between the rows, and nearly two feet from plant to plant in the rows. Keep the soil about the plants loose and open; and, in the process of cultivation, earth up the stems slightly in the manner of earthing pease. The pods will be fit for use in August and September.

It requires a long, warm season; and is most productive when started in a hot-bed, and grown in a warm, sheltered situation.

Use—The green pods are used while quite young, sliced in soups and similar dishes, to which they impart a thick, viscous, or gummy consistency. Thus served, they are esteemed not only healthful, but very nutritious.

The ripe seeds, roasted and ground, furnish a palatable substitute for coffee.

Varieties—

Buist's Dwarf Okra
A variety recently introduced by Mr. Robert Buist, of Philadelphia. Height two feet; being about half that of the old variety. Its superiority consists in its greater productiveness, and the little

space required for its development; while the fruit is of larger size and superior quality. It is said to produce pods at every joint.

Dwarf Okra.

Stem two feet and a half high, sometimes branched at the top, but generally undivided; leaves large, and, as in all varieties, five-lobed; flowers yellow, purple at the centre; pods erect, obtusely pointed, nearly as large in diameter as those of the Giant, but generally about five inches in length.

It is the earliest of the Okras, and the best variety for cultivation in the Northern and Eastern States.

Between this and the Tall, or Giant, there are numerous sub-varieties; the result both of cultivation and climate. The Tall sorts become dwarfish and earlier if long cultivated at the North; and the Dwarfs, on the contrary, increase in height, and grow later, if long grown in tropical climates.

The seeds of all the sorts are similar in size, form, and color.

Pendent-Podded

The plants of this variety differ slightly, if at all, from those of the Common or Dwarf Okra. It is principally, if not solely, distinguished by the pendulous or drooping character of its pods; those of all other sorts being erect.

Tall or Giant Okra

(White-Podded)

Stem five to six feet in height; pods erect, sharply tapering to a point, eight to ten inches in length, and about an inch and a half in diameter near the stem or at the broadest part.

With the exception of its larger size, it is similar to the Dwarf; and, if long cultivated under the influence of short and cool seasons, would probably prove identical.

It yields abundantly, but is best adapted to the climate of the Middle and Southern States.

Pods and flowers of red okra

166

Pepper/Capsicum (Capsicum annuum)

Of the Capsicum there are many species, both annual and perennial; some of the latter being of a shrubby or woody character, and from four to six feet in height. As they are mostly tropical, and consequently tender, none but the annual species can be successfully grown in open culture in the Middle States or New England.

The *Capsicum annuum,* or Common Garden-pepper, is a native of India.[1] The stalks vary in height from a foot to nearly three feet; the flowers are generally white or purple; the pods differ in a remarkable degree in size, form, colour, and acridness; the seeds are yellow, nearly circular, flattened, and, like the flesh or rind of the fruit, remarkable for their intense piquancy,—nearly forty-five hundred are contained in an ounce, and their vitality is retained five years.

***Propagation and Cultivation*—**The plants are always propagated from seeds. Early in April, sow in a hot-bed, in shallow drills six inches apart, and transplant to the open ground when summer weather has commenced. The plants should be set in warm, mellow soil, in rows sixteen inches apart, and about the same distance apart in the rows; or, in ordinary seasons, the following simple method may be adopted for a small garden, and will afford an abundant supply of peppers for family use: When all danger from frost is past, and the soil is warm and settled, sow the seeds in the open ground, in drills three-fourths of an inch deep, and fourteen inches apart; and, while young, thin out the plants to ten inches apart in the rows. Cultivate in the usual manner, and the crop will be fit for use early in September.

Use—

"The pod, or fruit, is much used in pickles, seasonings, and made dishes; as both the pod and seeds yield a warm, acrid oil, the heat of which, being imparted to the stomach, promotes digestion, and corrects the flatulency of vegetable aliments. The larger and more common sorts are raised in great quantities, by market gardeners in the vicinity of populous towns, for the supply of pickle-warehouses."

Species and Varieties—

Bell-Pepper

(Large Bell; Bull-Nose)

Bell-pepper

Plant two feet and upwards in height, stocky and branching, the stem and branches often stained or clouded with purple; leaves large, on long stems, smaller, smoother, and less sharply pointed, than those of the Squash-pepper; flowers white, sometimes measuring nearly an inch and a half in diameter.

The pods, which are remarkably large, and often measure nearly four inches deep and three inches in diameter, are pendent, broadest at the stem, slightly tapering, and generally terminate in four obtuse, cone-like points. At maturity, the fruit changes to brilliant, glossy, coral red.

The Bell-pepper is early, sweet and pleasant to the taste, and much less acrid or pungent than most of the other sorts. In many places, it is preferred to the Squash-pepper for pickling, not only because of its mildness, but for its thick, fleshy, and tender rind.

In open culture, sow in May, in drills sixteen inches apart, and thin the plants to twelve inches in the drills.

In England, they are pickled as follows: The pods are plucked while green, slit down on one side, and, after the seeds are taken out, immersed in salt and water for twenty-four hours; changing the water at the end of the first twelve. After soaking the full time, they are laid to drain an hour or two; put into bottles or jars; and boiled vinegar, after being allowed to cool, poured over them till they are entirely covered. The jars are then closely stopped for a few weeks, when the pods will be fit for use. In this form, they have been pronounced the best and most wholesome of all pickles.

Bird-Pepper

Stem fifteen to eighteen inches high; leaves very small; flowers white, about two-thirds of an inch in diameter; pods erect, sharply conical, an inch and three-quarters long, about half an inch in diameter, and of a brilliant coral-red when ripe.

The variety is late. If sown in the open ground, some of the pods, if the season be favorable, will be fit for use before the plants are destroyed by frost; but few will be fully perfected unless the plants are started under glass.

The Bird-pepper is one of the most piquant of all varieties, and is less valuable as a green pickle than many milder and thicker-fleshed sorts. It is cultivated in rows fourteen inches apart, and ten or twelve inches asunder in the rows. If sown in the open ground, make the rows the same distance apart, and thin the young plants to the same space in the rows.

The "Cayenne Pepper-pot" of commerce is prepared from Bird-pepper in the following manner:

"Dry ripe peppers well in the sun, pack them in earthen or stone pots, mixing common flour between every layer of pods, and put all into an oven after the baking of bread, that they may be thoroughly dried; after which, they must be well cleansed from the flour, and reduced to a fine powder. To every ounce of this, add a pound of wheat-flour, and as much leaven as is sufficient for the quantity intended. After this has been properly mixed and wrought, it should be made into small cakes, and baked in the same manner as common cakes of the same size; then cut them into small parts, and bake them again, that they may be as dry and hard as biscuit, which, being powdered and sifted, is to be kept for use."

Cayenne Pepper

The pods of this variety are quite small, cone-shaped, coral-red when ripe, intensely acrid, and furnish the Cayenne Pepper of commerce. Like the other species of the family, it is of tropical origin; and being a perennial, and of a shrubby character, will not succeed in open culture at the North.

Both the green and ripe pods are used as pickles, and also for making Chili vinegar or pepper-sauce; which is done by simply putting a handful of the pods in a bottle, afterwards filled with the best vinegar, and stopping it closely. In a few weeks, it will be fit for use.

The process of preparing Cayenne Pepper is as follows. The pods are gathered when fully ripe. "In India, they are dried in the sun; but in cooler climates they should be dried on a slow hot-plate, or in a moderately heated oven: they are then pulverized, and sifted through a fine sieve, mixed with salt, and, when dried, put into close, corked bottles, for the purpose of excluding the air. This article is subject to great adulteration, flour being often mixed with it; and, still worse, red lead, which is much of the same colour, and greatly increases the weight.

"A better method is to dry the pods in a slow oven, split them open, extract the seeds, and then pulverize them (the pods) to a fine powder, sifting the powder through a thin muslin sieve, and pulverizing the parts that do not pass through, and sifting again, until the whole is reduced to the finest possible state. Place the powder in air-tight glass bottles; but add no salt or other ingredient whatever."[2]

The pods of either of the long-fruited sorts, or those of the Cherry-pepper, prepared as above, will furnish a quality of "Cayenne" Pepper greatly superior to that ordinarily sold by grocers, or even by apothecaries and druggists.

The larger and milder kinds, powdered in the same manner, make a wholesome and pleasant grade of pepper of sufficient pungency for a majority of palates.

Cherry-Pepper

(Capsicum cerasiforme)

Stem twelve to fifteen inches high, strong and branching; leaves comparatively small, long, narrow, and sharply pointed; flowers white, three-fourths of an inch in diameter; pod, or fruit, erect, nearly globular or cherry-form, and, at maturity, of a deep, rich, glossy, scarlet colour. It is remarkable for its intense piquancy; exceeding in this respect nearly all the annual varieties.

Cherry-pepper

It is not so early as some of the larger sorts; but in favorable seasons will perfect a sufficient portion of its crop in the open ground, both for seed and pickling. For the latter purpose, the peppers should be plucked while still green, put into a common jar or wide-mouthed bottle, and vinegar added to fill the vessel. In a few weeks, they will be fit for use.

When in perfection, the plants are very ornamental; the glossy, coral-red of the numerous pods presenting a fine contrast with the deep-green foliage by which they are surrounded.

A variety occurs with larger, more conical, and pendent pods. The plant is also much larger, and quite distinct in its general character.

Yellow Cherry-Pepper

(Yellow-Fruited)

This is a variety of the Red Cherry. The plants have the same general habit, require the same treatment, and perfect their fruit at the same season. There is little real difference between the sorts, with the exception of the colour of the fruit; this being clear yellow.

To preserve either of these varieties for use in the dry state, all that is necessary is to cut off the plants close to the roots when the fruit is ripe, and hang them, with the fruit attached, in any warm and dry situation. They will retain their piquancy for years.

Chili Pepper

Pods pendent, sharply conical, nearly two inches in length, half an inch in diameter, of a brilliant scarlet when ripe, and exceedingly piquant; plant about eighteen inches high; leaves numerous, of small size, and sharply pointed; flowers white, nearly three-fourths of an inch in diameter.

Sow in a hot-bed in April, and transplant to the open ground in May, about fourteen inches apart in each direction. Requires a long, warm season.

Long Red Pepper

Fruit brilliant, coral-red, generally pendulous, sometimes erect, conical, often curved towards the end, nearly four inches in length, and from an inch to an inch and a half in diameter; skin, or flesh, quite thin, and exceedingly piquant.

Stalk about two feet high; foliage of medium size, blistered and wrinkled; flowers an inch in diameter, white.

The variety yields abundantly, but attains its greatest perfection when started in a hot-bed. The ripe pods, dried and pulverized as directed for Cayenne Pepper, make an excellent substitute for that article.

The plants, with ripe fruit, are very ornamental.

Long Yellow

Pods pendent, long, and tapering, three to four inches in length, and about an inch in their greatest diameter. At maturity, they assume a lively, rich, glossy yellow; and the plants are then showy and ornamental.

Stem two feet and upwards in height, slightly coloured with purple at the intersection of the branches and insertion of the leaf-stems; leaves of medium size, smaller and paler than those of the Long Red; flowers white, nearly an inch in diameter. Like the last named, it is very piquant. It is also late; and, to obtain the variety in perfection, the seed should be started in a hot-bed in April.

Purple or Blue Podded

(Black-Podded)

Fruit erect, on long stems, bluntly cone-shaped, two inches and a half in length, and a half or three-fourths of an inch in diameter at the broadest part. Before maturity, the skin is green or reddish-green, clouded or stained with black or purplish-brown; but, when ripe, changes to rich, deep, indigo-blue.

Plant two feet or upwards in height, more erect and less branched than other varieties, and much stained with purple at the intersection of the branches and at the insertion of the leaf-stems; leaves of medium size, or small, long, and sharply pointed; leaf-stems long, deep-green; flowers white, tipped with purple, about three-fourths of an inch in diameter; flower-stems long, purple.

A rare, richly coloured, and beautiful pepper, but not cultivated or of much value as an esculent. For its full perfection, a long, warm season is requisite. The plants should be started in a hot-bed in March or April, and transplanted in May to the open ground, fifteen inches apart.

Quince-Pepper

(Piment cydoniforme)

This variety is similar to the Sweet Spanish; but the fruit is rather longer, and its season of maturity is somewhat later. Its flavour is comparatively mild and pleasant; but, like the Sweet Spanish, it is not generally thick-fleshed. At maturity, the fruit is a brilliant coral-red.

Round or Large Red Cherry-Pepper

(Rond; Cerise grosse)

This is but a sub-variety of the Common Red Cherry-pepper, differing only in its larger size.

It is quite late, and should be started in a hot-bed.

Squash-Pepper

(Tomato-Shaped)

Fruit compressed, more or less ribbed, about two inches and three-quarters in diameter, and two inches in depth; skin smooth and glossy,—when ripe, of a brilliant coral-red; flesh thick, mild and pleasant to the taste, though possessing more piquancy than the Large Bell or Sweet Spanish.

Plant about two feet high, stout and branching; leaves broad and large; flowers white, an inch and a quarter in diameter; fruit drooping, the fruit-stem short and thick.

The Squash-pepper is extensively grown for the market, and is most in use in the pickle warehouses of the Eastern and Middle States. In field-culture, the plants are started in hot-beds in April, and, after the beginning of summer weather, transplanted to the open ground, fourteen to eighteen inches apart, according to the quality of the soil. The fruit is generally sold by weight; and an acre of land, in a fair state of cul-

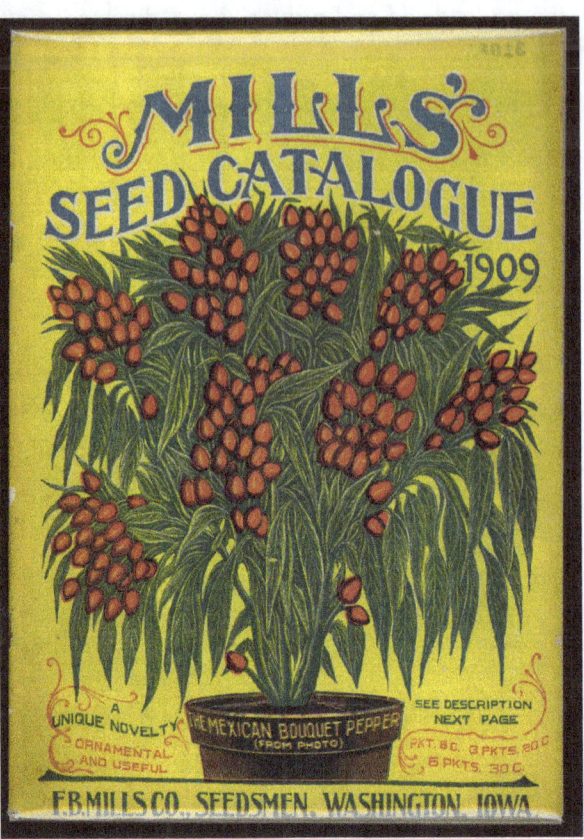

Mill's Seed Catalogue (1909), F.B. Mills Co., Seedsmen

tivation, will yield about three tons,—a bushel of the thick-fleshed sort weighing nearly thirty-two pounds. An excellent pickle may be made by preparing the peppers in the manner directed for the Bell variety.

As grown by different market-men and gardeners, there are several sub-varieties of the Squash-pepper, differing both in form and in the thickness of the flesh; the latter quality, however, being considered of the greater importance, as the thick-fleshed sorts not only yield a greater weight to the acre, but are more esteemed for the table.

The Squash-pepper succeeds well when sown in the open ground in May, in drills fourteen inches apart. The plants should be ten or twelve inches apart in the rows; for, when grown too closely, they are liable to draw up, making a weakly, slender growth, and yield much less than when allowed sufficient space for their full development. Low-growing, stocky, and branching plants are the most productive.

Sweet Mountain Pepper

This variety resembles the Large Bell, if it is not identical. The Sweet Mountain may be somewhat larger; but, aside from this, there is no perceptible difference in the varieties.

Sweet Spanish

(Piment monstreux)

Fruit obtusely conical, often four inches in length, and nearly three inches in diameter,—brilliant glossy scarlet at maturity; stem strong and sturdy, two feet or more in height; leaves large, but narrower than those of the Large Bell; flowers white, and of large size,—usually an inch and a half in diameter; fruit sometimes erect, but generally drooping.

Though one of the largest varieties, the Sweet Spanish is also one of the earliest. The flesh is sweet, mild, and pleasant; and the variety is much esteemed by those to whom the more pungent kinds are objectionable. When prepared in the same form, it makes a pickle equally as fine as the Large Bell.

The Sweet Spanish Pepper succeeds well if sown in the open ground in May. Make the rows sixteen inches apart, and thin the plants to a foot apart in the rows.

Yellow Squash-Pepper

(Yellow Tomato-Formed)

Fruit similar in form to the Squash-pepper, but of smaller size, erect or pendulous; orange-yellow at maturity.

The variety is later than the last named; much less productive; and, for pickling, is comparatively not worthy of cultivation.

1. This species is in fact native to the Americas.

2. The Book of the Garden. By Charles M'Intosh. 2 vols. Edinburgh and London, 1855.

Garden and Farm Manual (1906), Johnson & Stokes

167

Rhubarb (Rheum sp. et var)

(Pie-Plant)

This is a hardy, perennial plant, cultivated almost exclusively for its leaf-stalks. Its general character may be described as follows: Root-leaves large, round-heart-shaped, deep-green, and more or less prominently blistered; leaf-stems large, succulent, furrowed, pale-green, often stained or finely spotted with red, varying from two to three inches in diameter at the broadest part, and from a foot to three feet in length. The

Rhubarb stalks

flower-stalk is put forth in June, and is from five to seven feet in height, according to the variety; the flowers are red or reddish-white, in erect, loose, terminal spikes; the seeds are brown, triangular, membranous at the corners, and retain their germinative properties three years.

Soil and Cultivation—Rhubarb succeeds best in deep, somewhat retentive soil: the richer its condition, and the deeper it is stirred, the better; as it is scarcely possible to cultivate too deeply, or to manure too highly.

It may be propagated by seeds, or by a division of the roots; the latter being the usual method. When grown from seeds, the plants not only differ greatly in size and quality, but are much longer in attaining a growth suitable for cutting.

"Whether grown from seed, or increased by a division of the roots, a deep, rich soil, trenched to the depth of two or even three feet, is required to insure the full development of the leaf-stalks; for upon their size, rapidity of growth, and consequent tenderness of fibre, much of their merit depends. The seed should be sown in April, in drills a foot asunder; thinning the plants, when a few inches high, to nine inches apart. In the autumn or spring following, they will be fit for transplanting in rows three feet asunder, and the plants set three feet apart. If propagated by dividing the roots, it may be done either in autumn or spring; the same distance being given to the sets that is allowed for seedling plants. As, however, some of the varieties grow to a much larger size than others, a corresponding distance should be accorded them, extending to five feet between the rows, and three feet from plant to plant.

"The plants should be set out singly, and not in threes, as is so often done. For the first year, the ground between the rows may be cropped with lettuce, turnips, beans, or similar low-growing crops; but, after the second year, the leaves will cover the whole space, and require it also for their full development."[1]

After-Culture—This consists in keeping the soil well enriched, open, and clear of weeds; and in breaking over the flower-stalks, that they may not weaken the roots, and consequently reduce the size and impair the quality of the leaf-stalks.

Gathering the Crop—

"This is usually done in spring; commencing as soon as the stalks have attained a serviceable size. No leaves, however, should be plucked the first year, and only a few of the largest and first formed during the second; and this plucking should not be made too early in the season, because, in that case, the plants would be weakened. From the third year, as long as the roots or plantations last, it may be gathered with freedom. A plantation in good soil, and not overmuch deprived of its foliage, will last from ten to fifteen years.

"When the leaves are about half expanded, they may be plucked for use; but, when the largest returns are expected (as in the case of market-gardens), they should be allowed to attain their full size. In removing them, they should be pulled off close to the base, and not cut, to prevent an unnecessary escape of sap, which, in all succulent plants, flows more copiously from a clean cut than from one slightly lacerated or torn. The footstalks should then be separated from the leaves, and tied up in bundles of suitable size for market."[2]

Rhubarb is sometimes blanched. This may be effected without removing the plants, by means of sea-kale pots, or by empty casks open at the top, put over the crowns in March. It can, however, be more perfectly done by taking up the roots, and placing them in some dark place, with a temperature of 55° or 60°F; where they should be slightly covered with soil to prevent them from drying. When so treated, they are much more tender, crisp, and delicate than when grown exposed to the sun and air: but the quality is greatly impaired; the pulp, though somewhat acid, being generally comparatively flavourless.

Use—As before remarked, it is cultivated for its leaf-stalks; which are used early in the season, as a substitute for fruit, in pies, tarts, and similar culinary preparations. When fully grown, the expressed juice forms a tolerably palatable wine, though, with reference to health, of doubtful properties. As an article of commercial importance in the vegetable markets, it is of very recent date. In 1810, Mr. Joseph Myatts, of Deptford, England, long known for his successful culture of this plant, sent his two sons to the borough-market with five bunches of Rhubarb-stalks, of which they could sell but three. It is now disposed of by the ton, and many acres in the vicinity of nearly all large towns and cities are devoted exclusively to its cultivation.

Varieties—These are very numerous, as they are readily produced from the seed; but the number really deserving of cultivation is comparatively limited. Old kinds are constantly giving place to new, either on account of superior earliness, size, productiveness, or quality. The following are the prominent sorts cultivated:

Cahoon

Leaves remarkably large, often broader than long, and more rounded than those of most varieties; stalk short and thick,—if well grown, measuring from twelve to sixteen inches in length, and three inches or more in diameter; skin thick, uniformly green.

Its remarkable size is its principal recommendation. The texture is coarse, the flavour is harsh and strong, and it is rarely employed for culinary purposes.

In some localities, it is cultivated to a limited extent for the manufacture of wine; the juice being expressed from the stalks, and sugar added in the ratio of three pounds and a half to a gallon. This wine, though quite palatable, has little of the fine aroma of that made from the grape; and, if not actually deleterious, is much less safe and healthful. Any of the other varieties may be used for the same purpose; the principal superiority of the Cahoon consisting in its larger stalks, and consequently its greater product of juice.

Downing's Colossal

A large variety, nearly of the size of Myatt's Victoria. It is described as being less acid than the last named, and of a fine, rich, aromatic flavour.

Early Prince Imperial

Stalks of medium size; recommended by D. T. Curtis, Esq., Chairman of the Vegetable Committee of the Massachusetts Horticultural Society, as in all respects the best flavoured of any variety ever tested; and commended for general cultivation, as particularly adapted to the wants of the family, if not to the wishes of the gardener, to whom size and productiveness are more than flavour. It invariably turns red in cooking, which makes it preferable for the table as a sauce. When cooked, it is of the color of currant-jelly, and remarkably fine flavoured.

In 1862, it received the first prize of the Massachusetts Horticultural Society, as the best for family use.

Elford

(Buck's Rhubarb)

An early sort, well adapted for forcing. The stalks are rather slender, covered with a thin skin of a bright-scarlet color; and their substance throughout is of a fine red, which they retain when cooked, if not peeled,—a process which, owing to the thinness of the skin, is not considered necessary. Even when grown in the dark, the stalks still preserve the crimson tinge. It was raised from the seed of *Rheum undulatum*.

Hawke's Champagne

A new variety, said to equal the Prince Albert in earliness, and also to be of a deeper and finer color, and much more productive. It forces remarkably well; is hardy in open culture; and commands the highest market prices, both from its great size, and fine, rich color.

Mitchell's Royal Albert

Stalks large, red, and of excellent flavour. Early and prolific.

Red cabbage, rhubarb, and orange
Charles Demuth, 1929

Myatt's Linnæus

(*Linnæus*)

A medium-sized or comparatively small variety, recently introduced. "Besides being the earliest of all, and remarkably productive as well as high flavoured, and possessing little[Pg 631]acidity, it has a skin so thin, that removing it is hardly necessary; and its pulp, when stewed, has the uniform consistence of baked Rhode-Island Greenings; and it continues equally crisp and tender throughout the summer and early autumn." One of the best sorts for a small garden or for family use.

Myatt's Victoria

(*Victoria*)

Leaves large, broader than long, deep-green, blistered on the surface, and much waved or undulated on the borders. Leaf-stalks very large, varying from two inches and a half to three inches in their broadest diameter, and frequently measuring upwards of two feet and a half in length: the weight of a well-developed stalk, divested of the leaf, is about two pounds. They are stained with red at their base, and are often reddish, or finely spotted with red, to the nerves of the leaf.

It has rather a thick skin, is more acid than many other varieties, and not particularly high flavoured: but no kind is more productive; and this, in connection with its extraordinary size, makes it not only the most salable, but one of the most profitable, kinds for growing for the market.

It requires a deep, highly-manured soil; and the roots should be divided and reset once in four or five years. It is about a fortnight later than the Linnæus.

Nepal
(Rheum Australe; Rheum Emodi)

The leaf-stalks attain an immense size, but are unfit for use on account of their strongly purgative properties: but the leaves, which are frequently a yard in diameter, are useful in covering baskets containing vegetables or fruit; and for these the plant is sometimes cultivated.

Tobolsk Rhubarb
(Early Red Tobolsk)

Leaves comparatively small; leaf-stalks below medium size, stained with red at the base. It is perceptibly less acid than most varieties, and remarkable for fineness of texture and delicacy of flavour.

1,2. The Book of the Garden. By Charles M'Intosh. 2 vols. Edinburgh and London, 1855.

168

Sunflower (Helianthus annuus)

Tall Sunflower

Stem from five to eight feet or more in height; leaves heart-shaped, rough, three-nerved; flowers very large, terminal, nodding; the seeds are large, ovoid, angular, or compressed, nearly black, sometimes striped with white, and retain their germinative properties five years.

The plant is a native of South America.

Dwarf Sunflower

This species, which was introduced from Egypt, differs from the last principally in its more dwarfish habit of growth, but also in being less branched. The flowers are much smaller, and generally of a lighter color.

Sunflower (Helianthus annuus)
From the Flowers series for Old Judge Cigarettes MET

Soil and Cultivation—The Sunflower will thrive in almost any soil or situation, but succeeds best on land adapted to the growth of Indian Corn. It is always grown from seed, which should be sown in April, or the beginning of May, in drills three feet apart. When the plants are well up, they should be thinned to a foot asunder, and afterwards cultivated in the usual manner; stirring the ground occasionally, and keeping the plants free from weeds. The flowers appear in July, and the seeds ripen in August and September. The central flower is first developed; attains a larger size than any that succeed it; and ripens its seeds in advance of those on the side-branches. The heads of seeds should be cut as they successively mature, and spread in a dry, airy situation for three or four weeks; when the seeds will become dry and hard, and can be easily rubbed or threshed out.

Use—

"The seeds of both species yield an oil little inferior to that of the Olive for domestic purposes, and which is also well adapted for burning. In Portugal, the seeds are made into bread, and also into a kind of meal. They are also sometimes roasted, and used as a substitute for coffee; but the purpose for which they seem best adapted is the feeding of domestic fowls, pheasants and other game. The greatest objection to its culture is, that it is a most impoverishing crop, particularly the Large or Common Tall species."[1]

1. The Book of the Garden. By Charles M'Intosh. 2 vols. Edinburgh and London, 1855.

169

Tabacco (Nicotiana, sp.)

All the species and varieties of Tobacco in common cultivation are annuals; and most, if not all, are natives of this continent.

Connecticut Seed-leaf.

"Like other annual plants, it may be grown in almost every country and climate, because every country has a summer; and that is the season of life for all annual plants. In hot, dry, and short summers, like the northern summers of Europe or America, Tobacco-plants will not attain a large size; but the Tobacco produced will be of delicate quality and good flavor. In long, moist, and not very warm summers, the plants will attain a large size,—perhaps as much so as in Virginia; but the Tobacco produced will not have that superior flavor, which can only be given by abundance of clear sunshine, and free, dry air. By a skilful manufacture, and probably by mixing the Tobacco of cold countries with that[Pg 634] of hot countries, by using different species, and perhaps by selecting particular varieties of the different species, the defects in flavor arising from climate may, it is likely, be greatly remedied."

The species and varieties are as follow:—

Connecticut Seed-Leaf

(Peach-Leaf; Virginia Tobacco; Nicotiana tabacum)
Leaves oblong, regularly tapering, stemless and clasping, eighteen inches to two feet long, and from nine to twelve inches in diameter. When fully developed, the stem of the plant is erect and strong, five feet high, and separates near the top into numerous, somewhat open, spreading branches; the flowers are large, tubular, rose-colored, and quite showy and ornamental; the capsules are ovoid, or somewhat conical, and, if well grown, nearly half an inch in their greatest diameter; the seeds, which are produced in great abundance, are quite small, of a brownish color, and retain their germinative properties four years.

This species is extensively cultivated throughout the Middle and Southern States, and also in the milder portions of New England. In the State of Connecticut, and on the banks of the Connecticut River in Massachusetts, it is a staple product; and in some towns the value of the crop exceeds that of Indian Corn, and even that of all the cereals combined.

Guatemala Tobacco

A variety with white flowers. In other respects, similar to the foregoing.

Numerous other sorts occur, many of which are local, and differ principally, if not solely, in the size or form of the leaves. One of the most prominent of these is the Broad-leaved, which is considered not only earlier and more productive, but the best for manufacturing.

Propagation—It is propagated by seeds sown annually. Select a warm, rich locality in the garden; spade it thoroughly over; pulverize the surface well; and the last of April, or beginning of May, sow the seeds thinly, broadcast; cover with a little fresh mould, and press it well upon them either by the hoe or back of the spade. As they are exceedingly minute, much care is requisite in sowing, especially that they should not be too deeply covered. When the plants appear, keep them clear of weeds, and thin them out sufficiently to allow a free growth. A bed of seedlings nine or ten feet square will be sufficient for an acre of land. If preferred, the plants may be raised in drills eight inches apart, slightly covering the seeds, and pressing the earth firmly over them, as above directed. When the seedlings are four or five inches high, they are ready for transplanting.

Soil and Cultivation—Tobacco requires a warm, rich soil, not too dry or wet; and, though it will succeed well on recently turned sward or clover-turf, it gives a greater yield on land that has been cultivated the year previous, as it is less liable to be infested by worms, which sometimes destroy the plants in the early stages of their growth. The land should be twice ploughed in the spring; first as soon as the frost will permit, and again just previous to setting. Pulverize the surface thoroughly by repeated harrowing and rolling, and it will be ready to receive the young plants. The time for transplanting is from the 1st to the 20th of June; taking advantage of a damp day, or setting them immediately after a rain. If the ground is not moist at the time of transplanting, it will be necessary to water the plants as they are set.

"The ground should be marked in straight rows three feet apart, and slight hills made on these marks two feet and a half apart; then set the plants, taking care to press the earth firmly around the roots. As soon as they are well established, and have commenced growing, run a cultivator or horse-hoe between the rows, and follow with the hand-hoe; resetting where the plants are missing. The crop should be hoed at least three times, at proper intervals; taking care to stir the soil all over.

When the plants begin to flower, the flower-stem should be broken or cut off; removing also the suckers, if any appear; leaving from twelve to sixteen leaves to be matured.

Harvesting and Curing—In ordinary seasons, the crop will be ready for harvesting about the beginning of September; and should all be secured by the 20th of the month, or before the occurrence of frost. The stalks must be cut at the surface of the ground, and exposed long enough to the sun to wilt them sufficiently to prevent breaking in handling. They should then be suspended in a dry, airy shed or building, on poles, in such a manner as to keep each plant entirely separate from the others, to prevent mouldiness, and to facilitate the drying by permitting a free circulation of the air. Thirty or forty plants may be allowed to each twelve feet of pole. The poles may be laid across the beams, about sixteen inches apart.

> "When erected for the purpose, the sheds are built of sufficient height to hang three or four tiers; the beams being about four feet apart, up and down. In this way, a building forty feet by twenty-two will cure an acre and a half of Tobacco. The drying-shed should be provided with several doors on either side, for the free admission of air."

When the stalk is well dried (which is about the last of November or beginning of December), select a damp day, remove the plants from the poles, strip off the leaves from the stalk, and form them into small bunches, or hanks, by tying the leaves of two or three plants together, winding a leaf about them near the ends of the stems; then pack down while still damp, lapping the tips of the hanks, or bunches, on each other, about a third of their length, forming a stack with the buts, or ends, of the leaf-stems outward; cover the top of the stack, but leave the ends or outside of the mass exposed to the air. In cold weather, or by mid-winter, it will be ready for market; for which it is generally packed in damp weather, in boxes containing from two to four hundred pounds.

A fair average yield per acre is from fourteen to eighteen hundred pounds.

To save Seed—Allow a few of the best plants to stand without removing the flowering-shoots. In July and August, they will have a fine appearance; and, if the season be favourable, each plant will produce as much seed as will sow a quarter of an acre by the drill system, or stock half a dozen acres by transplanting. A single capsule, or seed-pod, contains about a thousand seeds.

Green Tobacco.

(Turkish Tobacco; Nicotiana rustica)

Leaves oval, from seven to ten inches long, and six or seven inches broad, produced on long petioles. Compared with the preceding species, they are much smaller, deeper coloured, more glossy, thicker, and more succulent. When fully grown, the plant is of a pyramidal form, and about three feet in height. The flowers are numerous, greenish-yellow, tubular, and nearly entire on the borders; the seed-vessels are ovoid, more depressed at the top than those of the Connecticut Seed-leaf, and much more prolific; seeds small, brownish.

The Green Tobacco is early, and remarkably hardy, but not generally considered worthy of cultivation in localities where the Connecticut Seed-leaf can be successfully grown. It is well adapted to the northern parts of New England and the Canadas; where it will almost invariably yield an abundance of foliage, and perfect its seeds.

It is very generally cultivated, almost to the exclusion of the other species, in the north of Germany, Russia, and Sweden, where almost every cottager grows his own Tobacco for smoking. It also seems to be the principal sort grown in Ireland.

There are several varieties, among which may be mentioned the Oronoco and the Negro-head, both of which have the hardiness and productiveness common to the species, but are not considered remarkably well flavored.

The plants should be started in spring, and transplanted as directed for the Connecticut Seed-leaf; but, on account of its smaller size and habit, two feet, or even twenty inches, between the plants, will be all the space required.

Green Tobacco.

170

Tomato (Solanum lycopersicum)

(Love-Apple)

The Tomato is a native of South America. It is a half-hardy annual, and is said to have been introduced into England as early as 1596. For a long period, it was very little used; and the peculiar, specific term, *lycopersicum,* derived from *lykos,* "wolf," and *persicon,* "a peach" (referring to the beautiful but deceptive appearance of the fruit), more than intimates the kind of estimation in which it was held.

It first began to be generally used in Italy, subsequently in France, and finally in England. In this country, its cultivation and use may be said to have increased fourfold within the last twenty years; and it is now so universally relished, that it is furnished to the table, in one form or another, through every season of the year. To a majority of tastes, its flavor is not at first particularly agreeable; but, by those accustomed to its use, it is esteemed one of the best, as it is also reputed to be one of the most healthful, of all garden vegetables.

When fully grown, the Tomato-plant is from four to seven feet and upwards in height or length, with a branching, irregular, recumbent stem, and dense foliage. The flowers are yellow, in branching groups, or clusters; the fruit is red, white, or yellow, and exceedingly variable in size and form; the seeds are lens-shaped, yellowish-white, or pale-gray,—twenty-one thousand are contained in an ounce, and they retain their vitality five years.

Propagation—The Tomato is raised from seeds, which should be sown in a hot-bed in March, or in the open ground as soon as the frost will permit. As the plants, even in the most favorable seasons, seldom perfectly mature their full crop, they should be started as early and forwarded as rapidly as possible, whether by hot-bed or open-air culture. If the seeds are sown in a hot-bed, the drills should be made five inches apart, and half an inch deep. When the plants are[Pg 640] two inches high, they should be removed to another part of the bed, and pricked out four or five inches apart, or removed into small pots, allowing a single plant to a pot. They are sometimes twice transplanted, allowing more space or a larger pot at each removal; by which process, the plants are rendered more sturdy and branching than they become by being but once transplanted.

As early in May as the weather is suitable, the plants may be set in the open ground where they are to remain, and should be three feet apart in each direction; or, if against a wall or trellis, three feet from plant to plant. Water freely at the time of transplanting, shelter from the sun for a few days or until they are well established, and cultivate in the usual form during summer.

If sown in the open ground, select a sheltered situation, pulverize the soil finely, and sow a few seeds in drills, as directed for the hot-bed. This may be done in November (just before the closing-up of the ground), or the last of March, or first of April. In May, when the plants are three or four inches high, transplant to where they are to remain, as before directed.

In gardens where tomatoes have been cultivated, young plants often spring up abundantly from the seeds of the decayed fruit of the preceding season. These, if transplanted, will succeed as well, and frequently produce fruit as early, as plants from the hot-bed or nursery-bed.

Sufficient plants for the garden of a small family may be started with little trouble by sowing a few seeds in a garden-pan or large flower-pot, and placing it in a sunny window of the sitting-room or kitchen. If the seed is sown in this manner about the middle or 20th of March, the plants will be of good size for setting by the time the weather will be suitable for their removal.

Forcing the Crop—The ripening of the fruit may be hastened by setting the plants against a south wall or close fence. As the plants increase in size, they must be nailed or otherwise attached to the wall or fence; and, if the weather be dry, liberally watered. When the two first trusses of bloom have expanded over each shoot, the shoot should be stopped by pinching off the portion which is beyond the leaf above the second truss, and no more lateral shoots should be suffered to grow; but the leaves must be carefully preserved, especially those near the trusses of bloom. The number of shoots on each plant will vary according to the strength and vigor of the particular plant; but three or four will be quite enough, leaving about half a dozen trusses of fruit.

Hoop-training of the tomato

As the fruit ripens, it must be well exposed to the sun. There will be nothing gained by allowing a great many fruit to ripen. The number above given will be sufficient, and the tomatoes will be much earlier and larger than if they were more numerous.

Culture and Training—A convenient, simple, and economical support for the plants may be made from three narrow hoops,—one twelve, another fifteen, and the third eighteen or twenty inches in diameter,—and attaching them a foot from each other to three stakes about four feet in length; placing the lower hoop so that it may be about ten inches from the surface of the ground after the stakes are driven. The adjoining figure illustrates this method of training. It secures abundance of light, free access of air, and, in skilful hands, may be made quite ornamental.

Hoop-training of the Tomato

Or a trellis may be cheaply formed by setting common stakes, four feet in length, four feet apart, on a line with the plants, and nailing laths, or narrow strips of deal, from stake to stake, nine inches apart on the stakes; afterwards attaching the plants by means of bass, or other soft, fibrous material, to the trellis, in the manner of grape-vines or other climbing plants. By either of these methods, the plants not only present a neater appearance, but the ripening of the fruit is facilitated, and the crop much more conveniently gathered when required for use.

Trellis-training

The French mode of raising tomatoes is as follows:

"As soon as a cluster of flowers is visible, they top the stem down to the cluster, so that the flowers terminate the stem. The effect is, that the sap is immediately impelled into the two buds next below the cluster, which soon push strongly, and produce another cluster of flowers each. When these are visible, the branch to which they belong is also topped down to their level; and this is done five times successively. By this means, the plants become stout, dwarf bushes, not above eighteen inches high. In order to prevent their falling over, sticks or strings are stretched horizontally along the rows, so as to keep the plants erect. In addition to this, all laterals that have no flowers, and, after the fifth topping, all laterals whatsoever, are nipped off. In this way, the ripe sap is directed into the fruit, which acquires a beauty, size, and excellence unattainable by other means."[1]

Trellis-training of the tomato.

Varieties—These are quite numerous. Some are merely nominal, many are variable or quite obscure, and a few appear to be distinct, and, in a degree, permanent. The principal are as follow:—

Apple-Tomato

(Apple-Shaped)

Fruit somewhat flattened, inclining to globular, depressed about the stem, but smooth and regular in its general outline. The size is quite variable; but, if well grown, the average diameter is about two inches and a half, and the depth two inches. Skin deep, rich crimson; flesh bright-pink, or rose-color,—the rind being thick and hard, and not readily reduced to a pulp when cooked.

Apple tomato

The Apple-tomato is early, hardy, productive, keeps well, and, for salad and certain forms of cookery, is much esteemed; but it is more liable to be hollow-hearted than any other of the large varieties.

In form, as well as in the thick, tough character of its rind, it resembles the Bermuda.

Bermuda

This is a red or rose-colored, apple-formed sort, extensively imported from Bermuda into the Middle and Northern States in May and the early summer months.

Like the preceding variety, it varies considerably in size,—some specimens measuring little more than an inch in diameter; while others from the same plant, matured at nearly the same season, frequently exceed a diameter of two inches and a half.

It possesses a thick, rather tough rind, which rarely becomes pulpy in the process of cooking; and, besides, is quite light and hollow-hearted. In size and form, it somewhat resembles the Apple-tomato. When cultivated in New England or the Middle States, it has little merit, either for its productiveness or early maturity.

Fejee

Fruit quite large, red, often blushed or tinged with pinkish-crimson, flattened, sometimes ribbed, often smooth, well filled to the centre; flesh pink, or pale-red, firm, and well flavored; plant hardy, healthy, and a strong grower.

Seeds received from different reliable sources, and recommended as being strictly true, produced plants and fruit in no respects distinguishable from the Perfected.

Fig-Tomato

(Red Pear-Shaped Tomato)

A small, red, pyriform or pear-shaped sort, measuring from an inch and a quarter to an inch and a half in length, and nearly an inch in its broadest diameter. Flesh pale-red, or pink, very solid and compact, and generally completely filling the centre of the fruit.

Like the Plum-tomato, it is remarkably uniform in size, and also in shape; but it is little used except for preserving,—other larger varieties being considered more economical for stewing, making catchup, and like purposes.

Annual of True Blue Seeds (1894), A.W. Livingston's Sons

The variety is usually employed for making tomato-figs, which are thus prepared:

"Pour boiling water over the tomatoes, in order to remove the skin; after which, weigh, and place in a stone jar, with as much sugar as tomatoes, and let them stand two days; then pour off the sirup, and boil and skim it till no scum rises; pour it over the tomatoes, and let them stand two days as before; then boil, and skim again. After the third time, they are fit to dry, if the weather is good; if not, let them stand in the sirup until drying weather. Then place them on large earthen plates, or dishes, and put them in the sun to dry, which will take about a week; after which, pack them down in small wooden boxes, with fine, white sugar between every layer. Tomatoes prepared in this manner will keep for years."[2]

Giant Tomato

(Mammoth)

An improved variety of the Common Large Red, attaining a much larger size. Fruit comparatively solid, bright-red, sometimes smooth, but generally ribbed, and often exceedingly irregular; some of the larger specimens seemingly composed of two or more united together. The fruit is frequently produced in masses or large clusters, which clasp about the stem, and rest so closely in the axils of the branches as to admit of being detached only by the rending asunder of the fruit itself; flesh pale-pink, and well flavored.

Like most of the other varieties, the amount of product is in a great degree dependent on soil, culture, and season. Under favorable conditions, twenty-five pounds to a single plant is not an unusual yield; single specimens of the fruit sometimes weighing four and even five or six pounds.

The Giant Tomato is not early, and, for the garden, perhaps not superior to many other kinds; but for field-culture, for market, for making catchup in quantities, or for the use of pickle-warehouses, it is recommended as one of the best of all the sorts now cultivated.

Grape or Cluster Tomato

(Solanum sp.)

This variety, or more properly species, differs essentially in the character of its foliage, and manner of fructification, from the Garden Tomato. The leaves are much smoother, thinner in texture, and have little of the musky odor peculiar to the Common Tomato-plant. The fruit is nearly globular, quite small, about half an inch in diameter, of a bright-scarlet color, and produced in leafless, simple, or compound clusters, six or eight inches in length, containing from twenty to sixty berries, or tomatoes; the whole having an appearance not unlike a large cluster, or bunch of currants.

The plants usually grow about three feet in height or length; and, in cultivation, should be treated in all respects like those of other varieties. Flowers yellow, and comparatively small. Early.

Though quite ornamental, it is of little value in domestic economy, on account of its diminutive size.

Large Red Tomato

Fruit sometimes smooth, often irregular, flattened, more or less ribbed; size large, but varied much by soil and cultivation,—well-grown specimens are from three to four inches in diameter, two inches and a half in depth, and weigh from eight to twelve ounces; skin smooth, glossy, and, when ripe, of a fine red color; flesh pale-red, or rose-color,— the interior of the fruit being comparatively well filled; flavor good.

Large red tomato

Not early, but one of the most productive of all the varieties; the plants, when properly treated, producing from twelve to fifteen pounds each.

From the time of the introduction of the Tomato to its general use in this country, the Large Red was almost the only kind cultivated, or even commonly known. The numerous excellent sorts now almost everywhere disseminated, including the Large Red, Oval, Fejee, Seedless, Giant, and Lester's Perfected, are but improved sub-varieties, obtained from the Common Large Red by cultivation and selection.

Large Red Oval-fruited Tomato

A sub-variety of the Large Red. Fruit oval, flattened, much less ribbed, more symmetrical, and more uniform in size, than the last named; well-grown specimens measure about four inches in one direction, three inches in the opposite, and two inches in depth; skin fine, deep-red, smooth and shining; flesh paler, the interior of the fruit well filled with pulp, and, when cooked, yielding a large product in proportion to the bulk. Prolific and well flavored, but not early; ripening at the time of the Large Red.

The variety is exceedingly liable to degenerate, constantly tending towards the Large Red; and can only be maintained in its purity by exclusive cultivation, and a continued use of seeds selected from the fairest, smoothest, best ripened tomatoes, having the peculiar oval form by which the variety is distinguished.

Large Yellow

Plant, in its general character, not distinguishable from the Large Red. The fruit also is quite similar in form and size; the principal mark of distinction being its color, which is a fine, clear, semi-transparent yellow. Flesh yellow, well filling the centre, and perhaps a little sweeter or milder than the Red; though generally not distinguishable when stewed or otherwise prepared for the table.

The variety is hardy, yields abundantly, and comes to per[Pg 648]fection with the Large Red. It is, however, not generally cultivated; the Red descriptions being more commonly used, and consequently better adapted for cultivation for the market.

Mexican

Fruit large, comparatively smooth, frequently of an oval form, bright-red, often tinted with rose or bright-pink; flesh pink, solid, filling the fruit to the centre.

It is similar to, if not identical with, the Perfected.

Perfected

(Lester's Perfected; Pomo d'Oro Lesteriano)

A recently introduced and comparatively distinct variety. Plant remarkably healthy and vigorous, often attaining a height or length of six or eight feet, and, in strong soil, of more than ten feet; fruit pinkish-red, or rose-red, of large size, comparatively smooth and regular, flattened, remarkably solid and well filled to the centre, and, when cooked, yielding a large return in proportion to its bulk; flesh firm, well flavored, with comparatively few seeds intermixed. In this last respect, not unlike the Seedless.

When started at the same time, it ripens two weeks after the early varieties, and continues to yield in great abundance until the plants are destroyed by frost. It is considered one of the best sorts for cultivation for the market, and by many is preferred to all others for the garden.

On the authority of a recent writer, the variety has already, to some extent, degenerated. Impure seed, or the influence of some peculiar locality, may have furnished grounds for the statement; but if the variety is genuine or unmixed, it will, in almost any soil or exposure, commend itself by its hardiness, solidity, and great productiveness.

Red Cherry-Tomato

A small, red Tomato, nearly spherical, and about half an inch in diameter. The fruit is produced in great profusion, in large bunches, or clusters; but is comparatively of little value, on account of its small size. It is sometimes used as a preserve, and by some is esteemed for pickling.

Red Plum-Tomato

Fruit bright-red or scarlet, oval, solid, an inch and a quarter or an inch and a half in depth, and about an inch in diameter; flesh pink, or rose-red, mild and well flavored; seeds comparatively few.

The variety is remarkable for its symmetry and for its uniform size. When ripe, the fruit is not easily distinguished from some varieties of scarlet plums. It is hardy, early, and yields abundantly: but the fruit is employed principally for pickling and preserving; its small size rendering it of little value for stewing or for catchup.

Mixed with the Yellow, they make a fine garnish, and are excellent for salad.

Round Red

A small, round, red variety, measuring about an inch in diameter. It is one of the earliest of all the cultivated sorts, but of little value except for pickling or preserving.

Round Yellow

Of the size and form of the foregoing, differing only in colour.

Seedless

Very similar to, if not identical with, the Perfected. Fruit almost rose-red, solid, and with comparatively few seeds.

Tree-Tomato

(New Upright; Tomate de Laye)

A new variety, raised from seed by Grenier, gardener to M. de Fleurieux, at a place in France called Chateau de Laye (whence the name), and introduced by M. Vilmorin of Paris.

It is distinct from all others; rising quite erect to the height of two feet or upwards, with a stem of remarkable size and strength. The branches are not numerous, and comparatively short, usually eight or ten inches in length,—thus requiring no heading-in; leaves not abundant, rather curled, much wrinkled, very firm, closely placed on the sturdy branches, and of a remarkably deep, shining-green color; fruit bright-red, of large size, comparatively smooth, and well filled to the centre,—in many respects, resembling the Perfected, though more regular in form.

From the peculiar, tree-like character of the plants, the variety is remarkably well adapted for cultivation in pots; but its late maturity greatly impairs its value as a variety for forcing. It is a slow grower, tardy in forming and perfecting its fruit, and, for ordinary garden culture, cannot be recommended as being preferable to the Perfected and other earlier and much more prolific varieties. It has been described as strictly self-supporting: but, though the fruit is produced in a remarkably close and almost clasping manner about the sturdy stem and branches, its weight often brings the plants to the ground; and consequently, in exposed situations, it will be necessary to provide stakes, or some similar means of support; though the plants never exhibit the rambling, recumbent character of the Common Tomato.

White Tomato

Plant similar in habit to the Large Red; fruit large, generally ribbed, often irregular, but sometimes comparatively smooth. Its distinguishing characteristic is its color, which, if the fruit be screened by foliage or if grown in the shade, is almost clear white; if much exposed to the sun, it assumes a yellowish tinge, much paler, however, than the Large Yellow. Flesh yellowish, more watery than that of the Large Red, and of a somewhat peculiar flavor, much esteemed by some, and unpalatable to others.

The variety is hardy, remarkably productive, as early as the Large Red, and equally large and solid: but its color, before and after being cooked, is unattractive; and it is rarely seen in the markets, and seldom cultivated for family use.

White's Extra Early

(Early Red; Extra Early)

A medium-sized Red variety, generally round, but frequently of an oval form, flattened, sometimes ribbed, but comparatively smooth, and, when fully matured, of a deeper color than the later Red sorts. Average specimens measure about two inches and a half in diameter, and an inch and a half in depth. The plants are moderately vigorous, and readily distinguished by their peculiar curled and apparently withering foliage.

Flesh pale-red, quite firm, mild, not very seedy, and well filling the fruit, which is considerably heavier than the Apple-shaped. When cooked, it yields a much greater product, in proportion to its size, than the last-named and similar hollow-hearted varieties. Productive, and of good quality.

Planted at the same time with the Common Red varieties, it will ripen about two weeks earlier. An excellent sort for the garden, and recommended for general cultivation.

In order to retain this or any other early variety in its purity, seed for planting should be saved from the smoothest, best formed, and earliest ripened fruit. Few of the numerous kinds now cultivated possess much permanency of character; and rapidly degenerate, if raised from seed taken from the scattered, irregular, and comparatively immature tomatoes remaining upon the plants at the close of the season.

Yellow Cherry-Tomato

A yellow variety of the Red Cherry-tomato,—differing only in color.
Quite showy, but of little value for culinary purposes.

Yellow Pear-Shaped Tomato

(Yellow Fig-Tomato)

A sub-variety of the Red Pear-shaped, with a clear, semi-transparent, yellow skin and yellow flesh. Like the preceding, it is little used except for preserving and pickling.

Yellow Plum-Tomato

A variety of the Red Plum, of the same size and form, and equally symmetrical; distinguished only by the color of its skin, which is a fine, clear, transparent yellow. It is used principally for preserving; its small size rendering it comparatively valueless for use in any other form.

When the two varieties are intermixed, the colors present a fine contrast; and a basket of the fruit is quite a beautiful object.

---

1. The Gardener's Chronicle. Weekly. By Prof. Lindley. 1844 to the present time.

2. Mrs. Eliza Marsh, in The Magazine of Horticulture, Botany, and Rural Affairs. By C. M. Hovey. Boston. Monthly. 1834 to the present time.

Annual of True Blue Seeds (1897), A.W. Livingston's Sons

www.ingramcontent.com/pod-product-compliance
Lightning Source LLC
Chambersburg PA
CBHW060521010526
44107CB00060B/2642